METHODS
FOR
GENETIC
RISK ASSESSMENT

METHODS FOR GENETIC RISK ASSESSMENT

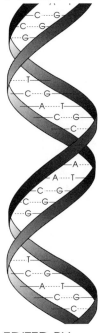

EDITED BY
DAVID J. BRUSICK

RA 1224.3
M.48
1994

LEWIS PUBLISHERS
Boca Raton   Ann Arbor   London   Tokyo

Library of Congress Cataloging-in-Publication Data

Methods for genetic risk assessment / edited by David J. Brusick.
   p.   cm.
   Includes bibliographical references and index.
   ISBN 1-56670-039-6
   1. Genetic toxicology.   2. Health risk assessment.   I. Brusick,
David.
   RA1224.3.M48   1994
   616'.042—dc20   93-5520
                                                                              CIP

This book contains information obtained from authentic and highly regarded sources. Reprinted material is quoted with permission, and sources are indicated. A wide variety of references are listed. Reasonable efforts have been made to publish reliable data and information, but the editor, authors, and the publisher cannot assume responsibility for the validity of all materials or for the consequences of their use.

Neither this book nor any part may be reproduced or transmitted in any form or by any means, electronic or mechanical, including photocopying, microfilming, and recording, or by any information storage or retrieval system, without prior permission in writing from the publisher.

All rights reserved. Authorization to photocopy items for internal or personal use, or the personal or internal use of specific clients, may be granted by CRC Press, Inc., provided that $.50 per page photocopied is paid directly to Copyright Clearance Center, 27 Congress Street, Salem, MA 01970 USA. The fee code for users of the Transactional Reporting Service is ISBN 1-56670-039-6/94/$0.00+$.50. The fee is subject to change without notice. For organizations that have been granted a photocopy license by the CCC, a separate system of payment has been arranged.

CRC Press, Inc.'s consent does not extend to copying for general distribution, for promotion, for creating new works, or for resale. Specific permission must be obtained from CRC Press for such copying.

Direct all inquiries to CRC Press, Inc., 2000 Corporate Blvd., N.W., Boca Raton, Florida 33431.

© 1994 by CRC Press, Inc.
Lewis Publishers is an imprint of CRC Press

No claim to original U.S. Government works
International Standard Book Number 1-56670-039-6
Library of Congress Card Number 93-5520
Printed in the United States of America 1 2 3 4 5 6 7 8 9 0
Printed on acid-free paper

PREFACE

The subjects of this book are the strategies and methods for genetic risk assessment. Genetic risk assessment is a science that is beginning to emerge as a consequence of fundamental advances made by research groups within the field of genetic toxicology, coupled with technological breakthroughs in basic genetics and biotechnology. As more is learned about the genetic composition of humans through data generated from research, epitomized by the Human Genome Initiative, it will become possible to assess the nature and possible consequences of genetic alterations induced by environmental agents. With the availability of this new technology and the concern raised by the increasing numbers of chemicals in the environment, more intensity will be focused on the potential for heritable damage to the human germ line.

The concerns identified above led to the initiation of a joint program between the United Nations Environmental Program (UNEP) and the International Commission for Protection against Environmental Mutagens and Carcinogens (ICPEMC). The goal of this joint program was to develop a document describing state-of-the-art activity in genetic risk assessment. This volume was intended to provide the foundation for future developments and refinements in the processes of risk assessment and risk extrapolation.

The initial outlines and objectives for the joint program were prepared at a UNEP-ICPEMC meeting in St. Petersburg, Russia in 1989. The Steering Committee for the project consisted of M.L. Mendelsohn (Chairman), J. Ashby, and P.H.M. Lohman. Five working groups were established to prepare authoritative reviews. The working groups met regularly from 1990 through 1992 and the results of these five working groups are provided in the five chapters of this volume.

ICPEMC was founded in 1977 and is associated with the International Association of Environmental Mutagen Societies. Its objectives are to identify and promote scientific principles and make recommendations that may serve as the basis for regulations designed to minimize genetic toxicity.

David Brusick, Chairman
ICPEMC

EDITOR

David J. Brusick, Ph.D., is the Scientific Director for Hazleton Washington. In 1970, he received a Ph.D. degree in genetics from Illinois State University and held a postdoctoral research position with the Food and Drug Administration. He was also a member of the faculty at the Howard University School of Medicine in Washington, D.C. Dr. Brusick is currently Adjunct Associate Professor at Howard University School of Medicine and at George Washington University. Dr. Brusick's research activities have focused on the areas of method development, validation, and data assessment in genetic toxicology. His professional activities include being a past President of the Environmental Mutagen Society and Chairman of the International Commission for Protection against Environmental Mutagens and Carcinogens. He is the author of more than 100 scientific publications and a textbook, *Principles of Genetic Toxicology* (second edition). Dr. Brusick has served on OTA and NRC/NAS committees and served on the NRC/NAS Safe Drinking Water Committee. He is currently the Editor of *In Vitro Toxicology,* a journal of cellular and molecular toxicology. Dr. Brusick is a Fellow of the Academy of Toxicological Sciences (ATS).

CONTRIBUTORS

John Ashby, Ph.D.
Central Toxicology Lab
Imperial Chemical Industries
Alderley Park, England

David J. Brusick, Ph.D.
Hazleton Washington
Vienna, Virginia

Larry D. Claxton, Ph.D.
Health Effects Research
  Laboratory
U.S. Environmental Protection
  Agency
Research Triangle Park, North
  Carolina

David M. DeMarini, Ph.D.
Health Effects Research
  Laboratory
U.S. Environmental Protection
  Agency
Research Triangle Park, North
  Carolina

Udo H. Ehling, Ph.D.
GSF Institut für Säugetiergenetik
Oberschleißheim, Germany

Jack Favor, Ph.D.
GSF Institut für Säugetiergenetik
Oberschleißheim, Germany

J.G. Filser, Ph.D.
GSF Institut für Toxikologie
Oberschleißheim, Germany

Robert H. Haynes, Ph.D.
Annual Reviews, Inc.
Palo Alto, California

Ronald Jensen, Ph.D.
University of California
San Francisco, California

John P. Knezovich, Ph.D.
Division of Environmental Sciences
Lawrence Livermore National
  Laboratory
Livermore, California

David W. Layton, Ph.D.
Division of Environmental Sciences
Lawrence Livermore National
  Laboratory
Livermore, California

Joellen Lewtas, Ph.D.
Health Effects Research Laboratory
U.S. Environmental Protection
  Agency
Research Triangle Park, North
  Carolina

Paul H.M. Lohman, Ph.D.
MGC-Department of Radiation
  Genetics and Chemical
  Mutagenesis
State University of Leiden
Leiden, The Netherlands

William R. Lower, Ph.D.
Environmental Trace Substance
  Research Center
University of Missouri
Columbia, Missouri

James T. MacGregor, Ph.D.
SRI International
Menlo Park, California

Thomas E. McKone, Ph.D.
Environmental Sciences Division
Lawrence Livermore National
  Laboratory
Livermore, California

**Mortimer L. Mendelsohn,
Ph.D., M.D.**
Biomedical and Environmental
  Research Program
Lawrence Livermore National
  Laboratory
Livermore, California
and
Radiation Effects Research
  Foundation
Hiroshima City, Japan

George G. Pesch, Ph.D.
Environmental Research Laboratory
U.S. Environmental Protection
  Agency
Narragansett, Rhode Island

Ann M. Richard, Ph.D.
Health Effects Research
  Laboratory
Genetic Toxicology Division
U.S. Environmental Protection
  Agency
Research Triangle Park, North
  Carolina

Michael D. Shelby, Ph.D.
National Institute of Environmental
  Health Sciences
Research Triangle Park, North
  Carolina

Eckerhardt W. Vogel, Ph.D.
Department of Radiation, Genetics,
  and Chemical Mutagenesis
State University of Leiden
Leiden, The Netherlands

Michael D. Waters, Ph.D.
Health Effects Research Laboratory
U.S. Environmental Protection
  Agency
Research Triangle Park, North
  Carolina

Jeffrey J. Wong, Ph.D.
State of California Department of
  Public Health Services
Toxic Substance Control Program
Sacramento, California

Albert A. van Zeeland, Ph.D.
Laboratory of Radiation, Genetics,
  and Chemical Mutagenesis
State University of Leiden
Leiden, The Netherlands

CONTENTS

Introduction

Chapter 1. Hazard Identification 1
D.M. DeMarini, A.M. Richard, M.D. Shelby, and M.D. Waters

Chapter 2. Assessment of Exposures to Genotoxic Substances 29
D.W. Layton, T.E. McKone, J.P. Knezovich, and J.J. Wong

Chapter 3. Methods for Dose and Effect Assessment 65
J. Favor, E.W. Vogel, A.A. van Zeeland, J.G. Filser, M.L. Mendelsohn, and U.H. Ehling

Chapter 4. Risk Characterization Strategies for Genotoxic Environmental Agents 125
J. Lewtas, D.M. DeMarini, J. Favor, D.W. Layton, J.T. MacGregor, J. Ashby, P.H.M. Lohman, R.H. Haynes, and M.L. Mendelsohn

Chapter 5. Monitoring Environmental Genotoxicants 171
J.T. MacGregor, J. Lewtas, G.G. Pesch, R. Jensen, L.D. Claxton, and W.R. Lower

Index 245

ACKNOWLEDGMENTS

This book is the result of a joint program between the United Nations Environmental Program (UNEP) and the International Commission for Protection Against Environmental Mutagens and Carcinogens (ICPEMC), to review methods and strategies for genetic risk assessment. ICPEMC and the editor would like to thank Jan Huismans of UNEP and Andrew Sors of the Environmental Programme of the European Community (EEC-DGX11) for their technical and financial contributions to this project.

Introduction

M. L. Mendelsohn, J. Ashby, and P. H. M. Lohman

The induction of genetic damage in the form of gene or chromosomal mutations to the germline of living organisms can be a serious environmental risk. The consequences of mutations include death of the organism or alterations transmitted by sexual or asexual reproduction to subsequent generations. Mutations are associated with both adverse health and survival outcomes in most species and have been linked to several somatic cell diseases, including cancer.

Spontaneous mutations occur in all species. The origin of these mutations is not known, but probably includes infidelity in DNA replication as well as effects of naturally occurring ultraviolet, ionizing radiations, free radicals, and other active oxygen species formed during normal metabolic processes. Background mutations are an important substrate for natural selection and species evolution; however, if the frequency of mutations occurring over a short time span is too great, a species' reproductive performance and ultimate survival may be in jeopardy. Spontaneous mutations also play a role in maintenance of the equilibrium for most genetic-based diseases.

Man-made mutagens released into the environment are believed to impose an additional genetic burden by adding new damage to the background mutation load and, consequently, contributing to the incidence of diseases and reproductive disorders. The degree to which these agents increase the mutation load is determined by exposure rates, metabolic detoxification processes, and the DNA repair capacities of the target species, and by the intrinsic potency of the mutagenic agent.

Chemical substances and their by-products are being released into the environment, on a worldwide basis, at increasing levels. It is estimated that up to 90% of these agents have not been adequately evaluated for their mutagenic activity toward somatic or germ lines of mammalian species. Because the pool of genes that will form all future generations of a species is held by the existing individuals, it is essential that their exposure to mutagens be minimized. Key to this objective is the ability to identify mutagens and aneugens among the array of man-made and naturally occurring substances in the milieu in which living species exist and to measure their intrinsic biological potency.

Once these agents have been identified, it becomes essential to determine their genetic risk to the genomes of somatic and germ cells of the exposed species. This process is called risk assessment and requires knowledge of (1) the abundance and distribution of the material and the exposure milieu, (2) its stability in that environment, and (3) its bioavailability to contiguous human and nonhuman biota. The integration of agent classification, exposure assessment, dose-effect extrapolation, and risk characterization allows one to make either absolute (quantitative) or comparative evaluations of genetic risk and to take the appropriate measures to manage that risk. Among issues relevant to the management process are risk perceptions and risk communication for human populations as well as the availability of resources and technologies to control the release and/or exposure to agents representing the most serious risk.

At the request of the United Nations Environmental Programme (UNEP), the International Commission for Protection Against Environmental Mutagens and Carcinogens has evaluated the current state-of-the-art knowledge in genetic risk assessment, characterization, and management for humans and nonhuman biota associated with exposure to genetically active substances in the environment.

Genetic risk assessment for humans has not been standardized to date because of serious information gaps in human epidemiology and risk characterization, comparative germline sensitivities, accurate dose-effect extrapolations, and the tremendous diversity among reproductive processes found across the range of plant and animal species.

The following chapters include a summary of the data requirements, assumptions, and methodologies currently employed to evaluate genetic risk. They also identify those aspects of the process where additional knowledge and better technologies are needed.

Identification of the full range of mutagens to which humans and the biota are exposed is an enormous task, even if the focus is restricted to chemicals generated from human activities. Environmental elements as basic as the diet must be considered as a possible source of mutagens. Hence, the identification process tends to be selective and proceeds along different lines, depending on the source of the chemical and the cause for concern.

From a risk management viewpoint, the simplest strategy is to prevent the introduction of new, potentially hazardous chemicals into commerce and the environment. Hence, screening of new chemicals for mutagenic activity plays an important role in limiting exposures to mutagens. Potentially hazardous chemicals can be identified early in the screening process on the basis of genotoxicity tests, structure-activity considerations, and exposure estimates. Such evidence may warrant additional testing or motivate industry to withdraw a chemical from further consideration before significant resources are committed or environmental exposures occur.

The second strategy is to focus on identifying mutagens among known chemicals resulting from established industrial processes or other human activities. The cause for concern could be based on significant population exposure, preliminary mutagenicity data, or association with an adverse health effect. Initial evaluation could lead to further testing and perhaps remedial action to eliminate or restrict exposure. Due to the high social and political costs of remedial action, this strategy tends to require greater investment of resources and burden of proof than does prevention.

The third strategy is to identify existing environmental mutagenic hazards, either through *in situ* monitoring or the collection of environmental samples such as air, water, or soil. Such an approach could be part of a prevention strategy whereby "sentinel" systems, typically sensitive to a broad range of genetic effects, are used as monitors of the genetic activity within an environment.

Finally, observation of an adverse health effect within a population may give rise to a suspicion that exposure to an environmental mutagen has occurred. If epidemiological data support a causal association between exposure and observed health effect, attempts may be made to identify and limit exposure to the specific chemical agents and/or their sources. This strategy and the previous strategy involving the use of sentinel systems are apt to deal with problems involving complex mixtures, exposure assessment, and/or environmental monitoring. These topics are treated more extensively in subsequent chapters of this volume.

Risk assessment attempts to define the nature and magnitude of the risk of a process, situation, or environment. In the context of human health, the risk has usually been expressed as the probability of a disease outcome or mortality. The assessment process is usually divided into four stages: (1) hazard identification, in which the harmfulness of an agent and its availability in the environment is suggested, (2) dose-response assessment, in which the dose of the agent and the disease outcome are linked through extrapolation methods, (3) exposure assessment, in which the agent is located in the environment and its transport and access to the target species is estimated, and (4) risk characterization, in which exposure assessment and dose-response (potency) assessment are combined to give the expected risk or disease outcome.

Environmental or human monitoring methods for genotoxic agents can be used at each stage of risk assessment to achieve different objectives. These objectives include (1) the identification of genotoxic hazards in a population or environment of interest, (2) the direct measurement of damage in a population, (3) dosimetry and determination of dose-response relationships, (4) evaluation of the relationship between exposure, genetic effect, and biological consequence, and (5) risk confirmation through association and/or intervention.

The development of a philosophical position related to the need for risk assessment immediately reveals an important issue: studies of human populations exposed to several mutagenic agents (including ionizing radiation and cancer chemotherapeutic agents) have failed to reveal any significant increase in genetic damage at the specific loci sampled. The results may simply be due to the inability of the selective techniques to detect small increases in genetic damage or an indication that human germ cells are quite resistant to the effects of mutagenic agents.

Most individuals familiar with the induction of mutations in experimental animals believe that there is sufficient similarity between rodent and human germ cells that a prudent philosophical position would be to assume that human germ cells are mutable by chemical mutagens and ionizing radiation.

In addition, the association between somatic-cell mutagens and germ-cell mutagens in mammals is sufficient to argue that agents known to induce gene or chromosomal damage in human somatic cells may also damage a portion of the germ cells. The quantitative association between these two cell populations is unknown for most species but would be expected to be compound specific, thus eliminating any attempts to express the relationship as a constant.

The relationship between mutations induced in mammalian somatic cells and other types of disease states has been moderately well documented for cancer. Induction of mutations at specific codons in some protooncogenes and tumor suppressor genes resulting in cell transformation provides a mechanistic explanation for the strong association. Individuals with DNA instability, manifested as chromosome aberrations, due to inherited conditions or environmental exposures, are also characterized by high risk for selected malignancies. As a consequence of the reasoning cited above, it appears to be safe to conclude that mutagenic agents should be considered hazardous and that their presence in the environment should be restricted or controlled to the greatest extent feasible.

For other agents, which have substantial human exposures and yet must be tolerated in the environment, some type of genetic risk assessment is desirable. The methods reviewed in this report are applicable for this purpose. The risk-benefit ratios obtained from evaluations, including a formal risk assessment, would be valuable in determining the probable health impact and are necessary for proper communication of the risks. Resources required for the production of a genetic risk assessment are considerable, limiting the approach to only a few circumstances.

One of the agents extensively studied in animals for genetic risk assessment which also has relevant human epidemiology data is ionizing radiation. Data from these studies are valuable to the development of a better understanding of the design and interpretation of results from animal models; however, data produced in risk assessment studies using chemicals cannot be directly compared to the radiation results at this time due to factors unique to the molecular dosimetry of chemicals. Exposure and dose are not related as a constant with chemicals. Metabolism, distribution, and excretion affect chemical assessments but are not important for assessing radiation effects. Because of these significant differences, the relative risks from radiation and chemicals cannot be accurately calculated. Clearly, this will be a fruitful area for continued study in the future.

In summary, the proper management of genetic risks associated with agents identified as gene or chromosomal mutagens should be decided by what is known about the probability of the agent or metabolite to reach somatic or germ-cell DNA and the benefits and associated costs required to control exposure of target species to the agent.

Risk management has been receiving increased attention in recent years. An important component of risk management is that of risk communication, especially to members of the public. Failure to communicate effectively the results of a risk assessment can lead to undesirable results, such as the incorrect interpretation of the nature and magnitude of the predicted risks, inappropriate risk-management decisions, and in some cases, public outrage. These outcomes are often unexpected because assessors usually presume that their published risk estimates and supporting analyses are sufficient to support subsequent actions or meet the information needs of the users. To avoid the pitfalls associated with such thinking, it is crucial to recognize that risk communication is not a one-way process (i.e., risk assessor to user), but a two-way process of information exchange between assessor and user. For those cases where a risk assessment is used solely by a regulatory body, with little or no public involvement, the risk assessor(s) and manager(s) should determine a mutually acceptable content and format — one that can support subsequent risk-management actions. So, for example, an agency responsible for regulating toxic substances in water should receive risk estimates that are expressed on the basis of the concentration of a substance in water, adjusted for appropriate water-based exposures (e.g., ingestion, inhalation, and dermal contacts). In this regulatory context, there are generally no issues raised regarding the credibility or neutrality of the assessors. However, when the public is actively involved as a stakeholder in the risk-management actions connected with the risk assessment, the credibility and neutrality of the organization performing the risk assessment can become critical factors influencing the transfer and interpretation of risk-assessment information.

Neutrality, for example, can be compromised by using self-serving risk comparisons designed to influence the reader's acceptance of the assessed risks. The acceptability of the health risk of a given toxic substance is a complex function of perceptional and cultural factors, economic and risk-risk trade offs, and the knowledge and education of the readers. Issues pertaining to risk acceptance are better left to the risk managers. In cases where the public is intimately involved in the risk-management process, risk assessors are well advised to interact with the interested public to define their information needs. As a result of such interactions, it may be necessary to place a much greater emphasis on the definition of commonly accepted assumptions and parameters. Additional discussions may also be needed of the inferences, uncertainties, and limitations of the analysis. Workshops can also be held to facilitate exchanges of information between assessors and the public. These are more helpful and constructive when held early in the risk-assessment process — not after the assessment is completed. Such actions will improve the credibility of the performing organization or assessment team, thereby improving (or at least not hindering) the communication of information. These observations suggest that risk-communication strategies should play a more important role in the preparation or risk assessments. In some instances it may be prudent to develop communication strategies at the same time that the risk-assessment approaches are being devised.

In rare situations, the recognition of a genetic hazard can lead directly to control measures. An example would be the decision to cease development of a candidate antiasthmatic agent shown to be a germ-cell mutagen in rodent studies. However, a more common sequence of events would be the recognition of a hazard, estimation of risk in the exposure situation anticipated, and initiation of appropriate control measures to reduce the risk to acceptable levels. The control measures selected in a given situation will be influenced by a range of subjective and often competing factors; it is, therefore, not possible to discuss control measures in general terms. It is, however, possible to illustrate the interplay of the subjective and objective components of risk management in the following example of two cancer risks.

A risk estimation will usually involve the calculation of a projected incidence of a given genetic disease in a particular exposed population. The incidence projected will be for a given time frame that itself will be determined by the disease condition. For example, exposure to a leukemogen results in the induction of leukemia over a ~10-year period, while bladder carcinogens normally have a latent/developmental period of ~20 years. Two identical examples of risk estimations could be as follows:

* An incidence of 5 in 10^4 skin cancer cases among 50 miners exposed to white arsenic
* An incidence of 5 in 10^4 leukemia cases among 2 million people exposed to an uncontrolled source of radiation

These two objective risk estimations will be subject to a range of arbitrary influences that will in turn affect the control measures instituted. The first of these is the extent to which the projected disease is encountered in the human population. For example, surveillance of the 50 miners is unlikely to yield any cases of skin cancer, while leukemia will be perceived as commonplace among the population exposed to radiation. There will, in fact, be 0.25 and 10,000 cases of cancer in these respective situations, despite the *individual* risk of succumbing to cancer being identical in each situation. Cancer is an easily recognized condition, but in the case of less well defined morbidity, its social perception will be reduced. The problem of perception of induced disease is then subject to the secondary influence of whose responsibility it is to institute control measures. For example, a government authority would probably sponsor control measures for the source of radiation as a priority, perhaps deciding to neglect the arsenic mine. In contrast, the mine owner would be faced with the single task of improving hygiene in the mine. Nonetheless, the example of the arsenic mine also has a political dimension because the risk estimation for the miners may not have been commissioned in the absence of appropriate legislation.

A further subjective complication is provided by the concept of risk-benefit analysis. The most basic problem is that the risk and the benefit often accrue to different individuals or groups of people. Where different groups are involved, the possibility of legal intervention gives rise to concepts such as negligence, acceptable risk, and relative risk. These terms are of obvious relevance to control measures, but they are subjective and are derived from socioeconomic backgrounds. When the risk and the benefit accrue to the same person or group, the idea of personal freedom intervenes. Thus, a community of people may expend great effort in measuring ppb levels of a pesticide in their environment, while accepting the personal freedom to smoke tobacco. The current interest in passive smoking illustrates how small changes in risk perception can have profound effects on risk management — thus, control measures are instituted for the lesser hazard of passive smoking while leaving the major hazard of primary smoking uncontrolled.

The final component of risk management is the extent to which the institution of control measures is practical. This ranges from (1) the simple decision to curtail development of a potential drug on its being found to be mutagenic and for which adequate nonmutagenic analogs already exist to (2) the difficulty faced on finding that a natural constituent of a staple diet is mutagenic. Thus, two similar risk estimations may lead to different risk management control measures.

The above considerations suggest that risk estimations are best made in the context of a well-developed perception of risk/benefit as it applies in the given socioeconomic environment. The above examples were derived from consideration of risk management in an economically developed community. However, in cases where a country is still concerned with improving life expectancy or reducing famine, more complex decisions on risk management may be required. As an example, a pesticide may dramatically enhance a subsistence crop and thereby lead to the saving of, e.g., 10^5 lives. However, that same pesticide may be shown by a risk assessment to entail a genetic disease risk of 1 in 10^5 when present as a residue in the crop. In a developed country, a heritable genetic risk of 1 in 10^5 may be considered unacceptable, but when the 10^5 people at risk have actually been kept alive by the crop, the matter becomes complex.

The best control option for reducing exposure to mutagens is to prevent them from entering the environment. The existing regulatory procedures in several industrialized countries (U.S., Japan) and larger geographic regions (European Commission) have the capability to identify most chemical mutagens. Thus, ideally, actions could be taken to prevent exposure of populations to new chemicals that could result in a health hazard. However, such regulatory actions are not universally practiced, especially in the less-developed countries of the world where societal priorities may result in different priorities being mandated. The expansion of premanufacture and premarket (preimport) testing and regulatory action can be effective tools for reducing exposures to chemical mutagens. UNEP could be a resource in disseminating health-related information already available on chemicals in commerce and industry as well as in providing guidance with respect to the implementation of programs for testing and regulating chemicals.

CHAPTER 1

Hazard Identification*

D. M. DeMarini, A. M. Richard, M. D. Shelby, and M. D. Waters

TABLE OF CONTENTS

I. Introduction 2
   A. Basis for Concern and Definition of Terms 2
   B. Nature of Mutation 3
   C. Germ Cells 3
   D. Incidence of Genetic Disease 4
   E. Somatic Cells 4

II. Identification of Mutagens 6
   A. Information Sources 7
   B. Information on Chemical and Physical Properties
      of Chemicals 14
   C. Computational Methods and Structure-Activity Studies 14
   D. Generation of New Data 16

III. Priority of Mutagenic Hazards 18

IV. Future Research 20

References 22

---

* An ICPEMC project sponsored by U.N. Environmental Program Working Group I. Hazard Identification.

I. INTRODUCTION

A. Basis for Concern and Definition of Terms

DNA is life's most basic and vital chemical component. Encoded in these informational macromolecules are complete instructions for the formation, function, and reproduction of virtually all forms of life. Concern for the health hazards resulting from exposure to genotoxic chemicals and ionizing radiation is centered on the detrimental effects that can result from induced alterations in the DNA of an exposed individual or on the progeny of that individual. Chemically or radiation-induced changes in DNA can result in malformation, malfunction, sterility, or death when transmitted to subsequent generations. These hazards are not limited to any biological sex, species, subgroup, or kingdom.

Life exists within a milieu of chemicals. Some are beneficial or essential to life, some are toxic, and some are without significant effect. Most are naturally occurring, i.e., are not the result of human activities, whereas others exist as the result of human activities. Similarly, ultraviolet and ionizing radiations are naturally occurring phenomena, but opportunities for exposure have increased as a result of man's activities.

Within the context of this document, environmental agents are considered to be the sum total of chemicals and radiations existing within the environment. Environmental mutagens comprise an important subset of environmental agents. Environmental mutagens (either chemicals or radiations) are agents that are capable of directly or indirectly interacting with DNA, resulting in changes in the sequence of bases and thus altering the informational content of the genetic material.

The environment contains naturally occurring mutagens that occur as a consequence of normal processes. For example, sunlight is mutagenic and causes skin cancer, especially among certain populations. Many plants consumed as food contain mutagens, as well as antimutagens. Many soil, water, and air samples contain naturally occurring mutagens. Consequently, most organisms have evolved some ability to repair at least some of the DNA damage caused by mutagens.

Although environmental mutagens occur independently of human activity, the focus of concern here is the detrimental effects of mutagens introduced into the environment by human activity. Some primary sources of human (anthropogenic) mutagens in the environment are agricultural chemicals such as pesticides, emissions from the burning of fossil fuels and other organic materials, effluents from industrial processes, and disposal of by-products from a wide range of domestic, industrial, and agricultural activities. In addition, mutagens are also released into the environment by chemical and nuclear accidents as well as by the deliberate release of mutagenic chemicals and radioactive material. No major component of the environment is free of anthropogenic mutagens; air, fresh and salt water, and soil have been contaminated to varying degrees as a result of human activities, and all life is potentially exposed to these agents. This chapter focuses on these anthropogenic mutagens.

HAZARD IDENTIFICATION

Detrimental health effects resulting from exposure to environmental mutagens may be placed in two broad categories. The first results from genetic damage to germ cells; the second from genetic damage to somatic cells.

B. Nature of Mutation

The genetic material of eukaryotic organisms, mammals included, is organized into chromosomes that consist of DNA and associated proteins. The organization of the genetic material can be used as a means for categorizing, in a simple fashion, the nature of the mutational events that are the basis for genetic disease.

The primary informational content of DNA is contained within the linear sequence of its nucleotide bases. Alterations in this sequence may result from the addition or deletion of nucleotides or from the substitution of one nucleotide for another during DNA replication or repair. Any one such molecular change may exert its effect by changing the function of a single gene and is referred to as a *gene or point mutation*.

In eukaryotic organisms, genes are arranged into discrete physical structures called chromosomes. At the level of chromosome structure, mutagens may induce breaks that may produce chromosome fragments or rearrangements. Such events can lead to the loss or gain of genetic material or chromosome rearrangements. These types of genetic damage are referred to as *chromosomal mutations*.

A third category of genetic damage involves changes in the number of chromosomes. Each species is characterized by a specific number of chromosomes; any departure from that number affects the genetic integrity of an organism. When that departure involves one or a few chromosomes, either too few or too many, the condition is referred to as *aneuploidy*.

All three types of mutations are known to be associated with human disease. Short-term tests (genotoxicity tests) are available in eukaryotic organisms, including mammals, to detect the induction by mutagens of all three categories of genetic damage. It should be noted that these three categories of genetic damage subsume a wide range of other genetic changes such as might result from sister chromatid exchange (SCE), genetic recombination, gene conversion, multilocus deletions, and gene amplification.

C. Germ Cells

The concern for mutations in the DNA of the germ cells extends beyond the exposed individual. The health of subsequent generations is dependent, first and foremost, on the inheritance of a genome (the total genetic material of an organism) that is free of mutant genes that might lead to malformation, disease, or untimely death. Although, to date, no instance of an increase in human genetic disease resulting from radiation- or chemically induced mutations has been clearly demonstrated, such effects have been demonstrated repeatedly in experimental mammals and a wide range of other organisms. Current knowledge of genetics, mutagenesis, and genetically based diseases leave no room for doubt that there exists a risk of genetic disease resulting from human exposure to germ-cell mutagens.

Several factors have contributed to the present lack of evidence for human germ-cell mutagens. These include the difficulties in identifying and studying suitable populations exposed to substantial levels of mutagens, the rarity of individual genetic diseases or marker genes, and the relatively large increase in mutations that would necessarily have to be detected in an epidemiology study. New methods using molecular techniques to detect induced genetic changes promise to increase the likelihood of detecting and studying the induction of mutations in human germ cells.

D. Incidence of Genetic Disease

It is clear that genetic diseases, both those originating in preexisting mutations and those resulting from newly formed mutations, contribute substantially to the human disease burden. McKusick (1990), for example, now lists more than 5000 different human genetic traits. A proportion of these traits are associated with diseases and are occasionally treatable, presently not curable, and may be transmitted to subsequent generations.

The incidence of genetic diseases among newborns is estimated to be 10 cases per 100 liveborn children (NRC, 1983; ICPEMC, 1983). Some of these disease traits appear to be due to mutations in a single gene; others appear to be due to mutations in several genes. One complication in arriving at a clear understanding of the incidence and origins of genetic disease is the uncertainty of how much genetic disease results from new mutations compared to the amount attributable to preexisting mutations within the gene pool.

Our knowledge of the true incidence and origins of human genetic disorders remains limited, and our mammalian models and other methods for studying germ-cell mutations and their impact on health require further development and refinement. Consequently, many assumptions are necessary to estimate genetic risk. Nonetheless, the well-documented existence of mutagens that cause germ cell mutations in nonhuman species, and the manifest existence of human genetic disease, clearly point to a potential human health risk associated with exposure to germ-cell mutagens. Hence, the identification of such hazards and the minimization of genetic risks to future generations are important public safety goals with a strong scientific basis.

E. Somatic Cells

Mutational damage induced in somatic cells can result in a range of effects from cell death through changes in metabolism and other phenotypic characters to a change or loss of growth control mechanisms. Such damage may also be

HAZARD IDENTIFICATION   5

a factor in the development of aging-related diseases, including cancer. Disruption of normal restraints on growth can lead to tumors and, ultimately, metastatic cancers. Consequently, the primary health concern for exposure of somatic tissues to mutagens is the risk of induced cancer (Crawford, 1985).

A causal association between genetic changes and cancer was proposed at the beginning of this century, and the first identified carcinogens, both radiation and chemicals, have subsequently been shown to be mutagens. Of the chemicals that have been established to be carcinogenic in humans, many are organic compounds that are mutagenic to a wide variety of organisms, both prokaryotic and eukaryotic, including mammals. (However, many environmental mutagens are not mutagenic per se and are converted into mutagens by metabolic activation.) In contrast to the high proportion of human carcinogens that are mutagenic, a substantial proportion of organic chemicals that show evidence of carcinogenicity in long-term rodent studies show little or no evidence of genetic toxicity. Whether one considers human carcinogens or rodent carcinogens, not all chemical carcinogens are genotoxic, and, therefore, tests for genetic toxicity are not adequate to detect all carcinogens. Thus, some carcinogens are operationally nongenotoxic.

Although our knowledge of the molecular events associated with the induction of cancer is increasing rapidly, the exact mechanisms by which carcinogens induce cancer remain unknown, but mutations are clearly involved in the process. Research on the mutational basis for cancer provides the greatest promise for understanding a mechanism by which human exposure to a carcinogen leads to cancer. Recent discoveries of specific genes that, when mutated, result in cancer have added substantially to the case for a mutational basis for cancer induction. Two general categories of cancer genes are oncogenes and tumor suppressor genes (or anti-oncogenes) (Pimentel, 1989; Cooper, 1990; Barrett, 1991).

Oncogenes are altered forms of normal genes called protooncogenes that are involved in the control of cell growth and differentiation. Genetic alteration (mutation) of protooncogenes can result in change in their function or expression, thus giving rise to cell division that is regulated less adequately than normal, leading to cancer. More than 50 oncogenes have been identified to date, and a variety of genetic changes, including base-pair substitutions, chromosomal aberrations, and gene amplification, have been shown to result in their carcinogenic properties (Bishop, 1991). *In vitro* studies have provided evidence that the effect of oncogenes is dominant, i.e., when oncogenes are introduced into normal cells in culture, the normal cells exhibit the characteristics of cancer cells even though the cells retain the normal counterparts (the protooncogenes) of the oncogenes.

In contrast to the dominant effect of oncogenes, genes have been identified that result in cancer when their function is lost. These genes are referred to as tumor suppressor genes. Normally functioning tumor suppressor genes play a role in constraining cell division, and it is in their absence or inactive state (due to mutational damage) that inadequately controlled cell growth and cancer results. In this case, the tumor phenotype is a recessive characteristic because the presence of a normal gene results in a normal (noncancerous) phenotype even when a mutant gene is present in the cell. The most thoroughly studied tumor suppressor gene is the gene responsible for retinoblastoma, tumors of the eye in humans (see Cooper, 1990). Most tumors, however, appear to be due to mutations in several genes, some of which are protooncogenes and some of which are tumor suppressor genes (Fearon and Voglestein, 1990).

The complexity of the carcinogenic process, the diversity of agents that induce cancer, including many that are not overtly mutagenic, and the multitude of diseases that are referred to as cancer make it seem highly unlikely that either mutagenicity in general or specific mutations in oncogenes and tumor suppressor genes are sufficient to explain completely the induction of all cancers by environmental agents. Indeed, arguments have been advanced regarding the importance of nongenotoxic mechanisms (see Clayson and Clegg, 1991). Nonetheless, the scientific evidence for a mechanistic role of mutations and, thus, mutagens in the etiology of induced cancer point clearly to the hazards of mutagen exposures.

Beyond the cancer hazard, mutations in somatic cells have been implicated in other health problems. Perhaps most notable is the emerging evidence for a mutational etiology in the induction of atherosclerotic plaques associated with cardiovascular diseases (ICPEMC, 1990). Other diseases that may result from somatic-cell mutations due to mutagen exposures include the phenomenon of aging, senile cataracts, and metaplasias of the stomach, intestine, and gall bladder (Hartman, 1983). Although the evidence for a mutational basis for these diseases is still limited, additional research may confirm a role for mutations in these and other noncancerous somatic-cell diseases.

In summary, the potential severity of the health hazards posed by mutagen exposures emphasizes the importance of identifying mutagens in the environment and, to the extent possible, eliminating them or limiting exposures of not only humans but of the entire biota.

II. IDENTIFICATION OF MUTAGENS

Identification of the full range of mutagens to which humans and the biota are exposed is an enormous task, even if the focus is restricted to chemicals generated from human activities. Even something as basic as the diet must be considered as a possible source of mutagens (Eckl et al., 1991). Hence, the identification process tends to be selective and proceeds along different lines, depending on the source of the chemical and the cause for concern.

From a risk-management viewpoint, the simplest strategy is to prevent the introduction of new, potentially hazardous chemicals into commerce and the environment. Hence, screening of new chemicals for mutagenic activity plays an important role in limiting exposures to mutagens. Potentially hazardous chemicals can be identified early in the screening process on the basis of genotoxicity tests, structure-activity considerations, and exposure estimates. Such evidence may warrant additional testing or motivate industry to withdraw a chemical from further consideration before significant resources are committed or environmental exposures occur.

The second strategy is to focus on identifying mutagens among known chemicals resulting from established industrial processes or other human activities. The cause for concern could be based on significant population exposure, preliminary mutagenicity data, or association with an adverse health effect. Initial evaluation could lead to further testing and perhaps remedial action to eliminate or restrict exposure. Due to the high social and political costs of remedial action, this strategy tends to require greater investment of resources and burden of proof than does prevention.

The third strategy is to identify existing environmental mutagenic hazards, either through *in situ* monitoring or the collection of environmental samples such as air, water, or soil. Such an approach could be part of a prevention strategy whereby "sentinel" systems, typically sensitive to a broad range of genetic effects, are used as monitors of the genetic activity within an environment. For example, feral rodents living on hazardous waste sites have been monitored for increases in genetic damage (Tice et al., 1987).

Finally, observation of an adverse health effect within a population may give rise to a suspicion that exposure to an environmental mutagen has occurred. If epidemiological data support a causal association between exposure and observed health effect, attempts may be made to identify and limit exposure to the specific chemical agents and/or their sources. This strategy and the previous strategy involving the use of sentinel systems are apt to deal with problems involving complex mixtures, exposure assessment, and/or environmental monitoring. These topics are treated more extensively in subsequent chapters of this book.

The following discussion focuses on the first two strategies outlined above dealing with preliminary screening and the identification of potential mutagenic hazards of individual chemicals. This identification process typically involves three steps: (1) review of existing sources of pertinent bioassay and structural information, (2) interpretation and extrapolation of existing data, and (3) generation of new data if existing data are judged insufficient for preliminary hazard identification.

A. Information Sources

A wide variety of information sources are available to aid in mutagenicity hazard identification, ranging from listings of physicochemical property data, to listings and profiles of bioassay test data, to detailed monographs on single chemicals or complex mixtures. A few of the more relevant and widely available data sources are mentioned below.

International organizations such as the World Health Organization (WHO) and the United Nations Environmental Programme (UNEP) are important sources of genotoxicity information for the international community. The International Register of Potentially Toxic Substances (IRPTC), administered under UNEP, provides information on use, regulatory action, and general toxic effects (including mutagenicity) of a select group of environmental chemicals of international significance. In addition, monographs on the genetic and carcinogenic hazards of environmental chemicals have been produced by the International Agency for Research on Cancer (IARC) and as Environmental Health Criteria (EHC) documents by the WHO International Programme for Chemical Safety (IPCS). These monographs provide in-depth surveys and analyses of existing data on single chemicals or small groups of chemicals, and they offer valuable guidance for developing more refined methods for carcinogen identification.

In addition, a number of countries have stores of information on environmental chemicals that can assist in efforts to identify mutagenic hazards. These include the European Community's Data Information Network (ECDIN), the European Chemical Industry Ecology and Toxicology Center (ECITOC), and Japan's Biological DataBase (BL-DB) (Hayashi et al., 1991). Within the U.S., widely used data sources of particular relevance to mutagenicity hazard identification have been generated within the Environmental Protection Agency (EPA), the National Institute of Occupational Safety and Health (NIOSH), and the National Institute of Environmental Health Sciences (NIEHS), among others. For a listing and discussion of additional toxicity and physicochemical property databases, see McCutcheon et al., 1990.

The largest collection of indexed publications on genetic toxicity resides at the Environmental Mutagen Information Center (EMIC).* EMIC's holdings include more than 74,000 publications from the open literature dealing with the genetic toxicity of approximately 25,000 chemical agents. Bibliographic and key experimental information from these publications are accessible on-line through the U.S. National Library of Medicine's Toxicology Data Network (TOXNET), with older information accessible through TOXLINE.

Gene-Tox committees were commissioned by EPA to review, interpret, and consolidate the genotoxicity information in EMIC (Waters and Auletta, 1981). The committees generated published reports and consensus protocols for each bioassay. In addition, the Gene-Tox database was established to provide a summary of the peer-reviewed literature on specific bioassays. The Gene-Tox database, in its current form, considers 73 different genotoxicity tests and includes over 4000 chemicals, at least 330 of which have been tested in five or more different types of genotoxicity tests (Table 1). Also included in the database, for each chemical tested in a given bioassay, are qualitative activities (positive or negative) that are assigned by Gene-Tox committees and updated

---

* Oak Ridge National Laboratory, Oak Ridge, TN.

HAZARD IDENTIFICATION

periodically on the basis of accumulated data. The Gene-Tox database, like EMIC, is accessible on-line through TOXNET.

A visual format for presenting the quantitative data contained in Gene-Tox for an array of genotoxicity tests has been developed by EPA in collaboration with the IARC (Garrett et al., 1984; IARC, 1987a; Waters et al., 1988a). A genetic activity profile (GAP) displays a range of genotoxicity test results for a single chemical in bar graph format according to the phylogenetic level of the test organism and genetic endpoint, with bar lengths corresponding to doses yielding positive or negative responses in each test (Waters et al., 1988a; Figure 1). GAPs and corresponding data listings have been reviewed and published for 187 compounds in IARC Monograph Supplement 6 (IARC, 1987a), and are currently available for approximately 370 chemicals tested in a subset of nearly 200 different genotoxicity tests (Table 1). Many of these chemicals have been evaluated in animal cancer tests and in epidemiological studies, as reported in IARC Monograph Supplement 7 (IARC, 1987b) and elsewhere (ATSDR, 1988; ATSDR, in press) upon request.

The Registry of Toxic Effects of Chemical Substances (RTECS), produced by NIOSH, is an additional source of mutagenicity data searchable on-line through TOXNET. Although RTECS contains a range of toxicity data on significantly more chemicals than Gene-Tox, RTECS data were obtained directly from unevaluated literature sources and, hence, are of more limited utility for the evaluation of mutagenic hazards.

The U.S. National Toxicology Program (NTP) administered within NIEHS has, through its testing program, accumulated a large database of *in vitro* and *in vivo* genetic toxicity test results on chemicals of environmental concern. Results have been generated using standardized test protocols and coded chemicals, providing improved internal consistency and comparability within and among tests. Although not publicly available in a database format, many of the test results have been published (see, e.g., Haseman et al., 1988; Ashby and Tennant, 1988, 1991; Ashby et al., 1989), others are being prepared for publication, and data on particular chemicals are available upon request.*

Although there are a large number of centralized sources of data on the toxic effects of chemicals, many of which include mutagenicity data, the Gene-Tox, GAP, and NTP databases provide the most extensive and widely accepted sources of mutagenicity data for use in hazard evaluation. When available, the NTP data are generally considered the most definitive due to the careful adherence to testing consistency and protocol.

Table 1 summarizes the distribution of chemicals among the various categories of test systems represented within the GAP, Gene-Tox, and NTP databases. The lack of consensus concerning the most informative bioassays for predicting genotoxicity or carcinogenicity has contributed to the large number

---

* National Toxicology Program, NIEHS, P.O. Box 12233, Research Triangle Park, NC 27709.

Table 1. Distribution of Tests in Genetic Activity Profile (GAP), Gene-Tox (GTX), and National Toxicology Program (NTP) Databases

Phylogenetic endpoint	GAP Chemicals[a]	GAP Entries[b]	GTX Chemicals	NTP Chemicals[c]
Prokaryotes				
DNA damage	115	309	676	
Gene mutation	282	3635	2694	1702 (SAL)
Lower eukaryotes				
DNA damage	18	22		
Recombination	99	266	506	
Gene mutation	101	276	352	
Chromosomal aberrations	1	1		
Aneuploidy	31	49	105	
Plants				
DNA damage	3	3		
Gene mutation	38	60	149	
SCE	7	12		
Micronuclei	11	13	8	
Chromosomal aberrations	40	105	223	
Insects				
Recombination	15	25		
Gene mutation	123	268	790	285 (DMX)
Chromosomal aberrations	28	49	61	
Aneuploidy	36	62	132	
Animals *in vitro*				
DNA damage	126	379	35	
Gene mutation	112	427	794	343 (G5T)
SCE	124	371	477	612 (SIC)
Micronuclei	23	27	9	
Chromosomal aberrations	118	334	197	616 (CIC)
Aneuploidy	23	39		
Cell transformation	94	278	550	
Human *in vitro*				
DNA damage	77	198	187	
Gene mutation	23	29		
SCE	93	244	701	
Micronuclei	5	6	10	
Chromosomal aberrations	92	271	11	
Aneuploidy	6	7		
Cell transformation	0	0		
Body fluids, host mediated	71	152	240	
Animals *in vivo*				
DNA damage	53	112	19	100 (UPR)
Gene mutation	20	31	94	
SCE	62	132	58	120 (SVA)
Micronuclei	103	252	426	40 (MVM)
Chromosomal aberrations	127	514	73	120 (CVA)

Table 1. (continued) Distribution of Tests in Genetic Activity Profile (GAP), Gene-Tox (GTX), and National Toxicology Program (NTP) Databases

Phylogenetic endpoint	GAP Chemicals[a]	GAP Entries[b]	GTX Chemicals	NTP Chemicals[c]
Aneuploidy	8	15		
Cell transformation	0	0		
Humans *in vivo*				
DNA damage	3	3		
SCE	37	88	1	
Micronuclei	3	4		
Chromosomal aberrations	52	145	26	
Aneuploidy	2	5		
Miscellaneous				
(Sperm morphology, DNA binding, intracellular communication, fish assays)	89	185	256	
Animal carcinogenicity	507	326		

[a] Number of chemicals for which test data are available.
[b] Number of individual results reported.
[c] SAL = *Salmonella typhimurium,* all strains; DMX = *Drosophila melanogaster,* sex-linked recessive lethal mutations; G5T = gene mutation, mouse lymphoma cells *in vitro;* SIC = sister chromatid exchange, Chinese hamster cells *in vitro;* CIC = chromosomal aberrations, Chinese hamster cells *in vitro;* UPR = unscheduled DNA synthesis, rat hepatocytes *in vivo;* SVA = sister-chromatid exchange, cells *in vivo;* MVM = micronucleus test, mice *in vivo;* CVA = chromosomal aberrations, other animal cells treated *in vivo.*

of genotoxicity tests represented. As discussed later, only 15 to 20 genotoxicity tests are in routine use currently, and even fewer are used for regulatory purposes. No one genotoxicity test is able to detect the full range of potential genotoxic damage induced by an agent, and the inherent limitations of each genotoxicity test must be taken into consideration when evaluating genotoxicity test data.

Even within a particular genotoxicity test, the predictive capability of the assay can vary a great deal, depending on the class of chemicals being tested. For example, the concordance between *Salmonella* assay results and rodent carcinogenicities is ~90% for polycyclic aromatic hydrocarbons (PAHs), but less than 20% for chlorinated organics (Claxton et al., 1988). Strategies for overcoming deficiencies in individual genotoxicity tests have focused on using a combination of genotoxicity tests in either a tier or battery approach to testing.

Table 1 indicates that the *Salmonella* test system, which provides a measure of gene mutation, has been used to analyze a substantial number of chemicals, and its behavior is well characterized for a variety of chemical classes. In contrast, other test systems have had only limited application and, thus, are less well characterized with respect to their predictive capabilities. In addition, the

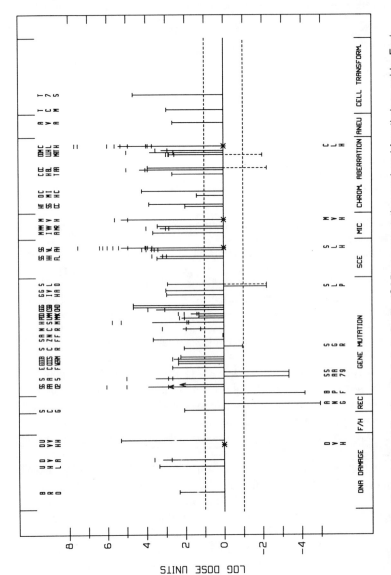

Figure 1. An example of the Genetic Activity Profile (GAP) representation produced for ethylene oxide. Each test response is shown as a line in the positive (up) or negative (down) direction with the length of the line determined by the lowest effective concentration for the positive responses or the highest ineffective concentration for negative agents.

HAZARD IDENTIFICATION 13

totals in Table 1 represent data from a wide variety of environmental sources, and include a large diversity of chemical structures and many possible mechanisms of genotoxic activity. As a result, although a substantial number of chemicals have been evaluated in genotoxicity tests, organization of the data according to genotoxicity test and chemical class results in groups of data that are too small to be informative or that contain significant data gaps (Ray et al., 1987). These data limitations have hampered efforts to use computational methods to predict the genotoxic activity of new chemicals.

A feature of most available genetic toxicity databases is the prevalence of compounds testing positive for mutagenicity. This feature reflects past testing priorities that focused resources on chemicals likely to exhibit genotoxic behavior. In recent years, this focus, particularly within the NTP program, has shifted to chemicals of potentially significant environmental or occupational exposure, irrespective of their mutagenicity. As a result, current data include a larger representation of nonmutagenic compounds, some of which are rodent carcinogens. Most genotoxicity tests in current use were designed specifically to detect genotoxic compounds, and it is generally recognized that two to three short-term tests (SSTs) can successfully detect the majority of such compounds. However, few available assays are capable of detecting nongenotoxic carcinogens, and little is known about the mechanism of action of such compounds. Clearly, future research in this area, with the possible development of new rodent-based genotoxicity tests, is required (Ashby and Morrod, 1991).

Another important feature of the NTP database is the availability of genotoxicity test data *and* rodent carcinogenicity results for a large number of chemicals. This has permitted a more critical evaluation of the ability of genetic toxicity tests to predict the potential rodent carcinogenicity of chemicals than had been possible previously. The results of such evaluations (e.g., Tennant et al., 1987; Zieger et al., 1990; Ashby and Tennant, 1991) have shown that genotoxicity tests are not as effective at predicting rodent carcinogens as once thought, largely due to the recent identification of a substantial number of rodent carcinogens that do not exhibit clear evidence of genetic toxicity. This has led to reassessment of the number and types of genotoxicity tests that are recommended for use in evaluating chemicals for potential carcinogenicity. In addition, a variety of nongenotoxic mechanisms have been implicated as the means by which some operationally nongenotoxic chemicals induce cancer in rodents (Clayson and Clegg, 1991).

In spite of ongoing debates on the relative utility and application of tests for genetic toxicity, available genotoxicity tests and databases provide valuable information for assisting efforts to identify the mutagenic hazards of a wide variety of chemicals. The challenges for the field of environmental mutagenesis are to identify and fill the most important data gaps, and to reach a consensus concerning the types and number of genotoxicity tests to be used in identifying or predicting mutagenic hazard. A timely discussion of these and other issues pertaining to available genotoxicity databases is presented in a series of articles from the Proceedings of the Symposium on Data Bases of Genotoxicity & Carcinogenicity and Their Usefulness for Hazard Evaluation (see, e.g., Parodi and Waters, 1991; Auletta et al., 1991; Ashby, 1991; Brusick, 1991).

B. Information on Chemical and Physical Properties of Chemicals

Existing data relevant to the identification of mutagenic hazard also include physical and chemical properties of a compound that are potentially discriminating with regard to its transport, metabolism, and reactivity with cellular components. Physical properties can be measured or calculated that relate to bioavailability and bioactivity, i.e., nonspecific processes such as transport and partitioning within the organism, and more specific processes relating to metabolism, chemical reactivity, and DNA adduct formation. A variety of sources are available that provide either data listings or a means for estimating physicochemical properties such as octanol/water partition coefficients, pK_as, melting points, etc. for a large variety of chemicals (McCutcheon et al., 1990).

Consideration of physical properties is factored into exposure modeling as well, where relevant properties include volatility, water solubility, and susceptibility to external environmental factors such as sunlight, soil bacteria, etc. Rough model estimates of these properties are often sufficient for the purposes of preliminary hazard identification; more sophisticated modeling of physical parameters associated with target population exposure, physiologically based pharmacokinetics, dose effect, and reactivity requirements could be developed for input into the final risk assessment equation (Reitz, 1990). Subsequent chapters of this volume consider the latter components of risk assessment in greater detail.

C. Computational Methods and Structure-Activity Studies

Computational tools have evolved in tandem with the growth of genetic toxicity databases, providing the ability to scan and access the large stores of available data. Computational methods also play a potentially important role in the organization, representation, and interpretation of such data (see, e.g., Klopman et al., 1985; Pet-Edwards et al., 1985; Garrett et al., 1986; Benigni, 1990), in attempts to design appropriate bioassay testing strategies (see, e.g., Changkong et al., 1985), and in the development of structure-activity relationship (SAR) models. It is in this last role that computational tools offer a way to extract information from existing genotoxicity data and apply the information to the preliminary identification of hazards among untested chemicals.

A number of structural classification schemes have been proposed for organizing existing databases and for use in SAR studies (Nesnow and Bergman, 1988; Wassam, 1980). Because information on mechanisms of action is usually lacking, chemical classifications tend to be based on common organic functionalities (e.g., ethers, esters, alcohols, etc.) or classes of chemicals known or suspected of being associated with particular types of biological activity (e.g., aromatic amines, polycyclic aromatic hydrocarbons, or nitroaromatics). Such classifications are subject to modification as additional chemicals are evaluated for mutagenicity and as knowledge of molecular mechanisms advances.

Some classes of chemicals, such as polycyclic aromatic hydrocarbons, aromatic amines, nitrosamines, and epoxides, contain a large representation of chemicals that are both mutagenic and carcinogenic. Thus, a preliminary screening might identify all members of these classes as initially suspect. This is the basis for Ashby's "structural alert" method in which one or more structural subunits alert the investigator to the potential mutagenicity and/or carcinogenicity of a compound (Ashby and Tennant, 1988). This rule-based method is appealing because of its (1) physical rationale, i.e., structural alerts are groups of atoms that are either electrophilic or capable of conversion to electrophiles, and (2) ease of use, i.e., the method requires no specialized computational tools. Further refinements (and added complexity) are being achieved through identification of structural features that modify the activity level within chemical classes and more accurately anticipate metabolism. A few attempts have been made to encapsulate a predetermined set of structure-based rules such as these into commercially available "expert" systems for broad-spectrum toxicity prediction, such as the HazardExpert program (CompuDrug, H-1136, Budapest, Hollán Ernö u5, Hungary). This type of program provides a potentially useful framework for developing screening strategies based on existing knowledge and can "learn" new rules as knowledge advances. The limitations of such methods are their reliance on current knowledge, difficulties encoding and extrapolating such knowledge, and lack of clear, well-established rules for the wide range of chemicals of interest.

Another approach to SAR modeling recognizes the limits in our current knowledge of the structural basis for mutagenicity, and attempts to infer relevant structural features (structural alerts) or physical property values from statistical analysis of existing data and, potentially, gain new insights into the mechanistic basis for such activity. A wide variety of statistically based SAR methods have been applied to the prediction and modeling of mutagenic and carcinogenic activity (see reviews in Goldberg, 1983; Frierson et al., 1986; Benigni et al., 1989; Richard et al., 1989; Karcher and Devillers, 1990). These studies differ with regard to the biological endpoint studied and the qualitative or quantitative nature of the activity measured, the number and diversity of chemical structures considered, the types of structural features or properties included, the types of statistical analyses or computational methods used, and the form of the final predictive model. They include Hansch-type quantitative SAR (QSAR) methods (Hansch, 1991), the CASE automated structural fragment identification program (Klopman et al., 1990), clustering analysis approaches such as SIMCA (Dunn and Wold, 1981), and automated linear regression or principle components analyses such as in TOPKAT and ADAPT, which utilize a wide variety of molecular descriptors (Enslein et al., 1986; Jurs et al., 1985).

QSAR methods such as the Hansch approach have been highly successful at modeling processes relating to the bioavailability and general toxicity of environmental chemicals (Kaiser, 1987; Karcher and Devillers, 1990). However, the limited availability of physical parameters for close structural analogs and the variation and complexities of the mechanisms involved have restricted the application of these methods for genotoxic and carcinogenic endpoints (Hansch, 1991). For similar reasons, relatively few theoretical modeling studies of chemical reactivity requirements pertaining to mutagenic or carcinogenic endpoints have been performed.

The major limitation of statistically based SAR methods is their reliance on a large data set within a test system and mechanistically distinct chemical class. As a result, most of the published studies utilizing these methods have analyzed the largest and most well-understood chemical classes (e.g., nitrosamines, PAHs, and aromatic amines) and test systems (e.g., *Salmonella*), and, thus, have made only limited contributions to furthering our understanding of the structural features important for mutagenicity. Attempts to overcome data limitations by analyzing data from structurally diverse chemicals within a test system have met with limited success (Blake et al., 1990; Rosenkranz and Klopman, 1990a, b). The validity of such an approach, however, has not been adequately demonstrated.

Many of the statistically based SAR computer programs, such as CASE, ADAPT, and TOPKAT, are commercially available or restricted in their availability. Although the commercial value of these programs has possibly accelerated their development, it also may have limited access to and hindered the objective analysis of their capabilities. Thus, their utility and the value of their predictions continue to be viewed with skepticism by many in the biological community, in large part due to their frequent dissociation from mechanistic explanation and experimental validation. However, gaining wider acceptance is the view that such methods may offer fresh insights or, at a minimum, a preliminary indication of potential mutagenic activity prior to testing.

Access to computerized databases and the use of SAR models can provide a valuable, relatively inexpensive resource to assist environmental risk-management efforts. Computational SAR methods clearly have not evolved to the stage where they can replace biologically based methods for evaluating the genotoxic effects of environmental agents on human or other populations, and they may never be capable of doing so. Provided the limitations of the methods are clearly understood, however, they can augment and guide such testing efforts and play an important role in preliminary hazard identification.

D. Generation of New Data

As discussed previously, a variety of genetic alterations are associated with genetically associated diseases (Hartman, 1983), including cancer (Bishop,

1991). In this regard, genotoxicity tests have been developed utilizing endpoints that reflect this variety of genetic damage (Ramel, 1986, 1988). Because of the diverse nature of these endpoints and the limitations of each genotoxicity test, no single genotoxicity test is capable of detecting all types of genetic damage. Thus, various schemes have been devised over the years whereby a limited number of genotoxicity tests are used in combination to identify an agent as a mutagen or nonmutagen.

The two main types of multitest schemes that have been employed to identify mutagens are the tier approach and the battery approach. The tier approach (Bridges, 1973, 1976) applies tests in a sequential manner and first uses inexpensive and rapid genotoxicity tests to identify mutagens. In general, these genotoxicity tests are conducted *in vitro* and err on the side of false positives, which are then minimized in the second tier, which uses more expensive, eukaryotic genotoxicity tests.

The second approach, battery testing, involves the construction of a test matrix or test battery that uses several genotoxicity tests in parallel, each of which is selected to compensate for the perceived limitations of the others in the battery (DHEW, 1977; UKEMS, 1983). In the battery approach, the response of a single test is not the sole determinant of genetic hazard, whereas a single test response can determine a "pass" or "fail" decision with the tier approach. Compared to the tier approach, the battery approach requires more initial resources but results in reduced testing time and broadened endpoint coverage. Battery testing is now the most common strategy for applying genotoxicity tests, even though it raises new problems regarding which tests should be included in the battery (Ashby and Morrod, 1991).

Selection of tests for inclusion in the battery is critical (IARC, 1986). However, there is currently no consensus regarding the types and numbers of genotoxicity tests that are adequate to declare a compound nonmutagenic or to identify an agent as a mutagen. Nonetheless, there is general agreement that genotoxicity tests should be included that detect two main classes of genetic damage, i.e., gene mutations and chromosomal mutations (Ashby, 1986; IPCS, 1990). The importance of aneuploidy, especially with regard to cancer and some genetic disease, is indisputable. (The most common genetic disease, Down syndrome, is associated with aneuploidy.) However, analysis for aneuploidy has not generally been used in routine screening of chemicals for genotoxic activity due primarily to the lack of simple, reliable assays for this endpoint. Nonetheless, the impact of this type of genetic damage on human health should encourage the development and incorporation of simple assays for this endpoint in routine use for screening environmentally important chemicals and mixtures for genotoxic activity.

A comparative study performed by the U.S. National Toxicology Program (NTP) has indicated that the inexpensive, simple, prokaryotic *Salmonella* (Ames) mutagenicity assay (Maron and Ames, 1983) is as informative in terms of identifying rodent carcinogens in a practical manner as are a variety of more expensive and time-consuming mammalian cell assays (Tennant et al., 1987). Although the NTP study does not provide justification for the inclusion of an additional assay, there is a theoretical basis for inclusion of a cytogenetic assay to detect chromosomal mutations. In its simplest form, an assay for micronuclei might be sufficient; however, a more detailed assay for chromosomal damage, such as one for chromosomal aberrations, could also be used. There is still disagreement as to whether the cytogenetic assay should be performed *in vitro* in cultured mammalian cells or *in vivo* in rodent (primarily mouse) bone marrow (Ashby and Morrod, 1991).

The value of using only the *Salmonella* mutagenicity assay and an *in vivo* cytogenetic assay in the initial test battery to detect carcinogens has been discussed by Ramel (1986, 1988), Shelby (1988), Shelby and Zeiger (1990), and Ashby et al. (1991), who have shown that all 31 organic chemicals (except benzene) that are human carcinogens, as confirmed by IARC, are genotoxic in both of these two genotoxicity tests. Although any combination of just two genotoxicity tests may fail to detect some proportion of mutagenic carcinogens, a battery composed of the *Salmonella* assay and an *in vivo* mouse bone marrow cytogenetic assay provides a practical approach to the initial identification of genotoxic carcinogen hazard. A slight variant of this approach would permit the use of either a mammalian cell *(in vitro)* cytogenetic assay or an *in vivo* cytogenetic assay, along with the *Salmonella* assay (DNHW, 1989).

Comparative mutagenesis studies using a variety of genotoxicity tests as well as comparison of genotoxicity test data to carcinogenicity data have been two main ways in which genotoxicity tests have been assessed for their ability to identify mutagens and/or carcinogens. Based on these studies, the use of the *Salmonella* assay and either an *in vivo* or *in vitro* cytogenetics assay is recommended as a reasonable means for identifying mutagenic hazards.

III. PRIORITY OF MUTAGENIC HAZARDS

Various approaches have been used to rank the relative hazards of a set of compounds based on genotoxicity test data derived from a battery of tests. These approaches have involved subjective assessment as well as computerized weight-of-evidence strategies, both of which rely on historical or statistical evidence and expert judgment (Brusick et al., 1986; Lohman et al., 1992). The use of several genotoxicity tests that encompass a variety of phyla and genetic endpoints would seem to offer the broadest and most sensitive means of detecting potential mutagenic hazard to any segment of the biota, including humans. In addition, *in vivo* mammalian assays are thought to have relevance to all mammals, including humans, due to the similar, although by no means identical, genetic apparatus, genetic processes, and metabolic capabilities among mammals. Thus, IARC (1989) has suggested that the ability of an agent to produce genotoxic effects in laboratory mammals should be regarded as having

HAZARD IDENTIFICATION

greater relevance to humans than results obtained in lower organisms. However, this suggestion must be taken with caution because some genotoxic endpoints cannot yet be measured as adequately in mammals as they can in other organisms (e.g., gene mutations in somatic cells).

Because of this, there may be limited value in ranking the genotoxic hazard of agents based on whether the agents induce genetic effects detectable in prokaryotic vs. eukaryotic organisms or in lower vs. higher eukaryotes. This is partly because concern for the biota includes all organisms and partly because of the intrinsic limitations of each genotoxicity test in terms of the range/types of genetic damage it can detect. Furthermore, there is little basis for ranking one type of genetic damage as more hazardous than another. In fact, for those human tumors for which the genetic basis is best understood, both gene/point mutations in oncogenes and deletions/chromosomal mutations in tumor suppressor genes are implicated (Fearon and Vogelstein, 1990; Bishop, 1991). Thus, a wide variety of types of genetic damage may play different but equally important roles in tumorigenesis.

With some exceptions, the potency of an agent in a genotoxicity test is not a reliable indicator of the mutagenic potency of the agent in another genotoxicity test or of its potential rodent or human carcinogenicity. This occurs because the level of exposure of the DNA to the electrophilic derivative of the applied carcinogen is seldom known, and correlation factors for the fixation and repair of chemically induced DNA lesions are seldom applied. The dose to which the organism is exposed may be quite difficult to determine in some assays. For example, many feeding studies in drosophila (fruit fly) involve nothing more than placing a small dish containing the mutagen or a piece of paper soaked in the mutagen into a bottle containing the flies. The concentration of mutagen in the solution in the dish or in the paper is usually known, but the dose to which the flies are exposed is quite difficult to determine. This raises the related problem of actual vs. effective or biological dose. The concentration of mutagen in the medium, air, or water is frequently known in laboratory experiments. However, the actual dose or amount of mutagen that enters the organism or reaches particular targets within the organism are rarely, if ever, known. This issue, which is referred to as dosimetry, is addressed elsewhere in this volume.

Another problem associated with determining mutagenic potency involves the actual calculation of mutagenic potency, which is complicated due to the various ways in which mutagenic potency can be expressed. Quite different mutagenic potencies can be obtained, depending on how potency is calculated. For example, mutant yield (mutants/treated population) vs. mutant frequency (mutants/survivors), expressed per dose or concentration of an agent, can give different views of the mutagenic potency of the same agent in the same test system (Eckardt and Haynes, 1981). Dose information and many other parameters have been incorporated into an automated system for the determination of mutagenic hazard (Lohman et al., 1992). This automated system attempts to model intuitive expert judgment into a rational, step-by-step analysis that can be performed by anyone with the appropriate computer system and software.

It is intuitively appealing to think that agents exhibiting high mutagenic potency (calculated by whatever means in whatever genotoxicity test) are more hazardous to humans than less potent mutagens. Some support for this idea was presented in a recent study by Rosenkranz and Ennever (1990), who analyzed a large number of compounds and concluded that the more potent carcinogens in rodents were more likely to be mutagens in *Salmonella*. However, for all of the reasons described above, the concept of using mutagenic potency to rank hazards is still controversial, and extrapolation of the mutagenic potency of an agent across genotoxicity tests or to humans is fraught with difficulties. Thus, the use of mutagenic potency in a ranking scheme for mutagenic hazards may be done, but should be approached with full knowledge of its limitations.

As discussed, genotoxicity tests identify mutagens in a variety of biological systems. The conclusions derived from a limited set of genotoxicity tests should be viewed as applicable to other animal, plant, or microbial species based on the universality of the genetic material. Even at the preliminary screening stage, identification of an agent as a mutagenic hazard implies some real or perceived risk to the target species. Hence, elements considered in more depth in the final risk assessment stage, such as production volume, exposure, target dose, etc., could be considered when selecting chemicals for mutagenicity screening and could influence the allocation of resources for preliminary mutagenicity testing.

Thus, all of the biological data, SAR data, and a crude estimate of production/exposure should be considered when ranking agents in terms of mutagenic hazard. However, the way in which to incorporate and integrate this variety of data follows no prescribed rules and will vary, depending on factors such as the cause for concern, the particular chemical agent(s) involved, and exposure routes. Although some formal schemes have been developed, the best approach at present is simply one in which knowledgeable people assess all of the available data and reach an informed judgment regarding the hazard posed by a particular agent.

IV. FUTURE RESEARCH

The ability to identify mutagens has improved remarkably since Muller (1927) detected the first known mutagen (X-rays) using fruit flies. More than 200 genotoxicity tests have been developed since then (Waters et al., 1988b), although probably less than 20 are in routine use or have been validated to any reasonable extent. The current ability to identify mutagens has been summarized in the proceedings of a recent international conference on environmental mutagenesis (Mendelsohn and Albertini, 1990). Although significant progress has been achieved in our ability to identify mutagens, many limitations persist.

HAZARD IDENTIFICATION

Although numerous mutagens have been identified that cause germ-cell mutations in laboratory mammals, no mutagens have yet been well documented that cause germ-cell mutations in humans (Allen et al., 1990; Neel and Lewis, 1990). However, because of the similarity of germ cells in mice and humans, the existence of mutagens that can cause germ-cell mutations in humans seems inevitable. Thus, the absence of confirmed human germ-cell mutagens is attributed primarily to the insensitivity of the methods currently available to determine an increase in germ-cell mutations in humans. Consequently, the development and application of methods to identify such mutagens are clearly needed. The recent application of DNA probes and the polymerase chain reaction (PCR) for the identification of mutations in human sperm, as well as the development and application of other molecular techniques (see Allen et al., 1990), should advance our ability to identify mutagens that cause germ-cell mutations in humans.

Molecular analysis of chemically induced germ-cell mutations in male mice indicates that the stage of the germ cell in which the mutation is induced, rather than the chemical composition of the mutagen itself, is the major determinant of the type of mutation induced by the mutagen (Russell et al., 1990). The distinction appears to be between premeiotic stages in which base-pair substitutions or small intragenic changes predominate and postmeiotic cell stages that yield mostly multilocus deletions. Thus, future research may permit the classification of germ-cell mutagens and the risks they pose according to the germ-cell stage that they affect.

With regard to the identification of chemical mutagens that cause somatic-cell mutations in humans, it now appears that the use of just two genotoxicity tests (the *Salmonella* mutagenicity assay and an *in vivo* mouse bone marrow cytogenetics assay) permits the detection of almost all genotoxic organic chemicals that are IARC-confirmed human carcinogens as well as the majority of genotoxic rodent carcinogens. It is presumed that these agents are carcinogenic via a mutagenic mechanism and, thus, may be considered to be human somatic-cell mutagens. Nonetheless, as discussed previously, the ability of genotoxicity tests to identify all potential human somatic-cell mutagens may not yet be satisfactory.

Perhaps the most important need at present is for an improved understanding of the mechanistic basis underlying somatic-cell diseases, especially cancer. Based on advances in these areas, improved genotoxicity tests should be developed. Steps in this direction are already occurring with molecular analysis of mutations in the *Salmonella* assay and various mammalian cell mutagenicity assays (see Mendelsohn and Albertini, 1990). The development of transgenic mice for mutagenesis studies also offers much promise (1) in identifying mutagens that are currently detected poorly or not at all and (2) for identifying species, sex, and tissue-specific mutagens in somatic as well as germ cells. The limited data available on the types of mutations induced by various mutagens in bacteria, yeast, fruit flies, mice, mammalian cells, and humans indicate that similar mutagenic mechanisms may account for certain classes or types of mutations obtained among all of these diverse organisms. Currently, chemical mutagens are classified according to chemical structure or, less commonly, by the type of mutation that the agent is presumed to induce. As work proceeds on the DNA sequencing of mutations in these organisms, it may be possible in the future to classify mutagens according to the mechanisms by which they induce mutations.

Recent advances in the ability to identify mutations in humans may lead to the identification of mutagens or exposures that increase the mutant and/or mutation frequency at certain genes in humans (see Mendelsohn and Albertini, 1990). This area of work should continue to progress as the Human Genome Project develops. It is likely that various techniques that emerge from the Human Genome Project may be applicable to the direct detection and analysis of mutations in humans.

In summary, an improved understanding of mutation mechanisms, and the resulting development of genotoxicity tests that detect mutagens based on these mechanisms, should improve our ability to identify mutagenic hazards. Further research on the possible mutational etiology of various somatic-cell diseases, such as cancer and atherosclerosis, are needed. Mutagen hazard identification for the protection of the biota requires additional development and refinement of genotoxicity tests and improved mutagen assessment strategies. All of these are contingent upon continued basic and applied research in mutagenesis.

Genotoxicity tests and procedures are available to identify effectively and efficiently the vast majority of environmental mutagens, including single chemicals and complex mixtures that may pose a hazard to public health or to the biota by virtue of their effect on DNA. Problems that remain to be resolved include (1) how many and which specific genotoxicity tests are sufficient to identify all genotoxic hazards, (2) how should nongenotoxicity be defined, and (3) how should mixed positive and negative genotoxicity test results be interpreted when an agent gives such results among multiple genotoxicity tests?

REFERENCES

Allen, J.W., B.A. Bridges, M.F. Lyon, M.J. Moses, and L.B. Russell (1990), *Biology of Mammalian Germ Cell Mutagenesis,* Banbury Rep. No. 34, Cold Spring Harbor Laboratory Press, Cold Spring Harbor, NY.

Ashby, J. (1986), The prospect for a simplified internationally harmonized approach to the detection of possible human carcinogens and mutagens, *Mutagenesis,* 1, 3–16.

Ashby, J. (1991), Aspects of database construction and interrogation of relevance to the accurate prediction of rodent carcinogenicity and mutagenicity, *Environ. Health Perspect.,* 96, 97–100.

Ashby, J. and R.S. Morrod (1991), Detection of human carcinogens, *Nature (London),* 352, 185–186.

Ashby, J. and R.W. Tennant (1988), Chemical structure, salmonella mutagenicity and extent of carcinogenicity as indicators of genotoxic carcinogenesis among 222 chemicals tested in rodents by the U.S. NCI/NTP, *Mutat. Res.,* 204, 17–115.

Ashby, J. and R.W. Tennant (1991), Definitive relationships among chemical structure, carcinogenicity and mutagenicity for 301 chemicals tested by the U.S. NTP, *Mutat. Res.,* 225, 229–306.

Ashby, J., R.W. Tennant, E. Zeiger, and S. Stasiewicz (1989), Classification according to chemical structure, mutagenicity to *Salmonella* and level of carcinogenicity of a further 42 chemicals tested for carcinogenicity by the U.S. National Toxicology Program, *Mutat. Res.,* 223, 73–103.

Ashby, J., H. Tinwell, R.D. Callander, and N. Clare (1991), Genetic activity of the human carcinogen sulphur mustard towards salmonella and the mouse bone marrow, *Mutat. Res.,* 257, 307–311.

ATSDR (1988), Toxicological Profile Series, Agency for Toxic Substances and Disease Registry, U.S. Public Health Service, Atlanta, GA.

Auletta, A.E., M. Brown, J.S. Wassom, and M.C. Cimino (1991), Current status of the Gene-Tox Program, *Environ. Health Perspect.,* 96, 33–36.

Barrett, J.C. (1991), The relationship between mutagenesis and carcinogenesis, in *Cold Spring Harbor Symposium on the Origins of Human Cancer,* Cold Spring Harbor Laboratory Press, Cold Spring Harbor, NY.

Benigni, R. (1990), Rodent tumor profiles, salmonella mutagenicity and risk assessment, *Mutat. Res.,* 224, 79–91.

Benigni, R., C. Andrioli, and A. Giuliani (1989), Quantitative structure-activity relationships: principles and applications to mutagenicity and carcinogenicity, *Mutat. Res.,* 221, 197–216.

Bishop, J.M. (1991), Molecular themes in oncogenesis, *Cell,* 64, 235–248.

Blake, B.W., K. Enslein, V.K. Gombar, and H.H. Borgstedt (1990), salmonella mutagenicity and rodent carcinogenicity: quantitative structure-activity relationships, *Mutat. Res.,* 241, 261–271.

Bridges, B.A. (1973), Some general principles of mutagenicity screening and a possible framework for testing procedures, *Environ. Health Perspect.,* 6, 221–227.

Bridges, B.A. (1976), Use of a three-tier protocol for evaluation of long-term toxic hazards, particularly mutagenicity and carcinogenicity, in *Screening Tests and Chemical Carcinogenesis,* R. Montesano, H. Bartsch, and L. Tomatis, Eds., Publ. No. 12, International Agency for Research on Cancer, Lyon, France, 549–559.

Brusick, D. (1991), A proposed method for assembly and interpretation of short-term test data, *Environ Health Perspect.,* 96, 101–111.

Brusick, D., J. Ashby, F. de Serres, P. Lohman, T. Matsushima, B. Matter, M.L. Mendelsohn, and M. Waters (1986), Weight-of-evidence scheme for evaluation and interpretation of short-term results, in *Genetic Toxicology of Environmental Chemicals, Part B: Genetic Effects and Applied Mutagenesis,* C. Ramel, B. Lambert, and J. Magnussen, Eds., Alan R. Liss, New York, 121–129.

Chankong, V., Y.Y. Haimes, H.S. Rosenkranz, and J. Pet-Edwards (1985), The carcinogenicity prediction and battery selection (CPBS) methods: a Bayesian approach, *Mutat. Res.,* 153, 135–166.

Claxton, L., A.G. Stead, and D. Walsh (1988), An analysis by chemical class of salmonella mutagenicity test as predictors of animal carcinogenicity, *Mutat. Res.,* 205, 197–225.

Clayson, D.B. and D.J. Clegg (1991), Classification of carcinogens: polemics, pendantics, or progress?, *Regul. Toxicol. Pharmacol.,* 14, 147–166.

Cooper, G.M. (1990), *Oncogenes,* Jones and Barlett, Boston.

Crawford, B.D. (1985), Perspectives on the somatic mutation model of carcinogenesis in *Advances in Modern Environmental Toxicology,* W.G. Flamm and R.J. Lorentzen, Eds., Princeton Scientific, Princeton, N.J.

DNHW (1989), The Use of Genotoxicity Tests in Regulation, Department of National Health and Welfare, Ottawa, Canada.

DHEW (1977), Approaches to determining the mutagenic properties of chemicals: risk to future generations, *J. Environ. Pathol. Toxicol.,* 1, 301–352.

Dunn III, W.J. and S. Wold (1981), The carcinogenicity of N-nitroso compounds: a SIMCA pattern recognition study, *Bioorg. Chem.,* 10, 29–45.

Eckardt, F. and R.H. Haynes (1981), Quantitative measures of induced mutagenesis, in *Short-Term Tests for Chemical Carcinogens,* H.F. Stich and R.H.C. San, Eds., Springer-Verlag, Berlin, 457–473.

Eckl, P.M., T. Alati, and R.L. Jirtle (1991), The effects of a purified diet on sister chromatid exchange frequencies and mitotic activity in adult rat hepatocytes, *Carcinogenesis,* 12, 643–646.

Enslein, K., G.W. Blake, M.E. Tomb, and H.H. Borgstedt (1986), Prediction of Ames test results by structure-activity relationships, *In Vitro Toxicol.,* 1, 33–44.

Fearon, E.R. and B. Vogelstein (1990), A genetic model for colorectal tumorigenesis, *Cell,* 61, 759–767.

Frierson, M.R., G. Klopman, and H.S. Rosenkranz (1986), Structure-activity relationships (SARs) among mutagens and carcinogens: a review, *Environ. Mutagen.,* 8, 283–327.

Garrett, N.E., H.F. Stack, M.R. Gross, and M.D. Waters (1984), An analysis of the spectra of genetic activity produced by known or suspected human carcinogens, *Mutat. Res.,* 134, 89–111.

Garrett, N.E., H.F. Stack, and M.D. Waters (1986), Evaluation of the genetic activity profiles of 65 pesticides, *Mutat. Res.,* 168, 301–325.

Goldberg, L. (1983), *Structure-Activity Correlation as a Predictive Tool in Toxicology. Fundamentals, Methods, and Applications,* Hemisphere, Washington, D.C.

Hansch, C. (1991), Structure-activity relationships of chemical mutagens and carcinogens, *Sci. Tot. Environ.,* 109/110, 17–29.

Haseman, J.K., J.E. Huff, E. Zeiger, and E.E. McConnell (1988), Comparative results of 327 chemical carcinogenicity studies, *Environ. Health Perspect.,* 74, 229–235.

Hartman, P.E. (1983), Mutagens: some possible health impacts beyond carcinogenesis, *Environ. Mutagen.,* 5, 139–152.

Hayashi, M.M. Nakadate, T. Osada, T. Ishibe, S. Tanaka, A. Maekawa, T. Sofuni, Y. Nakata, N. Kanoh, S. Hashiba, Y. Takenaka, and M. Ishidate, Jr. (1991), A fact database for toxicological data at the National Institute of Hygienic Sciences, Japan, *Environ. Health Perspect.,* 96, 57–60.

IARC (1986), *Long-Term and Short-Term Assays for Carcinogens: A Critical Approach,* IARC Sci. Publ. No. 83, International Agency for Research on Cancer, Lyon, France.

IARC (1987a), *IARC Monographs on the Evaluation of Carcinogenic Risks to Humans: Genetic and Related Effects,* An Update of Selected IARC Monographs from Volumes 1 to 42, Suppl. 6, International Agency for Research on Cancer, Lyon, France.

HAZARD IDENTIFICATION 25

IARC (1987b), *IARC Monographs on the Evaluation of Carcinogenic Risks to Humans: Carcinogenicity,* An Update of Selected IARC Monographs from Volumes 1 to 42, Suppl. 7, International Agency for Research on Cancer, Lyon, France.

IARC (1989), *IARC Monographs on the Evaluation of Carcinogenic Risks to Humans, Some Organic Solvents, Resin Monomers and Related Compounds, Pigments and Occupational Exposures in Paint Manufacture and Painting,* Vol. 47, International Agency for Research on Cancer, Lyon, France.

IPCS (1990), Summary Report on the Evaluation of Short-Term Tests for Carcinogens (Collaborative Study on In Vivo Tests). Environmental Health Criteria 109, World Health Organization, Geneva.

ICPEMC Committee 4 (1983), Estimation of genetic risks and increased incidence of genetic disease due to environmental mutagens, *Mutat. Res.,* 115, 255–291.

ICPEMC (1990), The possible involvement of somatic mutations in the development of atherosclerotic plaques, *Mutat. Res.,* 239, 143–148.

Jurs, P.C., T.R. Stouch, M. Czerwinski, and J.N. Narvaez (1985), Computer-assisted studies of molecular structure-biological activity relationships, *J. Chem. Inf. Comput. Sci.,* 25, 296–308.

Kaiser, K.L.E. (1987), *QSAR in Environmental Toxicology II,* D. Reidel, New York.

Klopman, G., R. Contreras, H.R. Rosenkranz, and M.D. Waters (1985), Structure-genotoxic activity relationships of pesticides: comparison of the results from several short-term assays, *Mutat. Res.,* 147, 343–356.

Klopman, G., M.R. Frierson, and H.S. Rosenkranz (1990), The structural basis of the mutagenicity of chemicals in *Salmonella typhimurium:* the Gene-Tox data base, *Mutat. Res.,* 228, 1–50.

Lohman, P.H.M., M.L. Mendelsohn, D.H. Moore II, M.D. Waters, D.J. Brusick, J. Ashby, and W.J.A. Lohman (1992), A method for comparing, combining and interpreting short-term genotoxicity data: the basic system, *Mutat. Res.,* 266, 7–25.

Maron, D.M. and B.N. Ames (1983), Revised methods for the salmonella mutagenicity test, *Mutat. Res.,* 113, 173–215.

McCutcheon, P., O. Nørager, W. Karcher, and J. Devillers (1990), *Practical Applications of Quantitive Structure-Activity Relationships (QSAR) in Environmental Chemistry and Toxicology,* Karcher, W. and J. Devillers, Eds., Kluwer, Boston, 13–24.

McKusick, V.A. (1990), *Mendelina Inheritance in Man,* 9th ed., Johns Hopkins University Press, Baltimore.

Mendelsohn, M.L. and R.J. Albertini (1990), *Mutation and the Environment,* Parts A–E. Wiley-Liss, New York.

Muller, H.J. (1927), Artificial transmutation of the gene, *Science,* 66, 84–87.

Neel, J.V. and S.E. Lewis (1990), The comparative radiation genetics of humans and mice, *Annu. Rev. Genet.,* 24, 327–362.

Nesnow, S. and H. Bergman (1988), An analysis of the Gene-Tox carcinogen data base, *Mutat. Res.,* 205, 237–253.

NRC (1983), *Identifying and Estimating the Genetic Impact of Chemical Mutagens.* National Academy Press, Washington, D.C.

Parodi S. and M.D. Waters, (1991), Introduction and summary. Genotoxicity and carcinogenicity databases: an assessment of the present situation, *Environ. Health Perspect.,* 96, 3–4.

Pet-Edwards, J., H.S. Rosenkranz, V. Chankong, and Y.Y. Haimes (1985), Cluster analysis in predicting the carcinogenicity of chemicals using short-term assays, *Mutat. Res.,* 153, 167–185.

Pimental, E. (1989), *Oncogenes,* 2nd ed., CRC Press, Boca Raton, FL.

Ramel, C. (1986), Deployment of short-term assays for the detection of carcinogens; genetic and molecular considerations, *Mutat. Res.,* 168, 327–342.

Ramel, C. (1988), Short-term testing — are we looking at wrong endpoints?, *Mutat. Res.,* 205, 13–24.

Ray, V.A., L.D. Kier, K.L. Kannan, R.T. Haas, A.E. Auletta, J.S. Wassom, S. Nesnow, and M.D. Waters (1987), An approach to identifying specialized batteries of bioassays for specific classes of chemicals: class analysis using mutagenicity and carcinogenicity relationships and phylogenetic concordance and discordance patterns. I. Composition and analysis of the overall data base, *Mutat. Res.,* 185, 197–241.

Reitz, R.H. (1990), Distribution, persistence, and elimination of toxic agents (pharmacokinetics), in *Progress in Predictive Toxicology,* D.B. Clayson, I.C. Munro, P. Shubik, and J.A. Swenberg, Eds., Elsevier, Amsterdam, 79–90.

Richard, A.M., J.R. Rabinowitz, and M.D. Waters (1989), Strategies for the use of computational SAR methods in assessing genotoxicity, *Mutat. Res.,* 221, 181–196.

Rosenkranz, H.S. and F.K. Ennever (1990), An association between mutagenicity and carcinogenic potency, *Mutat. Res.,* 244, 61–65.

Rosenkranz, H.S. and G. Klopman (1990a), The structural basis for the mutagenicity of chemicals in *Salmonella typhimurium:* the National Toxicology Program data base, *Mutat. Res.,* 228, 51–80.

Rosenkranz, H.S. and G. Klopman (1990b), Structural basis of carcinogenicity in rodents of genotoxicants and nongenotoxicants, *Mutat. Res.,* 228, 109–124.

Russell, L.B., W. L. Russell, E.M. Rinchik, and P.R. Hunsicker (1990), Factors affecting the nature of induced mutations, in *Biology of Mammalian Germ Cell Mutagenesis,* Allen, J.W., B.A. Bridge, M.F. Lyon, M.J. Moses, and L.B. Russell, Eds., Banbury Report No. 34, Cold Spring Harbor Laboratory, New York, 271–285.

Shelby, M.D. (1988), The genetic toxicity of human carcinogens and its implications, *Mutat. Res.,* 204, 3–15.

Shelby, M.D. and E. Zeiger (1990), Activity of human carcinogens in the salmonella and rodent bone-marrow cytogenetics tests, *Mutat. Res.,* 234, 257–261.

Tennant, R.W., B.H. Margolin, M.D. Shelby, E. Zeiger, J.K. Haseman, J. Spalding, W. Caspary, M. Resnick, S. Stasiewicz, B. Anderson, and R. Minor (1987), Prediction of chemical carcinogenicity in rodents from in vitro genetic toxicity assays, *Science,* 236, 933–941.

Tice, R.R., B.G. Ormiston, R. Boucher, C.A. Luke, and D.E. Paquette (1987), Environmental biomonitoring with feral rodent species, in *Short-Term Bioassays in the Analysis of Complex Environmental Mixtures V,* S.S. Sandhu, D.M. DeMarini, M.J. Mass, M.M. Moore, and J.L. Mumford, Eds., Plenum Press, New York, 175–179.

UKEMS (1983), Report of the UKEMS Sub-committee on Guidelines for Mutagenicity Testing. Part 1, United Kingdom Environmental Mutagen Society, London.

Wassom, J.S. (1980), The storage and retrieval of chemical mutagenesis information, in *Progress in Environmental Mutagenesis,* Marija Alacevic, Ed., Elsevier, Amsterdam, 313–330.

Waters, M.D. and A. Auletta (1981), The GENE-TOX program: genetic activity evaluation, *J. Chem. Inf. Comput. Sci.*, 21, 35–38.

Waters, M.D., H.F. Stack, A.L. Brady, P.H.M. Lohman, L. Haroun, and H. Vainio (1988a), Use of computerized data listings and activity profiles of genetic and related effects in the review of 195 compounds, *Mutat. Res.*, 205, 295–312.

Waters, M.D., H.F. Stack, J.R. Rabinowitz, and N.E. Garrett (1988b), Genetic activity profiles and pattern recognition in test battery selection, *Mutat. Res.*, 205, 119–138.

Zeiger, E., J.K. Haseman, M.D. Shelby, B.H. Margolin, and R.W. Tennant (1990), Evaluation of four in vitro genetic toxicity tests for predicting rodent carcinogenicity: confirmation of earlier results with 41 additional chemicals, *Environ. Mol. Mutagen.*, 16 (Suppl. 18), 1–14.

CHAPTER 2

Assessment of Exposures to Genotoxic Substances

D. W. Layton, T. E. McKone, J. P. Knezovich, and J. J. Wong

TABLE OF CONTENTS

I. Introduction 30

II. Role of Exposure Assessment in the Risk Assessment and
Management Process 31

III. Exposure and Risk 33

IV. Sources of Genotoxic Substances in Contact Media 34
    A. Releases to Indoor and Outdoor Air 35
    B. Source Terms for Water 37
    C. Releases to Soils 37
    D. Genotoxic Components of Foods, Medicines, and
       Consumer Products 37

V. Contaminant Concentrations in Contact Media: The Role of
Transport and Transformation Processes 38
    A. Overview of Transport Process within a
       Single Compartment 38
       1. Indoor and Outdoor Air 38
       2. Surface Waters 39
       3. Groundwater 40
       4. Soil 41
    B. Transformation Processes 42
       1. Photolysis 42
       2. Hydrolysis 43
       3. Oxidation and Reduction 43
       4. Microbial Transformation 44
    C. Cross-Media Transfers and Multimedia Models 44
    D. Environmental Landscape Properties
       and Physicochemical Properties 46

VI. Exposure Pathways for Biota 47
    A. Transfers to Crops/Plants 47
    B. Transfers to Aquatic and Terrestrial Species 48

VII. Exposure Pathways for Humans 49
    A. Ingestion..50
    B. Inhalation 51
    C. Dermal Uptake 52

VIII. Exposure Assessment Methods 53

IX. Quantifying and Reducing Uncertainties 55

X. Recommendations 56

Acknowledgment 58

References 58

I. INTRODUCTION

The potential effects of a genotoxic agent on humans and biota are controlled by a complex sequence of environmental and biological events, beginning with the genotoxicant's formation within or transfer to foods, medicines, consumer products, soils, air, and water. Subsequent contact by an organism with the genotoxicant in contaminated media sets the stage for toxicokinetic processes affecting the absorption or uptake into, distribution and metabolism within, and elimination of those substances from the organism. The ultimate expression of genotoxicity then becomes a reflection of a balance between the interaction of the genotoxicant (or its reactive metabolite[s]) with the cellular target (e.g., DNA), subsequent cellular damage, and biological repair, and tissue or organ reserve capacity to maintain physiological function. The probability or likelihood that a genotoxic agent causes a deleterious effect within an exposed population (i.e., its risk) is, therefore, a function of the sources of that genotoxic substance, its occurrence, concentrations, and bioavailability in contact media, toxicokinetic properties, and genotoxic potency as a function of effective dose.

The focus of this chapter is on the exposure component of the risk-assessment process for genotoxic agents. Specifically, it addresses pertinent topics dealing with the presence and/or formation of genotoxicants in environmental media, their movement within and between contaminated media, and the contacts of humans and biota with those media. As such, we discuss principles and methods, along with supporting data requirements, for quantifying the contact with such agents at the exchange boundaries of organisms (e.g., lungs, gastrointestinal track, gills, and skin) and the uncertainties associated with exposure assessments. In addition, we address exposure assessment techniques that involve measurements as well as modeling and the interplay between these two approaches for quantifying exposures. No attempt is made herein at an exhaustive presentation on the subject of exposure; rather, this chapter summarizes cogent methodologies and provides references to authoritative sources for those readers desiring further details.

II. ROLE OF EXPOSURE ASSESSMENT IN THE RISK ASSESSMENT AND MANAGEMENT PROCESS

The principal elements of the risk assessment process for toxic substances are depicted in Figure 1. This process is based on the scheme proposed by the National Research Council, U.S. National Academy of Sciences (NRC, 1983), consisting of a four-component process: hazard identification, exposure assessment, dose-response assessment, and risk characterization. We have specified three additional subtasks for the exposure assessment component, one dealing with the analysis of the source terms of genotoxicants, the second involving analysis of the transport and transformation of those substances in environmental media, and the third concerning assessment of the contacts with contaminated media. Hazard identification is essentially a qualitative evaluation of the potential genotoxic risk of a given substance, based on information regarding its genotoxicity in various bioassays and the magnitude of possible exposures (for additional details, see Chapter 1, this volume). A formal risk assessment on a genotoxic substance is begun only after the hazard evaluation step provides sufficient evidence indicating that adverse effects could occur in potentially exposed populations. The exposure assessment component of a risk assessment normally starts with source-term analyses that focus on the mechanism(s) by which a genotoxicant is formed as well as the rates at which it is released into the environment from a specific source or sources. This type of analysis is crucial to the subsequent components of an exposure assessment and the study of approaches for managing the genotoxic risk of the substance by source reduction. Fate and transport analyses address not only the movement or interaction of a genotoxicant within and between environmental compartments, but also potential chemical and biological transformation processes that may alter its bioavailability and/or toxicity. The basic goal of the transport and transformation analyses is to quantify the spatial and temporal variations in the concentrations of a substance in environmental media. Exposure assessment includes the identification of target populations and potential exposure pathways, and importantly, estimation of the rates at which contact is made with the genotoxicant in foods, air, water, etc. The goal of the dose-response assessment, which is the next important component of the overall risk-assessment

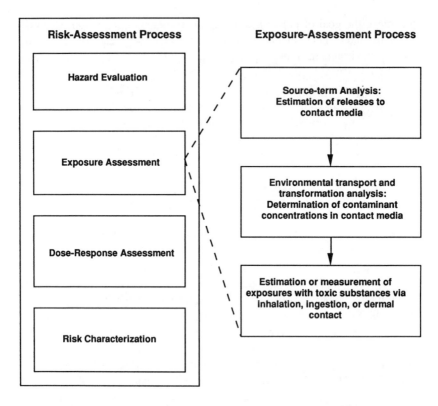

Figure 1. Overview of the risk- and exposure-assessment processes.

process, is to relate the predicted or measured exposures to genotoxic endpoints of concern. The final task is the characterization of the risk posed by a genotoxic agent. This step is used to integrate information on the nature of the source(s) of the genotoxicant, the contamination of contact media, characteristics of the specific population(s) at risk, the magnitude and duration of contacts with the contaminated media, and the dose-response relationship between potential exposure levels and genotoxic endpoints.

Because scientific understanding concerning the activity of genotoxicants at the organ, tissue, or cellular level is incomplete, potential risk-management measures based on intervention at these levels is not presently feasible. Current risk management or risk reduction will have to be achieved by strategies relying upon source and/or exposure reduction. In concept, the risk-assessment process provides not only a mechanism for estimating the potential level of unacceptable health impact, but also a vehicle to "back calculate" from a level of acceptable risk to an associated allowable level of the genotoxic chemical in contact media such as soil, air, water, and foods. Thus, the results of risk reduction through management of the contaminant source and/or the exposure conditions can serve as the most feasible and therefore the most viable option to achieve the goal of public health and environmental protection. Exposure can be limited by (1) reducing the source of the active or precursor agent in the ambient environment through engineering controls or modifications of engineering practices, (2) limiting the contact of humans or target organisms with contaminants in one or more environmental media through modification of the behavior or activity of the exposed populations, and (3) some combination of 1 and 2. Thus, the risk-assessment process not only frames the risk-management problem, but also provides a process for understanding and choosing the potential solutions. Wong et al. (1991) noted that in developing risk-management strategies and making risk-reduction decisions, governmental regulatory institutions must often balance the public demands for absolute protection or "zero risk" against the real limits of public and private resources. The process of choosing the potential solution or risk-management option is intended not only to consider issues uncovered during the risk-assessment phase, but also to integrate considerations such as engineering feasibility, financial resources, sociopolitical needs, and real or potential benefit of the proposed risk-reduction solution. The demanding policy areas of such a process, by virtue of their vulnerability to public and political controversy, encompass (1) the assignment of an appropriate weight to each of these considerations and (2) the integration of these considerations into the final risk-management decision.

III. EXPOSURE AND RISK

Based on a definition of exposure as "contact with a chemical or physical agent" (U.S. EPA, 1985), the exposure assessment components in risk assessment are needed to translate contaminant sources into quantitative estimates of the amount of contaminant that reaches and ultimately crosses biological exchange boundaries, such as the lungs, the GI tract, and the skin. Where appropriate, further toxicokinetic modeling of the biological compartments can provide estimates of genotoxicants reaching cellular target sites. In current practice, exposure assessments rely either on the measurement of a contaminant (or its surrogates) or on concentration estimates from contaminant-transport models. Often implicit in these methods is the assumption that exposure can be linked by simple parameters to ambient concentrations of the substance in air, water, and soil. An exposure assessment can be made more valuable when it provides a comprehensive view of time and activity patterns, microenvironmental data, exposure routes, and major sources of uncertainty (Wallace, 1986; Spengler et al., 1981; McKone and Ryan, 1989). Yet, the common practice in exposure evaluations has often been to use single exposure routes and mean or point estimates for most parameters. A more realistic approach to exposure assessment requires that we address two important issues: (1) the multifactorial nature of exposure (e.g., spatial and temporal exposure patterns, array of potential exposure pathways, presence within the environmental ¡compartment of alternate biological sinks and degradation processes) and (2) the analysis and treatment of uncertainties associated with the development of specific exposure scenarios.

One can represent the overall process of predicting the population risk resulting from a genotoxicant released at a source S (in kilograms per year) as a function of five factors:

$$R = N \times f(S, E[S], D[E]), B(D[E])$$ (1)

where R is the risk expressed as the annual incidence of a genotoxic effect associated with the source S in a population of size N, E is the exposure function, which converts the source into a time-weighted exposure in the population, D is the dose factor, the fraction of contaminant delivered to the organism (or target tissue/cells) after contact, and B is the potency (i.e., genotoxic risk as a function of effective dose rate) associated with the predicted or measured exposure.

A key consideration in an exposure assessment is that it provide a quantitative expression of the contact with a genotoxicant that is consistent with the nature of the dose-response relationship for the substance being considered. If, for example, the critical metric for predicting a given genotoxic endpoint is the peak concentration within a target tissue during a critical time interval of cell sensitivity, then an exposure assessment would focus on those conditions producing such a concentration. Similarly, if the probability of genetic damage is a linear function of the exposure, then an exposure metric such as the average contact rate (e.g., milligrams of chemical ingested per kilogram of body weight per day) over an exposure period might be an appropriate metric. Other factors that must be considered in exposure assessments are nonlinearities as well as thresholds in dose-response relationships. Nonlinear responses between exposures and effects increase the complexity of an exposure assessment because effects may no longer be predicted from a simple linear scaling of exposure. A threshold (or "apparent" threshold) of response means that exposures below the "no-effect contact" rate with a contaminated medium would not induce a genotoxic effect, and thus an exposure assessment would necessarily focus on those situations leading to contacts exceeding the threshold. There is some evidence for the presence of thresholds in dose-response relationships for genotoxic chemicals as assessed by Ehling et al. (1983), but the prevailing expectation is for nonthreshold relationships.

IV. SOURCES OF GENOTOXIC SUBSTANCES IN CONTACT MEDIA

An initial step in an exposure assessment for a genotoxic substance that has already been identified as a potential hazard is the characterization of its sources in different contact media. Knowledge of the source-term characteristics of a genotoxic substance is important in order to develop measurement or modeling efforts to determine the substance's concentrations in contact media. Pertinent information includes the physical and chemical properties of the substance(s) of concern and attributes of the source (e.g., emission rate of gas or particles to indoor and outdoor air). Sources can be categorized in terms of space (e.g., area source vs. point source), time (e.g., transient vs. chronic release), and mode of formation (endogenous or exogenous substance). Source-term analyses are typically designed to determine key physical and chemical factors that produce the observed genotoxicity of effluents from a given process and to determine release rates to contact media.

A. Releases to Indoor and Outdoor Air

People spend most of their time indoors, and hence the releases of genotoxic substances to indoor air can be a controlling factor in determining the total inhalation exposure to such substances. Outdoor sources are important as well because of the potential for the contamination of local and regional atmospheres together with subsequent transfer to biota via deposition processes. As shown in Table 1, indoor sources include such things as volatile consumer products, tap-water contaminants, building materials, combustion products from smoking and heaters, and the infiltration of outdoor contaminants. The Ames/*salmonella* assay has been widely used to investigate possible sources of genotoxic substances in air. Van Houdt et al. (1984), for example, in a study of the mutagenic activity of suspended particles in indoor and outdoor air, reported that tobacco smoke was the most important source of mutagens in the indoor air of the 39 houses they surveyed. Suspended particles in kitchens also exhibited mutagenic activity, and cooking is a likely source.

Once the source or sources of a genotoxic substance have been identified, a follow-on task is to quantify the emission rate of the compound(s) to indoor air. One technique for determining emissions of gases and particles from indoor sources is to solve an indoor-transport model for the emission term (i.e., release rate) that provides the best agreement between measured and predicted concentrations of the contaminant. Traynor et al. (1987) used this approach to estimate emissions of benzo[a]pyrene from wood-burning stoves. McKone and Knezovich (1991) determined the volatilization rate of a volatile organic compound (VOC) from shower water by conducting a mass-balance experiment in which the volatilization was calculated by subtracting the discharge rate of the VOC in drain water (i.e., product of VOC concentration in water and water flow rate) from the discharge rate at the shower head. This experimental approach was used to verify a theoretical model for estimating volatilization rates of VOCs based solely on their physicochemical properties (McKone, 1987).

Table 1. Overview of Some Potential Sources of Genotoxic Substances in Indoor and Outdoor Environments

Air (indoor)	**Groundwater**
Emissions of volatile consumer products	Leaking underground tanks
	Waste-water injection wells
Degassing of building materials	Percolation pits
Combustion of heating fuels	Leach lines from septic tanks
Exhalation of soil gases to indoor air	Leachate from landfills
Volatilization of dissolved gases in tapwater	**Surface Water**
	Treated discharges from cities and factories
Pesticide applications indoors	Untreated discharges from cities and factories
Infiltration of airborne toxics	
Tracking of contaminated soil indoors	Nonpoint runoff from agricultural lands
Cigarette smoking	
Cooking	Nonpoint runoff from urban lands
Air (outdoor)	Disinfection byproducts
Combustion of transportation fuels	Accumulation in aquatic biota
Combustion of fuels in energy production	Accidental discharges to water
	Crops/Plants
Combustion of heating fuels	Pesticide/herbicide applications
Emissions of contaminants from soil and water	Endogenous substances
	Accumulation in terrestrial fauna
Applications of agricultural chemicals	Uptake of soil contaminants
Transient releases of industrial gases	**Foods**
Chronic releases of industrial gases	Pyrolysis products in cooked meats
Soils	
Applications of agricultural chemicals	Food additives
Applications of sewage sludge	Residues from packaging materials
Residues from transportation sources	

Gases and particles in the ambient (outdoor) atmosphere are produced from a variety of stationary sources such as power plants, home stove/furnaces, and industrial facilities. Automotive emissions from gasoline- and diesel-powered vehicles represent another important source of atmospheric contaminants containing mutagenic substances (Lewtas, 1985). Releases from stationary point sources are characterized by both an emission rate (mass/unit time) and other parameters (e.g., exit velocity, stack height, and exhaust gas temperature) that are needed as inputs to atmospheric dispersion models. Emission factors, which express emissions as the ratio of an emission rate and an associated fuel processing rate (e.g., grams of contaminant emitted per day divided by the kilograms of coal burned per day) can be used to determine releases of specific substances. DeMarini et al. (1991) developed emission factors for mutagens (revertants per kilogram of fuel burned) produced during the incineration of the pesticide Dinoseb.

METHODS FOR EXPOSURE TO GENOTOXIC AGENTS

B. Source Terms for Water

Ground and surface waters receive contaminants from many different sources. Domestic wastes constitute one of the largest sources of waste water entering surface streams and groundwater. Approximately 25% of the housing units in the U.S., for example, discharge waste water to septic tanks whose effluents can subsequently percolate to groundwater. Domestic sewage can contain a variety of contaminants associated with consumer products, including solvents and VOCs such as toluene and benzene (see Viraraghavan and Hashem, 1986). Discharges of liquid wastes from domestic or industrial waste-water treatment facilities are considered point sources of pollution because they occur at a specific location (outlet) along a surface-water body. Nonpoint sources of water contamination usually originate from runoff from large urban and agricultural areas and are harder to characterize because of their diffuse nature.

C. Release to Soils

Direct releases to soil occur in the form of pesticide, herbicide, and fertilizer applications, land treatment of domestic and industrial wastes, and chronic releases from vehicles, resulting from the wear of brakes and tires as well as oil leaks. Accidental discharges to the soil surface from storage tanks and miscellaneous spills during the transport of genotoxic substances can also occur. Releases to soil are normally quantified in terms of mass per unit area per unit time. Pesticide applications to agricultural fields, for example, can range from under 1 to over 20 kg/ha (see Wauchope, 1978). Discharges of mutagenic substances to soil in the form of sewage sludge amendments have been monitored using the Ames/*salmonella* assay (Fiedler et al., 1991). One potential artifact of such measurements is the presence of naturally occurring substances in soils that are mutagenic to *salmonella* bacteria. Knize et al. (1987) have shown that extracts of nonagricultural soils can exhibit a "potent mutagenic response".

D. Genotoxic Components of Foods, Medicines, and Consumer Products

Although a great deal of attention has been placed on the genotoxicity of various contaminant emissions into the environment from industrial and municipal sources, genotoxic substances have also been found in foods, medicines, and consumer products. Genotoxic substances in these types of contact media are of concern because of the potential for widespread population exposures. In this regard, it is relevant here to note that heterocyclic amines in cooked meats (pyrolysis products) are among the most potent mutagens that have been measured with the Ames/*salmonella* assay (Felton and Knize, 1990). Other examples of mutagen-bearing products include drugs (Kramers et al., 1991) and pesticides (Cid et al., 1990).

V. CONTAMINANT CONCENTRATIONS IN CONTACT MEDIA: THE ROLE OF TRANSPORT AND TRANSFORMATION PROCESSES

For the purposes of exposure assessment, a genotoxic substance that has been identified in one component of the environment must be characterized in terms of its transport and transformation within that compartment and its transport to other components of the environment. We are particularly interested in those components of the environment with which humans are most likely to have contact. In order to carry out this characterization, we view the environment as a series of interacting compartments. In this framework, one must determine whether a substance will (1) remain or accumulate within the compartment of its origin, (2) be transported by dispersion or advection within the compartment of its origin, (3) be physically, chemically, or biologically transformed within the compartment of its origin (i.e., by hydrolysis, oxidation, etc.), or (4) be transported to another compartment by cross-media transfer (i.e., volatilization, precipitation, etc.). The purpose of this section is to provide an overview of methods that can be used to determine how the competition among these processes results in the temporal and spatial variation of the concentrations of genotoxic contaminants in the environment.

In the sections below, we review methods for characterizing the environmental distribution of genotoxic substances using both measurements and models. We begin by considering transport processes that lead to the dispersion of substances within a single environmental medium. This is followed by an overview of transformation processes. We next consider cross-media transfers.

A. Overview of Transport Processes within a Single Compartment

Most of the transport models that have been developed for describing the behavior of contaminants in the environment have dealt with specific environmental compartments, such as indoor and outdoor air, surface water and sediments, groundwater, and soils. These single-compartment models operate at various levels of spatial and temporal detail, depending on the particular conditions being assessed. Our following discussions highlight some of the more commonly used methods for characterizing contaminant transport in environmental compartments. Additional information on transport modeling for use in exposure assessments can be found in Jones et al. (1991).

1. Indoor and Outdoor Air

Several approaches have been used to estimate expected indoor air pollutant concentrations (for a review, see Wadden and Scheff, 1983). These include deterministic models based on a pollutant mass balance around a particular indoor air volume, a variety of empirical approaches based on statistical evaluation of test data and (usually) a least-squares regression analysis, or a combination of both approaches — empirically fitting the parameters of a mass-balance model with values statistically derived from experimental measurements. All three approaches have advantages and weaknesses. The mass-balance models provide more generality in their application, but the results lack accuracy and precision. The empirical models, when applied within the range of measured conditions for which they were fitted, provide more accurate information. Mass-balance models include single and multiple compartment models. Often the component of the indoor-air mass balance models that is most difficult to represent is the role of indoor surfaces as sources or sinks for contaminants.

Substances in outdoor air are dispersed by atmospheric advection and diffusion. Meteorological parameters have an overwhelming influence on the behavior of contaminants in the lower atmosphere; among them, wind parameters (direction, velocity, and turbulence) and thermal properties (stability) are the most important. The standard models for estimating the time and spatial distribution of point sources of contamination in the atmosphere are the Gaussian plume models. These models are obtained from solution of the classical differential equation for time-dependent diffusion in three dimensions. Pasquill (1961) has discussed the physical basis, analytical solutions, and use of these equations. Turner (1970) has compiled a workbook on applications of these solutions to air pollution problems, including application of the Gaussian plume to area and line sources. Another approach to the dispersion of substances in the atmosphere is based on the application of a mass balance to a volume element, parcel, or box of air. This gives rise to the "box" models. In this approach, the region to be studied is divided into cells or boxes. The concentration in each box is assumed to be uniform and is a function of the box volume, the rate at which material is being imported, emission rates within the box, and the rate at which material is exported from the box. Gifford and Hanna (1973) have shown that in a simple box model, the yearly average concentration of pollutant within the box is proportional to the source strength in mass per unit area divided by the wind speed. Based on data from U.S. cities, they have developed proportionality coefficients for gases and particles that can be used to estimate long-term contaminant concentrations.

2. Surface Waters

The transport of genotoxicants in surface waters is determined by two factors: the rate of physical transport in the water system and chemical reactivity. Chemical reactions in surface waters are discussed below in Section V.B. Physical transport process are dependent to a large extent on the type of water body under consideration — oceans, seas, estuaries, lakes, rivers, or wetlands. Schnoor (1981) and Schnoor and MacAvoy (1981) have summarized important issues relating to surface water transport. At low concentration, contaminants in natural waters exist in both a dissolved and a sorbed phase. In slow-moving surface waters, both advection and dispersion are important. In rapidly moving water systems, advection controls mass transport, and dissolved substances move at essentially the same velocity as the bulk water in the water system. Contaminants that are sorbed to colloidal material and fine suspended solids can also be entrained in current, but they may undergo additional transport processes that increase their effective residence time in surface waters. Such processes include sedimentation and deposition, and scour and resuspension. Thus, determining the transport of contaminants in surface waters requires that we follow both water movement and sediment movement.

A water balance is the first step in assessing surface water transport. A water balance is established by equating gains and losses in a water system with storage. Water can be stored within estuaries, lakes, rivers, and wetlands by a change in elevation or stage. Water gains include inflows (both runoff and stream input) and direct precipitation. Water losses include outflows and evaporation.

3. Groundwater

In groundwater, the dilution of contaminants occurs at a much slower rate than it does in air and surface water. After precipitation water infiltrates the ground surface, it travels vertically down through the unsaturated zone to where it contacts the water table and then flows approximately horizontally. This horizontal movement is driven by the hydraulic gradient, which is the difference in hydraulic head at two points divided by the distance (along the flow path) between the points. Bear and Verruijt (1987) and Freeze and Cherry (1979) have compiled extensive reviews on the theory and modeling of groundwater flow and on the transport of contaminants in groundwater. The movement of contaminants in groundwater is described by two principal mechanisms: (1) gross fluid movement (advective flow) and (2) dispersion. Dispersion depends on both fluid mixing and molecular diffusion. The transport of many chemical species in groundwater is often slowed or "retarded" relative to the flow of the bulk fluid by sorption of the contaminant material to soil particles or rock. As pointed out by Bear and Verruijt (1987), there are multitudes of groundwater models available for assessing the transport of contaminants in the subsurface environment, ranging from simple, one-dimensional hand calculations to large, three-dimensional computer programs. The choice of an appropriate model for any situation will depend to a large extent on the information available, the type of information needed to carry out an exposure assessment, and the tolerance of the analyst for large, complex computer programs.

4. Soil

Soil, the thin outer zone of the earth's crust that supports rooted plants, is the product of climate and living organisms acting on rock. A true soil is a mixture of air, water, mineral, and organic components (Horne, 1978). The relative mix of these components determines to a large extent how a chemical will be transported and/or transformed within the soil. The movement of water and contaminants in soil is typically vertical as compared to horizontal transport in the groundwater (i.e., saturated) zone. A genotoxic compound in soil is partitioned between soil water, soil solids, and soil air. For example, the rate of volatilization of an organic compound from soil solids or from soil water depends on the partitioning of the compound into the soil air and on the porosity and permeability of the soil.

Soils are characteristically heterogeneous. A trench dug into the soil zone typically reveals several horizontal layers having different colors and textures. These multiple layers are often divided into three major horizons: (1) the A horizon, which encompasses the root zone and contains a high concentration of organic matter, (2) the B horizon, which is unsaturated, is below the roots of most plants and contains a much lower organic carbon content, and (3) the C horizon, which is the unsaturated zone of weathered parent rock consisting of bedrock, alluvial material, glacial material, and/or soil of an earlier geological period (Bowen, 1979). Contaminants in the A horizon (or upper soil) are transported horizontally by runoff, transported upward by diffusion, volatilization, root uptake, and the capillary motion of water, transported downward by diffusion and leaching, and transformed chemically by processes such as photolysis and biodegradation. Contaminants in the B horizon (or middle soil) are transported upward by diffusion and the capillary motion of water, transported downward by diffusion and leaching, and transformed chemically by biodegradation, but to a lesser degree than transformation in the upper soil. Contaminants in the C horizon (or lower soil) are transported upward by diffusion and downward by leaching into groundwater.

Models developed for assessing the behavior of contaminants in soils can be categorized in terms of the transport/transformation processes being modeled. Partition models such as the fugacity models of Mackay (1979) and Mackay and Paterson (1981, 1982) describe the distribution of a contaminant among the liquid, solid, and water phases of soils. Jury et al. (1983) have developed an analytical screening model that can be used to calculate the extent to which contaminants buried in soil evaporate in the atmosphere. The multimedia model GEOTOX (McKone and Layton, 1986) has been used to determine the inventory of chemical elements and organic compounds in soil layers following various contamination events. This model addresses volatilization to atmosphere, runoff to surface water, and leaching to groundwater and first-order chemical transformation processes.

B. Transformation Processes

The transformation of genotoxic substances in the environment can have a profound effect on their potential for accumulation by biota. Chemical transformations, which may occur as a result of biotic or abiotic processes, can significantly reduce the concentration of a substance or alter its structure in such a way as to enhance or diminish its genotoxicity. For example, nitrogenous compounds, which are largely represented by aliphatic and aromatic amines, are of particular interest due to their potential genotoxic activities. Transformation processes (e.g., phototransformation, oxidation, and reduction reactions) can lead to the interconversion of these compounds between their related condensation products (e.g., azo compounds) and oxidation products (e.g., primary amines). Such transformations may prolong the persistence of these compounds in the environment and determine their genotoxic potencies.

For organic chemicals, knowledge of a compound's half-life for any given transformation process provides a very useful index of persistence in environmental media. Because these processes determine the persistence and form of a chemical in the environment, they also determine the amount and type of substance that is available for the exposure of species of interest. Experimental methods (Howard et al., 1978) and estimation methods (Lyman et al., 1982) are available for defining these fate processes in a variety of media and should be applied to the determination of the environmental fate of genotoxicants of concern. Specific information on the rates and pathways of transformation for individual chemicals of concern must be obtained directly from experimental determinations or derived indirectly from information on chemicals that are structurally similar. Consequently, quantitative estimates are difficult to derive for classes of compounds for which empirical data are lacking.

Good qualitative estimates of transformation processes can be made based on a review of chemical's transport and fate in the environment. The extent to which an environmental mutagen is transformed will largely depend on the media into which it is distributed. For example, volatile chemicals that are released to the atmosphere are likely to undergo photolytic and hydrolytic transformations. Conversely, chemicals present in soil and water are less likely to undergo photochemical transformations due to light attenuation. Therefore, in the following discussion, the major processes that result in chemical transformations are discussed in relation to the environmental media in which they are most likely to occur. Exposure assessments must incorporate this information to determine the extent to which a chemical's persistence in given media influences exposure to a given population.

1. Photolysis

Most organic contaminants are capable of undergoing photolytic decomposition. Such decompositions can be partial, resulting in the formation of stable

METHODS FOR EXPOSURE TO GENOTOXIC AGENTS

byproducts, or complete, resulting in the destruction of the compound. The potential for such photochemical transformations can generally be predicted based on the molecule's ability to absorb radiant energy in the near-ultraviolet and visible light range (240 to 700 nm). Although solar radiation at the earth's surface is attenuated by the atmosphere, it is generally sufficient (i.e., >290 nm) to break bonds in many compounds. All compounds that contain aromatic rings absorb energy at environmental wavelengths, as do compounds that contain halogen atoms (e.g., Cl and Br) and unsaturated carbon chains (e.g., alkenes and alkynes). Phototransformation of a chemical may result in two effects that are relevant to its environmental fate: (1) fragmentation, in which a molecular bond is broken, forming two free radicals, or (2) rearrangement, such as conversion from *cis* to *trans* isomers. Of these two possible reactions, fragmentation is likely to play the greatest role in the fate of genotoxic contaminants due to its potential for the degradation of contaminants present in the environment. Such transformations may result in relatively short half-lives (e.g., hours to days) for contaminants such as pesticides that are applied directly to crops or surface soils. Photolytic breakdown of pesticides can substantially reduce the concentrations of chemicals that are applied to crops and thereby diminish potential exposures to farm workers and consumers.

2. Hydrolysis

Hydrolytic transformation of organic chemicals can be a significant fate process of genotoxic compounds that are present in aqueous environments. Hydrolysis is most important for chemicals that have functional groups (e.g., amides, esters, carbamates, and organophosphates) that can be rapidly altered (e.g., minutes to days) in the presence of water. For amides and carbamates, hydrolytic cleavage yields aromatic and aliphatic amines with an increased likelihood of mutagenic activity. Conversely, hydrolytic degradation of compounds that contain stable substituents (e.g., halogenated compounds such as carbon tetrachloride) can have half-lives of several thousand years. Because hydrolytic reactions are driven by the availability of hydrogen and hydroxide ions, the pH of the environment can have a dramatic influence on the rate of hydrolysis for any given compound. Hydrolytic transformations that are relatively slow at neutral pH can occur at rates that are several orders of magnitude greater under acidic or basic conditions (Tinsley, 1979). Therefore, the relative importance of hydrolysis to the environmental fate of a mutagenic compound will depend on the chemical structure of the compound as well as the pH of the environmental medium.

3. Oxidation and Reduction

Many inorganic and organic mutagens can undergo oxidation or reduction reactions in the environment. These reactions are important because they can influence the environmental fate and toxicological properties of the compound and are most significant in aqueous environments (see Stumm and Morgan, 1981). An index of a compound's ability to be oxidized or reduced is provided by a knowledge of its reduction potential (E^o), which is the voltage at which it is transformed to its reduced state. A similar measure of the environment's ability to reduce a compound is provided by the redox potential (pE), which is a measure of electron activity. Redox potentials are relatively high and positive in oxidized environments (e.g., surface waters), and low and negative in reduced environments (e.g., aquatic sediments and the terrestrial subsurface). These environmental conditions are especially important for inorganic mutagens that are rarely present in their elemental form in the environment. Arsenic, for example, exists primarily in its oxidized form (arsenate) in the atmosphere and surface waters and in its reduced form (arsenite) in sediments.

4. Microbial Transformation

The transformation of organic and inorganic compounds by microorganisms that are present in environmental media can have a profound influence on their persistence and toxicity. Due to their broad range of enzymatic capabilities, microorganisms are capable of transforming many inorganic and organic compounds. Such transformations can result in the partial degradation of a compound (e.g., conversion of DDT to DDE), mineralization (i.e., complete transformation to carbon dioxide and water), or synthesis of a stable product (e.g., formation of methyl arsenicals from arsenate). While these processes generally result in the detoxification of the parent compound, genotoxic products may also be formed. For example, the microbial metabolism of aromatic amines can result in the formation of mutagenic byproducts (Lyons et al., 1985; Yoneyama and Matsumura, 1984).

The susceptibility of many organic compounds to microbial transformation can be predicted based on a knowledge of chemical structure (Boethling and Sabljic, 1989). Predicting the genotoxicity of the transformation products, however, generally requires empirical information.

C. Cross-Media Transfers and Multimedia Models

Efforts to assess human exposure from multiple media date back to the 1950s, when the need to assess human exposure to global fallout led rapidly to a framework that included transport both through and among air, soil, surface water, vegetation, and food chains (Whicker and Kirchner, 1987). In contrast to the single-medium paradigm for assessing exposure, in a multimedia approach, we locate all points of release to the environment, characterize mass-balance relationships (e.g., between sources and sinks in the environment), trace contaminants through the entire environmental system, observing and recording changes in form as they occur, and identify where in this chain of events control efforts would be most appropriate. Efforts to apply such a framework to nonradioactive organic and inorganic toxic chemicals have been more recent and have not as yet achieved the level of sophistication extant in the radioecology field.

The first widely used multimedia compartment models for organic chemicals were the "fugacity" models proposed by Mackay (1979) and Mackay and Paterson (1981, 1982). Fugacity models have been used extensively for modeling the transport and transformation of nonionic organic chemicals in complex environmental systems (see Mackay, 1991). Fugacity is a way of representing chemical activity at low concentrations. Fugacity has units of pressure (pascal [Pa]) and can be regarded physically as the partial pressure exerted by one environmental phase or compartment on another. When two phases have the same fugacity, neither exerts more pressure to escape than the other, meaning they are in equilibrium. For phases in equilibrium, the escaping tendency (fugacity) of a chemical is the same in all phases. This characteristic of fugacity-based modeling often simplifies the mathematics involved. Fugacity models can also be used to represent a dynamic system in which the fugacities in two adjacent compartments are changing in time due to an imbalance of sources and losses or a dynamic system that has achieved steady state by balancing gains and losses but fugacities are not equal. In a multimedia model using such fugacity relations, major components of the environment are lumped into homogeneous subsystems or compartments that can exchange mass with other adjacent compartments. Quantities or concentrations within compartments are described by a set of linear, coupled, first-order differential equations. A compartment is described by its total mass, total volume, solid-phase mass, liquid-phase mass, and gas-phase mass. Mass flows among compartments include solid-phase flows, such as dust suspension or deposition, and liquid-phase flows, such as surface runoff and groundwater recharge. The transport of individual chemical species among compartments occurs by diffusion and advection at the compartment boundaries. Each chemical species is assumed to be in chemical equilibrium among the phases within a single compartment. However, there is no requirement for equilibrium between adjacent compartments. Decay and transformation processes (such as radioactive decay, photolysis, biodegradation, etc.) are treated as first-order, irreversible removals.

Cohen and co-workers introduced the concept of the multimedia compartment model as a screening tool with the MCM model (Cohen and Ryan, 1985) and more recently with the spatial multimedia compartment model (SMCM) (Cohen et al., 1990), which allows for nonuniformity in some compartments. Another multimedia screening model, called GEOTOX (McKone and Layton, 1986; McKone, 1987), was one of the earliest multimedia models to explicitly address human exposure. The compartment structure use for the GEOTOX analysis described here is illustrated in Figure 2.

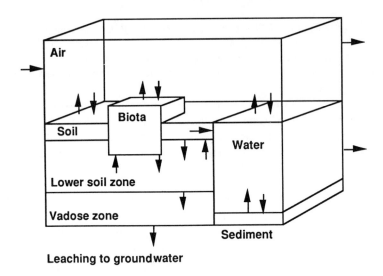

Figure 2. An illustration of mass exchange processes modeled in a seven-compartment environmental transport and transformation model such as GEOTOX.

D. Environmental Landscape Properties and Physicochemical Properties

Multimedia models require two types of input data, one providing properties of the environment or landscape receiving the contaminants and the other describing the properties of the contaminants. Because it is often impractical to develop detailed parameter sets for the landscapes surrounding a large number of facilities, it is more practical to develop landscape data sets that are representative of various regions of the world. The data needed to construct a landscape data set include meteorological data such as average annual wind speed, deposition velocities, air temperature, and depth of the mixing layer, hydrological data such as annual rainfall, runoff, water infiltration to soil, groundwater recharge, and surface-water depth and sediment loads, and soil properties such as bulk density, porosity, water content, erosion rates, and root zone depth.

Multimedia models also require that we measure or estimate solid/liquid phase partition coefficients between the soil and soil water, between groundwater and aquifer material, and between surface water and sediments, air/liquid partition coefficients, and the diffusion coefficients of substances in air and water. Lyman et al. (1982) have compiled methods for estimating basic chemical properties and parameters relevant to environmental transport between soil and air. Mackay (1979) has illustrated an approach for calculating mass transfer at the water/air and air/soil interfaces. The use of multimedia transport methods requires as a minimum that we determine the following set of chemical properties:

- Sorption partition factors, K_d, in the surface soil, vadose zone, saturated zone, and surface-water sediments
- Vapor pressure of the contaminant compound
- Solubility of the contaminant in water
- Diffusivities of the contaminant in pure air and water
- Transformation-rate constants for radioactive decay and chemical or biological degradation processes

For organic compounds, the sorption coefficient, K_d, is estimated as the product of the organic carbon partition coefficient and the fraction organic carbon in each compartment. Organic carbon partition coefficients are estimated as a function of the octanol-water partition coefficient (Karickhoff, 1985). The effective diffusivity of a substance in soil is estimated from the diffusivities in pure air and water and from soil properties using methods proposed and validated by Jury et al. (1983). Mass transfer at soil/air, air/water, and water/sediment interfaces are estimated using the two-resistance boundary layer model described in Lyman et al. (1982).

VI. EXPOSURE PATHWAYS FOR BIOTA

The factors that determine the accumulation of genotoxic substances in plants and animals must be understood so that species at risk can be identified and food-chain transfers to humans can be established. Although much information exists on the accumulation of environmental contaminants by plant and animal species, relatively little information is available concerning compounds that possess mutagenic activity. Therefore, in the following discussion, the general principles that determine chemical accumulation are defined and highlighted with examples of mutagenic compounds.

A. Transfers to Crops/Plants

Plants generally have contact with two environmental media — air and soil. Plant interactions with these media are not understood well enough to define an accurate method of predicting chemical uptake. The translocation of mutagenic chemicals from soils appears to be a relatively minor pathway for the accumulation of these compounds in plants (Fiedler e al., 1991). In the absence of experimental studies, there has been a reliance on simple bioconcentration factors (BCFs) that relate either a soil or air concentration to a plant concentration. The earliest use of vegetation BCFs was for assessing the effects of global fallout by relating concentrations of radionuclides in plants to concentrations in soil (Ng, 1982). More recently, vegetation BCFs have been proposed for organic chemicals (Travis and Arms, 1988). What is meant by the "concentration in a plant" is difficult to understand in larger plants, where roots, stems, and leaves may each have different chemical concentrations. It is now common to see BCFs relating chemical concentrations in leaves to concentrations in air, concentrations in roots to concentrations in soil, and concentrations in stems to concentrations in soil. Recent studies on bioconcentration of nonionic organic chemicals has focused on correlations between BCFs and known chemical properties such as Henry's law constants (H) and octanol-water partition coefficients (K_{ow}). This work has often shown that the logarithm of BCF has a positive correlation to the logarithm of K_{ow} and a negative correlation to the logarithm of H (Bacci et al., 1990; Briggs et al., 1982, 1983; Travis and Hattemer-Frey, 1988). Travis and Arms (1988), however, have found a negative correlation between the logarithm of soil-plant BCFs and the logarithm of K_{ow}, indicating that such statistical correlations may be unreliable. Ryan et al. (1988) and Riederer (1990) report the unreliability of simple plant BCFs. McFarlane et al. (1987) have demonstrated experimentally that chemicals with similar values of K_{ow} can have drastically different long-term partitioning, again point out a lack of understanding of the mechanisms determining plant uptake and partitioning.

The failure of simple BCF correlation models to predict reliably plant uptake has resulted in the publication of several more-detailed models of chemical uptake from soil and/or air by plants. For radioactive fallout studies, Whicker and Kirchner (1987) developed a model called PATHWAY, which is a dynamic model that includes air/plant and soil/plant uptake and includes treatment of plant growth and senescence. However, the use of only two plant compartments — plant tissues and plant surfaces — limits its ability to model translocation within the plant. Also, this model has not been used for organic chemicals. Calamari et al. (1987) proposed two fugacity-based equilibrium models of plant uptake. Riederer (1990) has proposed a fugacity-based equilibrium model that treats only the leaf-air interface, but includes much detail on the composition of leaves. Based on azalea leaf experiments with five chemicals, Bacci et al. (1990) have developed a correlation of leaf-air bioconcentration factors with air-water and octanol-water partition coefficients. Trapp et al. (1990) have used barley plants in growth chambers to develop a four-compartment — air, soil, roots, and shoots — steady-state fugacity model and found that, in most cases, chemicals in the roots of the barley plants reach equilibrium with the soil, whereas chemicals in the shoots are in equilibrium with the air. Paterson and Mackay (1989) have proposed a seven-compartment fugacity model for plants that, in addition to the leaves, stem, and roots included in the models described above, includes compartments for xylem, phloem, and cuticle. This model was validated against experimental results obtained by Bacci and Gaggi (1986), who found that foliage concentrations were not directly dependent on chemical concentration in soil, but rather on concentration in air.

B. Transfers to Aquatic and Terrestrial Species

The transfer of a contaminant in water, air, soil/sediment, and food to an aquatic or terrestrial organism is governed by three basic processes: uptake from these contact media, distribution to target tissue or products (e.g., eggs, milk, etc.), and elimination via excretion and/or metabolic transformation. Normally, biotransfers of contaminants are measured or estimated for steady-state conditions in which intakes are equal to losses. Under such conditions, a BCF is often determined as the ratio of the concentration of a substance in the target tissue to the concentration in a given environmental medium. For fish, the BCF is taken as the concentration of a xenobiotic substance in fish flesh (or lipids) to the contaminant's concentration in water (see Spacie and Hamelink, 1985). The BCF for neutral organic compounds can be estimated from regression equations based on selected physicochemical properties, particularly a compound's octanol-water partition coefficient or aqueous solubility. Regression equations for BCFs appear in Van Gestel et al. (1985) and Isnard and Lambert (1988). There are cases, though, in which the measured levels of organic or inorganic substances in fish are considerably higher than the values predicted from BCF equations. This occurs as a result of bioaccumulation, which refers to the accumulation of a substance in fish from both water (i.e., bioconcentration) and the consumption of contaminated foods. Thus, a substance that bioaccumulates will usually occur at its highest concentrations in fish occupying the top trophic levels in a food chain. Generally, only those substances that are lipophilic and persistent will bioaccumulate. Models for simulating these bioaccumulation processes have been presented by Thomann (1989) and Barber et al. (1991).

Much of the information available on the biotransfer of chemicals to terrestrial species deals with agricultural animals. Steady-state concentrations of a given organic or inorganic substance in meat, milk, or eggs are measured against the corresponding concentration in an animal's diet. The convention, though, for describing the biotransfer of contaminants to food-chain animals is to develop biotransfer factors (BTFs), which are ratios of the concentration in a food product to intake (the associated units are days per kilogram for a meat product and days per liter for milk) (Stevens, 1991; Ward and Johnson, 1986). These factors are multiplied by the daily dietary intake (feed and water) of a contaminant (i.e., mass per day) to obtain the steady-state concentration in target tissue or product. Travis and Arms (1988) have developed regression equations for predicting BTFs for organic contaminants accumulating in beef meat and cow milk. The equations are based on the octanol-water partition coefficient of the substance.

VII. EXPOSURE PATHWAYS FOR HUMANS

Human exposures to genotoxic substances can result from contacts with contaminated soils, water, air, and food as well as with drugs and consumer products. Exposures may be dominated by contacts with a single medium or may reflect concurrent contacts with multiple media. The nature and extent of such exposures depends largely on two things: (1) human factors and (2) the concentrations of a genotoxic substance in contact media. Human factors include all behavioral, sociological, and physiological characteristics of an individual that directly or indirectly affect his or her contact with the substances of concern. Important factors in this regard are contact rates with food, air, water, soils, drugs, etc. Activity patterns, which are defined by an individual's allocation of time spent at different activities and locations, are also significant because they directly affect the magnitude of inhalation exposure to substances present in different indoor and outdoor environments. From an exposure-assessment standpoint, the principle challenge is, therefore, to estimate or measure a person's exposure as a function of relevant human factors and measured and/or estimated concentrations in contact media. Table 2 gives a matrix showing the many interrelationships (or pathways) that can exist between contaminated media and the three possible routes of exposure. These complexities indicate that an integrated approach for assessing exposures to genotoxic substances is of paramount concern, as there are often multiple pathways and routes of exposure that need to be characterized.

The principal output of an exposure assessment is a quantitative measure or estimate of contact with a substance over a specified period of time. The exposure metric normally used is a rate, expressed as mass of chemical per day or per kilogram of body weight per day. In addition, the exposures should be estimated for specific routes (i.e., inhalation, ingestion, and dermal uptake) because metabolism (and hence metabolic activation of a substance) within the body is affected by route of contact. Route-specific uptake, distribution, and metabolism is accounted for in physiologically based pharmacokinetic models (Bogen and Hall, 1989) that could be used in risk assessments of genotoxic agents (see Rhomberg et al., 1990). Below, we discuss the basic physiologic inputs required for quantitative estimates of exposure by each route.

A. Ingestion

The intakes of food and beverages often constitute the primary input parameters for characterizing exposures that occur via ingestion. Hence, dietary information is needed for the population(s) that are or could be exposed to the substance(s) addressed in an exposure assessment. In the U.S., a stratified random sample of the population is conducted very 10 years to ascertain average dietary intakes for a 3-d period. Figure 3 depicts the daily average intakes for major food groups and beverages for the survey that was completed in 1977–1978 (from Yang and Nelson, 1986). Such data can be used to define intakes of all the food items in a given group or selected food items. Moreover, results from this type of survey can be used to identify representative foods for chemical monitoring and to prepare specialized statistical studies involving a specific food group (e.g., cooked meats) or type of beverage. The difficulties of completing large dietary surveys will represent a problem for many developing countries; nevertheless, in societies with

METHODS FOR EXPOSURE TO GENOTOXIC AGENTS

Table 2. Matrix of Environmental Media and Exposure Pathways

Pathways	Air	Soil	Water
Inhalation	Inhalation of ambient outdoor air Inhalation of indoor air	Inhalation of soil vapors that migrate to indoor air Inhalation of soil particles transferred to indoor air	Indoor inhalation of contaminants transferred from tap water
Ingestion	Ingestion of fruits, vegetables, and grains contaminated by atmospheric particle deposition Ingestion of fruits, vegetables, and grains contaminated by foliar uptake from the gas phase of the atmosphere Ingestion of meat, milk, and eggs contaminated through inhalation by animals	Human soil ingestion Ingestion of meat, milk, and eggs contaminated through soil ingestion by animals Ingestion of fruits, vegetables, and grains by transfer from soil Ingestion of meat, milk, and eggs contaminated by transfer from soil to plants to animals	Ingestion of tap water Ingestion of irrigated fruits, vegetables, and grains Ingestion of meat, milk, and eggs, from animals consuming contaminated water Ingestion of fish and sea food
Dermal contact		Dermal contact with soil	Dermal contact in baths and showers

uniform diets, small surveys may prove to be quite reliable. Ingestion of soil represents another possible exposure pathway to environmental contaminants (LaGoy, 1987). Issues related to the estimation of soil intakes are addressed by Stanek and Calabrese (1991) and Calabrese and Stanek (1991).

B. Inhalation

Inhalation exposures are often difficult to quantify because of the spatial and temporal variations in the concentrations of air contaminants. In fact, because the concentrations of many substances vary considerably between indoor and outdoor air, it is often crucial to determine the amounts of time that individuals spend in specific indoor and outdoor environments. The complexities of modeling accurately the concentrations of an airborne contaminant at multiple locations

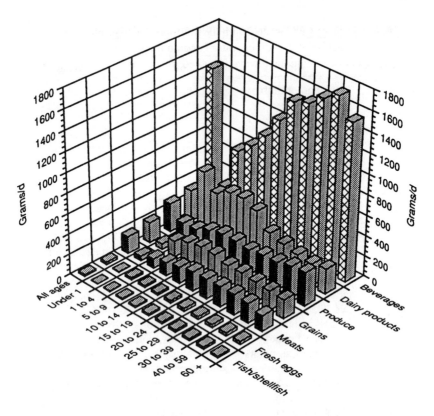

Figure 3. Daily intake of foods and beverages from a sample of the U.S. population over the years 1977–1978. (Data from Yang, Y. Y. and C. B. Nelson (1988), *Health Phys.*, 50, 245–257.)

has led some researchers to develop personal monitors that either actively or passively monitor the air an individual comes in contact with over the course of a day. Estimates of short-term inhalation exposures to contaminated particles and gases require as input the breathing rates associated with different physical activities. Much of the data on breathing rates have been summarized by the U.S. EPA (1989a). For exposures to airborne contaminants lasting weeks and months, inhalation rates should be directly related to the oxygen required to metabolize the fat, carbohydrate, and protein in diets. Recently, Lyon (1993) has developed a methodology for estimating age- and sex-dependent inhalation rates that are metabolically consistent with food-energy intakes.

C. Dermal Uptake

Quantitative estimates of dermal uptake exposure are frequently required for exposure assessments that address contaminants in dusts or soils and bath, shower, and swimming water. Often, these estimates include a rather large uncertainty because we must deal with the transport of chemicals within the skin layer, the interaction of the soil or water layer on the skin with the skin surface, and the dynamic conditions always involved in scenarios addressing soil and water contact with the skin. Dermal exposure to environmental contaminants can occur during a variety of activities and can be associated with several environmental media, e.g., contact with contaminated water during bathing, washing, or swimming, contact with contaminated soil during work, gardening, and recreation outdoors, and contact with sediment during wading and fishing. For contact with contaminated water containing volatile organic compounds, Brown et al. (1984) have demonstrated that dermal uptake could be comparable to ingestion exposure. Brown et al. estimated skin permeability under the assumptions that (1) dermal uptake of contaminants occurs mainly by passive diffusion through the stratum corneum, (2) resistance to diffusive flux through layers other than the stratum corneum is negligible, and (3) steady-state diffusive flux is proportional to the concentration difference between water on the skin surface and internal body water. They have determined that K_p is on the order of 1 cm/h for volatile organic compounds. Contact scenarios, exposure time, exposure frequency, and average time for bathing, showering, and swimming scenarios are discussed in U.S. EPA documents (1989a, b).

For contact with contaminated soil, McKone (1990) has proposed a general model for estimating the uptake of an organic compound from an air-soil-skin system. This model includes two important concepts: (1) tortuous diffusion pathways through soil and skin and (2) boundary layer effects at the soil-air interface. Central to the approach is a fugacity model that uses the physical and chemical properties of the compound, skin, and soil to estimate transport across the combined skin and soil layer, taking evaporation into account. Other inputs for this model include the daily mass loading of dirt on skin, skin surface area, and exposure duration. In its risk assessment for TCDD, the U.S. EPA (Schaum, 1984) used the work of Roels et al. (1980) to determine that some 0.5 mg of soil per square centimeter of skin surface could be deposited on the hands of children. Sedman (1989) reviewed the literature on measured soil contact to develop an estimate of lifetime-average dermal soil loading of roughly 0.45 g/d. Sedman (1989) also observed that a typical soil loading of 0.5 mg/cm^2 exists on exposed skin surfaces and that most skin surfaces would not contain this level of soil loading. Values for body surface area by age and sex have been compiled by the ICRP (1975) and the U.S. EPA (1989b). Contact scenarios, exposure time, exposure frequency, and average time for soil contact scenarios are discussed in U.S. EPA documents (U.S. EPA, 1989a, b).

VIII. EXPOSURE ASSESSMENT METHODS

A basic challenge of any exposure assessment is to determine in a cost-effective manner the nature and magnitude of human contacts with the substances of concern. This challenge confronts risk assessors/managers in both developing and industrialized countries. The complexities embodied in the various exposure pathways shown in Table 2 suggest that approaches that systematically link the concentrations of a genotoxic substance in contact media with human exposures offer the greatest opportunity for preparing assessments that are as realistic as possible. No generic methodologies are available that can address all substances, sources of contamination, and modes of exposure; however, approaches have been developed that do establish useful frameworks for evaluating exposures.

One of the earliest approaches for systematically assessing multipathway exposures to environmental contaminants is termed the exposure commitment method, developed by Bennett (1981). The basic objective of this approach is to calculate the steady-state concentration of a contaminant in the human body resulting from exposures to a contaminant present in a given environmental medium. Exposure commitments (i.e., contaminant concentration in human tissue) are calculated from transfer factors that are estimated as the ratios of the steady-state concentrations of a contaminant in adjoining compartments of an exposure pathway. An exposure commitment is determined by multiplying the transfer factors associated with the adjoining compartments of a given pathway of exposure, e.g., air → plants → livestock → diet. This method has been applied to organic chemicals (Bennett, 1983; Jones and Bennett, 1989) and metals (Bennett, 1982). The published applications of the exposure commitment methodology depend on measured concentrations of the substances in different compartments to estimate transfer factors. The retrospective nature of this approach limits its usefulness for predicting exposures to chemicals for which there are little or no monitoring data available.

McKone and Daniels (1991) have recently developed an approach that explicitly allows for the estimation of exposures to a substance based on either measured or predicted concentrations of the substance in contact media. Pathway exposure factors (PEFs) are used to link chemical concentrations in multiple environmental media to human exposure. Specifically, each PEF numerically translates a chemical concentration in each of the primary environmental media (air, water, and soil) into exposure rates (in milligrams per kilograms per day) for specific routes of exposure. Incorporated into each PEF is information on human physiology and lifestyle, as well as data describing pollutant behavior in food chains or in microenvironments, such as indoor air.

When environmental concentrations are constant in time, the average population exposure to contaminant concentration C_k (measured or modeled) in environmental medium "k" (i.e., air, water, or soil) is given by the product of the concentration and the pathway exposure factor, PEF (k→i), which is calculated as

$$\text{PEF}(k \rightarrow i) = \left[\frac{\text{CR}_i}{\text{BW}}\right] \times \frac{C_i}{C_k} \times \frac{\text{EF} \times \text{ED}}{\text{AT}} \qquad (2)$$

where $[CR_i/BW]$ is the contact rate per kilogram of body weight, such as kg in contact medium i (i.e., air, tap water, milk, soil, etc.) to the concentration (soil)/kg/d, l (milk)/kg/d, or m^3 (air)/kg/d, C_i/C_k is the ratio of the concentration in environmental medium k (air, water, soil) (units depend on the two media), EF is the exposure frequency in days per year, ED is the exposure duration in years, AT is the averaging time in days, and PEF (k→i) is the pathway exposure factor relating the concentration C_k in medium k to the chronic daily intake (in milligrams per kilograms per day) during the period ED.

McKone and Daniels have developed equations for determining PEFs associated with each of the pathways presented in Table 2. Because the exposure commitment approach of Bennett (1981) and the pathway exposure approach are based on similar principles, we compared the two approaches using PCBs and dioxin, which have already been assessed using the exposure commitment method (i.e., PCBs, Bennett, 1983, dioxin, Jones and Bennett, 1989). The results of that comparison, shown in Table 3, indicate that both approaches yield comparable results. This is encouraging because it demonstrates the congruence between an approach based on a retrospective analysis and one based on a prospective or predictive technique.

IX. QUANTIFYING AND REDUCING UNCERTAINTIES

At best, mathematical models only approximate real systems, and therefore their predictions are inherently uncertain. In evaluating the reliability of exposure or risk-assessment models, two questions must be asked: (1) how large is the uncertainty in the model predictions?, and (2) how much confidence can be placed in the results? To address these questions, exposure and risk should be presented as probability distributions so that uncertainty in risk and exposure can be characterized by expectation (mean) and spread (variance). When the magnitude of uncertainty in an exposure and risk assessment confounds the process of environmental management, decision makers must identify strategies for reducing uncertainty.

The need to address human exposure in a multimedia framework brings with it a need to characterize the uncertainty in human exposure models and the combined uncertainty in exposure and dose/response models. In characterizing uncertainty in exposure models, three key issues should be considered: (1) uncertainty in predicting the relationship between sources of contaminants and concentrations in the accessible environment, (2) uncertainty in quantifying exposure or transfer factors that relate environmental concentrations to levels of exposure, and (3) the important contributions to the combined uncertainty in environmental dispersion and pathway exposure factors.

Table 3. Comparison of Transfer Factors and Pathway Exposure Factors for TCDD and PCB

Chemical and Exposure Pathway	Exposure-to-Concentration Ratio	
TCDD	Transfer factors derived from measured data by Jones and Bennet (1989)	Pathway exposure factors calculated from chemical properties using methods from McKone and Daniels (1991)
Air/inhalation	0.28	0.31
Air/plant/diet (m^3/kg/d)	31	36
Air/milk/diet (m^3/l/d)	40	42
Air/meat/diet (m^3/l/d)	21	17
Air/food/diet (total) (m^3/l/d)	93	95
Water/fish/diet (l/kg/d)	17	12
Water/ingestion (l/kg/d)	0.017	0.029
Water/dermal uptake	n/a	0.037
PCB (60% Cl)	Transfer factors derived from measured data by Bennet (1992)	Pathway exposure factors calculated from chemical properties using methods from McKone and Daniels (1991)
Air/inhalation	0.31	0.39
Air/plant/diet (m^3/kg/d)	508	252
Air/milk/diet (m^3/l/d)	n/a	18,000
Air/meat/diet (m^3/l/d)	166	1,000
Water/fish/diet (l/kg/d)	350	320
Water/ingestion (l/kg/d)	0.017	0.029
Water/dermal uptake	n/a	0.037

Note: For TCDD, $K_{ow} = 4 \times 10^6$, BCF (fish) = 39,000; for PCB, $K_{ow} = 2.8 \times 10^6$, BCF (fish) = 1×10^6.

X. RECOMMENDATIONS

Measurement and modeling techniques are available to assess quantitatively exposures to genotoxic substances. The best analytical approach, however, will always depend on chemical- and site-specific considerations. In order to tailor an exposure assessment so that it fulfills its role in an integrated risk assessment of a genotoxic substance, the outputs must be consistent with the input requirements of the dose-response assessment. Risk assessors in developing countries can develop reasonable and flexible exposure-assessment strategies by using a judicious mix of modeling and monitoring approaches. In the following subsections, we outline some of the choices available for dealing with each of the components of the exposure assessment process.

Source terms — Assuming that the hazard identification component has demonstrated that a given genotoxic substance poses a potential health risk, literature reviews can be conducted to determine emission factors, release rates, etc. for appropriate sources. Alternatively, source-term models available in the literature can be used to estimate releases or, if necessary, can be developed from available data and/or physicochemical properties. If these approaches cannot be pursued, monitoring/measurement programs must be developed to acquire the data necessary to quantify the release rates needed as input to environmental transport models if they are to be used to estimate concentrations in contact media.

Concentrations in contact media — We strongly recommend that analyses be performed on the potential distributions of a substance in the environment using one of the various compartmental models described earlier. The input requirements of these models are modest and they are all available for use on personal computers, either as spreadsheet implementations or stand-alone codes. A key benefit derived from using such models is the early designation of the media that are potentially the most important contaminant reservoirs leading to exposure. That information can then be used in several ways, including (1) the design of monitoring approaches and potential control strategies that focus on the most important media, (2) the implementation of more precise environmental transport models that are specific to the primary media, and (3) the development of studies to address data gaps involving the behavior of a contaminant in environmental media. It is very important to examine the tradeoffs between the use of measurement/monitoring programs or modeling to determine concentrations of genotoxic substances in the environment. Of particular concern is the availability of trained personnel to run models or to operate sampling and analysis programs. Cost considerations must also be addressed.

Exposure assessment — Human exposures to genotoxic substances can occur via multiple environmental media and routes, and hence exposure assessment methodologies must be designed so that they (1) focus on the most important exposure modes and environmental media and (2) deemphasize or screen out unimportant pathways of exposure. Existing methodologies based on the exposure commitment approach or the pathway exposure factor methodology can be used in conjunction with information on the sources of a substance and its distribution in the environment. Nevertheless, although many exposure factors (e.g., activity patterns, food ingestion, housing properties, etc.) needed in such assessment methods have been developed for industrialized countries, similar factors are generally unavailable for less developed regions of the world and need to be quantified. When this is the case, we recommend that sensitivity analyses be performed to identify those parameters that are most important to quantify. Finally, exposure assessments must inevitably deal with various sources of uncertainty, ranging from ignorance regarding source terms, transport, and transformation phenomena to statistical variability of parameters. These sources of uncertainty should be dealt with either qualitatively or quantitatively. This will help ensure that the exposure assessors honestly confront the limitations of their assessments.

ACKNOWLEDGMENT

This work was performed under the auspices of the U.S. Department of Energy by the Lawrence Livermore National Laboratory under Contract W-7405-Eng-48.

REFERENCES

Bacci, E., D. Calamari, C. Gaggi, and M. Vighi (1990), Bioconcentration of organic chemical vapors in plant leaves: experimental measurements and correlation, *Environ. Sci. Technol.,* 24, 885–889.

Bacci, E. and C. Gaggi (1986), Chlorinated pesticides and plant foliage translation experiments, *Bull. Environ. Contam. Toxicol.,* 37, 850–857.

Barber, M. C., L. A. Suarez, and R. R. Lassiter (1991), Modelling bioaccumulation of organic pollutants in fish with an application to PCBs in Lake Ontario salmonids, *Can. J. Fish. Aquat. Sci.,* 48, 318–337.

Bear, J. and A. Verruijt (1987), *Modeling Groundwater Flow and Pollution,* D. Reidel, Dordrecht, Holland.

Bennett, B.G. (1981), The exposure commitment method in environmental pollutant assessment, *Environ. Monit. Assess.,* 1, 21–36.

Bennett, B. G. (1982), Exposure of man to environmental nickel — an exposure commitment assessment, *Sci. Total Environ.,* 22, 203–212.

Bennett, B. G. (1983), Exposure of man to environmental PCBs — an exposure commitment assessment, *Sci. Total Environ.,* 29, 101–111.

Boethling, R. S. and A. Sabljic (1989), Screening-level model for aerobic biodegradability based on a survey of expert knowledge, *Environ. Sci. Technol.,* 23, 672–679.

Bogen, K. T. and L. C. Hall (1989), Pharmacokinetics for regulatory risk analysis: the case of 1,1,1-trichloroethane (methyl chloroform), *Regul. Toxicol. Pharmacol.,* 10, 26–50.

Bowen, H. J. M. (1979), *Environmental Chemistry of the Elements,* Academic Press, London.

Briggs, G. G., R. H. Bromilow, and A. A. Evans (1982), Relationship between lipophilicity and root uptake and translocation of non-ionized chemicals by barley, *Pestic. Sci.,* 13, 495–504.

Briggs, G. G., R. H. Bromilow, A. A. Evans, and M. Williams (1983), Relationships between lipophilicity and the distribution of non-ionized chemicals in barley shoots following uptake by the roots, *Pestic. Sci.,* 14, 492–500.

Brown, H. S., D. R. Bishop, and C. R. Rowan (1984), The role of skin absorption as a route of exposure for volatile organic compounds (VOCs) in drinking water, *Am. J. Publ. Health,* 74, 479–484.

Calabrese, E. J. and E. J. Stanek (1991), A guide to interpreting soil ingestion studies. II. Qualitative and quantitative evidence of soil ingestion, *Regul. Toxicol. Pharm.,* 13, 278–292.

Calamari, D., M. Vighi, and E. Bacci (1987), The use of terrestrial plant biomass as a parameter in the fugacity model, *Chemosphere,* 16, 2359–2364.

Cid, M. G., D. Loria, and E. Matos (1990), Genotoxicity of the pesticide propoxur and its nitroso derivative, NO-propoxur, on human lymphocytes in vitro, *Mutat. Res.,* 232, 45–48.

Cohen, Y. and P. A. Ryan (1985), Multimedia modeling of environmental transport: trichloroethylene test case, *Environ. Sci. Technol.,* 9, 412–417.

Cohen, Y., W. Tsai, S. L. Chetty, and G. J. Mayer (1990), Dynamic partitioning of organic chemicals in regional environments: a multimedia screening-level approach, *Environ. Sci. Technol.,* 24, 1549–1558.

DeMarini, D. M., V. S. Houk, J. Lewtas, R. W. Williams, M. G. Nishioka, R. K. Srivastava, J. V. Ryan, J. A. McSorley, R. E. Hall, and W. P. Linak (1991), Measurement of mutagenic emissions from the incineration of the pesticide dinoseb during application of combustion modifications, *Environ. Sci. Technol.,* 25, 910–913.

Ehling, U. H., D. Averbeck, P. A. Cerutti, J. Friedman, H. Greim, A. C. Kolbye, and M. L. Mendelsohn (1983), Review of the evidence for the presence of thresholds in the induction of genetic effects by genotoxic chemicals, *Mutat. Res.,* 123, 281–341.

Felton, J. S. and M. G. Knize (1990), Heterocyclic-amine mutagens/carcinogens in foods, *Handb. Exp. Pharmacol.,* 94/I, 471–502.

Fiedler, D. A., K. W. Brown, J. C. Thomas, and K. C. Donnelly (1991), Mutagenic potential of plants grown on municipal sewage sludge-amended soil, *Arch. Environ. Contam. Toxicol.,* 20, 385–390.

Freeze, R. A. and J. A. Cherry (1979), *Groundwater,* Prentice-Hall, Englewood Cliffs, NJ.

Gifford, F. A. and S. R. Hanna (1973), Modeling urban air pollution, *Atmos. Environ.,* 7, 131–136.

Horne, R. A. (1978), *The Chemistry of Our Environment,* Wiley-Interscience, New York.

Howard, P. H., J. Saxena, and H. Sikka (1978), Determining the fate of chemicals, *Environ. Sci. Technol.,* 12, 398–407.

ICRP (1975), *Report of the Task Group on Reference Man,* International Commission on Radiological Protection, Pergamon Press, New York.

Isnard, P. and S. Lambert (1988), Estimating bioconcentration factors from octanol-water partition coefficient and aqueous solubility, *Chemosphere,* 17, 21–34.

Jones, K. C., T. Keating, P. Diage, and A. C. Chang (1991), Transport and food chain modeling and its role in assessing human exposures to organic chemicals, *J. Environ. Qual.,* 20, 317–329.

Jones, K. C. and B. G. Bennett (1989), Human exposure to environmental polychlorinated dibenzo-*p*-dioxins and dibenzofurans: an exposure commitment assessment for 2,3,7,8-TCDD, *Sci. Total Environ.,* 78, 99–116.

Jury, W. A., W. F. Spencer, and W. J. Farmer (1983), Behaviour assessment model for trace organics in soil. I. Model description, *J. Environ. Qual.,* 12, 558–564.

Karickhoff, S. W. (1985), Pollutant sorption in environmental systems, in *Environmental Exposure from Chemicals,* Vol. 1, Neely, W. B. and G. E. Blau, Eds., CRC Press, Boca Raton, FL.

Knize, M. G., B. T. Takemoto, P. R. Lewis, and J. S. Felton (1987), The characterization of the mutagenic activity of soil, *Mutat. Res.,* 192, 23–30.

Kramers, P. G. N., J. M. Gentile, B. J. A. Gryseels, P. Jordan, N. Katz, K. E. Mott, J. J. Mulvihill, J. L. Seed, and H. Frohberg (1991), Review of the genotoxicity and carcinogenicity of antischistosomal drugs: is there a case for a study of mutation epidemiology: report of a task group on mutagenic antischistosomals, *Mutat. Res.,* 257, 49–89.

La Goy, P. K. (1987), Estimated soil ingestion rates for use in risk assessment, *Risk Anal.,* 7, 355–359.

Layton, D. (1993), Metabolically consistent breathing rates for use in dose assessments, *Health Phys.,* 64, 23–36.

Lewtas, J. (1985), Comparative potency of complex mixtures: use of short-term genetic bioassays in cancer risk assessment, in *Short-Term Bioassays in the Analysis of Complex Environmental Mixtures IV,* Waters, M. D., S. S. Sandhu, J. Lewtas, L. Claxton, G. Strauss, and S. Nesnow, Eds., Plenum Press, New York, 363–375.

Lyman, W. J., W. F. Reehl, and D. H. Rosenblatt (1982), *Handbook of Chemical Property Estimation Methods. Environmental Behavior of Organic Compounds,* McGraw-Hill, New York.

Lyons, C. D., S. E. Katz, and R. Bartha (1985), Persistence and mutagenic potential of herbicide-derived aniline residues in pond water, *Bull. Contam. Toxicol.,* 35, 696–703.

Mackay, D. (1979), Finding fugacity feasible, *Environ. Sci. Technol.,* 13, 1218–1223.

Mackay, D. (1991), *Multimedia Environmental Models, The Fugacity Approach,* Lewis, Chelsea, MI.

Mackay, D. and S. Paterson (1981), Calculating fugacity, *Environ. Sci. Technol.,* 15, 1006–1014.

Mackay, D. and S. Paterson (1982), Fugacity revisited, *Environ. Sci. Technol.,* 16, 654–660.

McFarlane, J. C., T. Pfleeger, and J. Fletcher (1987), Transpiration effect on the uptake and distribution of bromacil, nitrobenzene, and phenol in soybean plants, *J. Environ. Qual.,* 16, 372–376.

McKone, T. E. (1990), Dermal uptake of organic chemicals from a soil matrix, *Risk Anal.,* 10, 407–419.

McKone, T. E. (1987), Human exposure to volatile organic compounds in household tap water: the indoor inhalation pathway, *Environ. Sci. Technol.,* 21, 1194–1201.

McKone, T. E. and J. I. Daniels (1991), Estimating human exposure through multiple pathways from air, water, and soil, *Regul. Toxicol. Pharmacol.,* 13, 36–61.

McKone, T. E. and J. P. Knezovich (1991), The transfer of trichloroethylene (TCE) from a shower to indoor air: experimental measurements and their implications, *J. Air Waste. Manage. Assoc.,* 41, 832–837.

McKone, T. E. and D. W. Layton (1986), Screening the potential risk of toxic substances using a multimedia compartment model: estimation of human exposure, *Regul. Toxicol. Pharmacol.,* 6, 359–380.

McKone, T. E. and P. B. Ryan (1989), Human exposures to chemicals through food chains: an uncertainty analysis, *Environ. Sci. Technol.,* 23, 1154–1163.

National Research Council (NRC) (1983), *Risk Assessment in the Federal Government: Managing the Process,* National Academy Press, Washington, D.C.

Ng, Y. C. (1982), A review of transfer factors for assessing the dose from radionuclides in agricultural products, *Nucl. Saf.,* 23, 57–71.

Pasquill, F. (1961), The estimation of the dispersion of windborne material, *Meteorol. Mag.,* 90, 33–49.

Paterson, S. and D. Mackay (1989), Modeling the uptake and distribution of organic chemicals in plants, in *Intermedia Pollutant Transport: Modeling and Field Measurements,* Allen, D. T., Y. Cohen, and I. R. Kaplan, Eds., Plenum Press, New York, 283–292.

Rhomberg, L., V. L. Dellarco, C. Siegel-Scott, K. L. Dearfield, and D. Jacobson-Kram (1990), Quantitative estimation of the genetic risk associated with the induction of heritable translocations at low-dose exposure: ethylene oxide as an example, *Environ. Mol. Mutagen.,* 16, 104–125.

Riederer, M. (1990), Estimating partitioning and transport of organic chemicals in the foliage/atmosphere system: discussion of a fugacity-based model, *Environ. Sci. Technol.,* 24, 829–837.

Roels, H. A., J. P. Buchet, R. R. Lauwerys, P. Bruaux, F. Claeys-Thoreau, A. Fontaine, and G. Verduyn (1980), Exposure to lead by the oral and pulmonary routes of children living in the vicinity of a primary lead smelter, *Environ. Res.,* 22, 81–94.

Ryan, J. A., R. M. Bell, J. M. Davison, and G. A. O'Connor (1988), Plant uptake of nonionic organic chemicals from soils, *Chemosphere,* 17, 2299–2324.

Schaum, J. J. (1984), Risk Analysis of TCDD Contaminated Soils, Office of Health and Environmental Assessment, U.S. Environmental Protection Agency, EPA/600/8-84/031.

Schnoor, J. L. (1981), Fate and transport of dieldrin in Coraville Reservoir: residues in fish and water following a pesticide ban, *Science,* 211, 840–842.

Schnoor, J. L. and D. C. MacAvoy (1981), A pesticide transport and bioconcentration model, *J. Environ. Eng.,* 107, 1229–1245.

Sedman, R. M. (1989), The development of applied action levels of soil contact: a scenario for the exposure of humans to soil in a residential setting, *Environ. Health Perspect.,* 79, 291–313.

Spacie, A. and J. L. Hamelink (1985), Bioaccumulation, in *Fundamentals of Aquatic Toxicology,* Rand, J. M. and S. R. Petrocelli, Eds., Hemisphere, Washington, D.C., Chap. 17.

Spengler, J. D., D. W. Docker, W. A. Turner, J. M. Wolfson, and B. G. Ferris (1981), Long-term measurements of respirable sulfates and particles inside and outside homes, *Atmos. Environ.,* 15, 23–30.

Stanek, E. J., III and E. J. Calabrese (1991), A guide to interpreting soil ingestion studies. I. Development of a model to estimate the soil ingestion detection level of soil ingestion studies, *Regul. Toxicol. Pharm.,* 13, 263–277.

Stevens, J. B. (1991), Disposition of toxic metals in the agricultural food chain. I. Steady-state bovine milk biotransfer factors, *Environ. Sci. Technol.*, 25, 1289–1294.

Stumm, W. and J. J. Morgan (1981), *Aquatic Chemistry, An Introduction Emphasizing Chemical Equilibria in Natural Waters,* 2nd ed., John Wiley & Sons, New York.

Thomann, R. V. (1989), Bioaccumulation model of organic chemical distribution in aquatic food chains, *Environ. Sci. Technol.*, 23, 699–707.

Tinsley, I. J. (1979), *Chemical Concepts in Pollutant Behavior,* John Wiley & Sons, New York.

Trapp, S., M. Matthies, I. Scheunert, and E. M. Topp (1990), Modeling the bioconcentration of organic chemicals in plants, *Environ. Sci. Technol.*, 24, 1246–1252.

Travis, C. C. and A. D. Arms (1988), Bioconcentration of organics in beef, milk, and vegetation, *Environ. Sci. Technol.*, 22, 271–274.

Travis, C. C. and H. A. Hattemer-Frey (1988), Uptake of organics by aerial plant parts: a call for research, *Chemosphere,* 17, 277–284.

Traynor, G. W., M. G. Apte, A. R. Carruthers, J. F. Dillworth, D. T. Grimsrud, and L. A. Gundel (1987), Indoor air pollution due to emissions from wood-burning stoves, *Environ. Sci. Technol.*, 21, 691–697.

Turner, D. B. (1970), A Workbook of Atmospheric Dispersion Estimates, U.S. Environmental Protection Agency, Washington, D.C., AP-26.

U.S. EPA (1985), Guidelines for exposure assessment, *Fed. Regist.*, 51(185), 34042–34053.

U.S. EPA (1989a), Exposure Factors Handbook, EPA/600/8-89/043, Exposure Assessment Group, Office of Health and Environmental Assessment, Office of Research and Development, Washington, D.C.

U.S. EPA (1989b), Risk Assessment Guidance for Superfund, Human Health Evaluation Manual, EPA/540/1-89/002, Office of Emergency and Remedial Response, U.S. Environmental Protection Agency, Washington, D.C.

Van Gestel, C. A. M., K. Otermann, and J. H. Canton (1985), Relation between water solubility, octanol/water partition coefficients, and bioconcentration of organic chemicals in fish: a review, *Regul. Toxicol. Pharmacol.*, 5, 422–431.

van Houdt, J. J., W. M. F. Jongen, G. M. Alink, and J. S. M. Boleij (1984), Mutagenic activity of airborne particles inside and outside homes, *Environ. Mutagen.*, 6, 861–869.

Viraraghavan, T. and S. Hashem (1986), Trace organics in septic tank effluent, *Water Air Soil Pollut.*, 28, 299–308.

Wadden, R. A. and P. A. Scheff (1983), *Indoor Air Pollution Characterization, Prediction and Control,* Wiley Interscience, New York.

Wallace, L. (1986), An Overview of the Total Exposure Assessment (TEAM) Study. Vol. 1, Final Report, U.S. Environmental Protection Agency, Washington, D.C.

Ward, G. M. and J. E. Johnson (1986), Validity of the term transfer coefficient, *Health Phys.*, 50, 411–414.

Wauchope, R. D. (1978), The pesticide content of surface water draining from agricultural fields — a review, *J. Environ. Qual.*, 7, 459–472.

Whicker, F. W. and T. B. Kirchner (1987), PATHWAY: a dynamic food-chain model to predict radionuclide ingestion after fallout deposition, *Health Phys.*, 52, 717–737.

Wong, J. J., G. M. Schum, E. G. Butler, and R. A. Becker (1991), Looking past soil cleanup numbers, in *Hydrocarbon Contaminated Soils and Groundwater,* Vol. 1, Kostecki, P. T., E. J. Calabrese, and C. E. Bell, Eds., Lewis Publishers, Chelsea, MI, 1–21.

Yang, Y. Y. and C. B. Nelson (1986), An estimation of daily food usage factors for assessing radionuclide intakes in the U.S. population, *Health Phys.,* 50, 245–257.

Yoneyama, K. and F. Matsumura (1984), Microbial metabolism of 4,4′-methylene-*bis*-(2-chloroaniline), *Arch. Environ. Contam. Toxicol.,* 13, 501–507.

CHAPTER 3

Methods for Dose and Effect Assessment

J. Favor, E. W. Vogel, A. A. van Zeeland, J. G. Filser, M. L. Mendelsohn, and U. H. Ehling

TABLE OF CONTENTS

I. Introduction 66

II. Pharmacokinetics 67
    A. Pharmacokinetic Principles and Models 68
        1. Basic Considerations 68
        2. Area Under the Curve (AUC) as Dose Surrogate 69
    B. Dose-Response Curves in Mutagenicity and
        Carcinogenicity Studies 72
        1. Vinyl Chloride and Vinyl Bromide 72
        2. 2-Nitropropane 72
        3. Ethylene and Ethylene Oxide 75
    C. Pharmacokinetical Considerations 76

III. Molecular Dosimetry 77
    A. Methodology 77
    B. DNA Adduct Formation 78
    C. DNA Adduct Removal by Repair Mechanisms 79
    D. Exposure vs. Mutation Fixation 81
        1. Organ Specificity 81
        2. Correlation of DNA Adduct Formation with
            Genetic Effects 81
        3. Molecular Mutation Spectra 83
    E. Molecular Dosimetry Considerations 84

IV. Somatic Cell Mutagenesis 84
    A. Short-Term Assays 84
    B. Other Laboratory-Based Mutation Tests 86
    C. Human Mutation Testing 86

V. Germ Cell Mutagenesis 88
   A. Mammalian Germ Cell Mutagenesis 88
      1. Molecular Characterization of Ethylnitrosourea
         (ENU)-Induced Mutations 88
      2. Test Systems 89
   B. Dose Response 91
   C. Relevant Genetic Endpoints 92
   D. Genetic Risk Estimation 92
   E. Female Germ Cells 94
   F. Germ Cell Mutagenesis Considerations 95

VI. Comparative Mutagenesis 95
   A. Experimental Approaches: Qualitative vs.
      Quantitative Comparisons 95
      1. Qualitative Parameters 95
      2. Quantitative Comparisons 96
         2.1 Efficiency 96
         2.2 Relative Efficiency 97
   B. Carcinogenic Potency Ranking: Interspecific
      Comparisons in Rodents 97
   C. Relative Carcinogenic Potency: Relationship to
      Genetic Activity Profiles 100
   D. Comparative Mutagenesis Considerations 104

VII. Overall Conclusions 107

VIII. Perspectives 108

References 109

I. INTRODUCTION

Humans are exposed to naturally occurring mutagenic compounds as well as compounds which are only present due to industrial, technological, and medical developments. The source of exposure of a human population to the class of mutagenically active chemicals which are artificially present in the environment may range from an unplanned release following an industrial accident to the planned introduction into the market of a chemical with some beneficial uses. An assessment of the dose and the mutagenic effect associated with human exposure is in both cases imperative. For the former, an estimate of the dose and mutagenic effect is required for a full assessment of the impact of the industrial accident. For the latter, an estimate of the dose and mutagenic effect is required in the risk/benefit evaluation, important for the decision if the

METHODS FOR DOSE AND EFFECT ASSESSMENT

planned human exposure is justifiable. Confronted with an actual or potential exposure of a human population to a mutagenically active compound, the assessment of dose and mutagenic effect is important if risk-management efforts are to be intelligently pursued. First, it is apparent that society cannot blindly accept the deleterious effects associated with industrial progress only to employ remedial action when the risks become identified. A more intelligent approach is to incorporate an evaluation of any risks as an integral part of the processes involved in industrial progress. Second, due to the heterogeneous nature of human populations and the current status of testing methods, one would not expect epidemiological studies or laboratory studies of human populations to demonstrate *de novo* mutagenic effects. Further, such an epidemiological approach contradicts the preferred policy of protecting human populations from harm.

We present the experimental assessment of dose and mutagenic effect in a human population due to exposure to a chemical in a series of independent but not mutually exclusive steps: pharmacokinetics, molecular dosimetry, somatic cell mutagenesis, and *in vivo* mammalian germ-cell mutagenesis. Ideally, an evaluation of a compound would include all of the above steps, and thus results would represent a mechanistic understanding of the mutagenic activity of a chemical (pharmacokinetics, molecular dosimetry) as well as the biological complexities associated with mutation fixation in the living organism. The extrapolation procedure breaks down if comparable laboratory results are not available. An evaluation of dose and mutagenic effect for such chemicals may be forced to rely upon less directly related mutagenicity test data. Alternatively, knowledge of the structure-activity relationships of chemicals may be useful in predicting mutagenic potency. Finally, this large discrepancy between the desired *in vivo* mutagenicity data required for an adequate genetic risk assessment and the data actually available may be reconciled by recent developments in mutagenicity test systems based upon transgenic mice. Such tests combine the speed and precision of microbial mutagenicity assays with the complexities of the mammalian *in vivo* situation. If these methods prove to be accurate in reflecting the *in vivo* situation, it may be reasonable to survey large numbers of chemicals either prospectively or retrospectively for a genetic risk estimation.

II. PHARMACOKINETICS

The biological response to a mutagenic exposure is best understood in terms of the delivery of the mutagenically active species of the substance to the relevant targets. The target dose of the active species is determined by its concentration and the time it remains at the site of action. Both parameters depend on the rates at which the xenobiotic is absorbed, distributed, and biotransformed to active metabolites or to products of reduced effectiveness and on the rate of excretion from the body. Pharmacokinetics describe these processes quantitatively. The biotransformation and catabolism of xenobiotics are mostly catalyzed enzymatically and therefore subject to saturation kinetics (e.g., according to Michaelis and Menten). Consequently, the target dose of the active metabolite is often not a linear function of the amount of xenobiotic administered. Thus, knowledge of the pharmacokinetics of a substance is indispensable for understanding dose-response relationships observed *in vitro* and in animal studies, and for extrapolation to the human situation.

In the following, a short introduction to pharmacokinetic procedures and models is given, and the importance of pharmacokinetics in toxicology is demonstrated for vinyl chloride, vinyl bromide, and 2-nitropropane as well as ethylene and ethylene oxide. *In vitro* and *in vivo*, various studies on prokaryotic and eukaryotic systems have been performed with vinyl chloride, ethylene oxide, and 2-nitropropane to detect mutagenic and genotoxic effects (summarized for vinyl chloride: ECETOC, 1988; for ethylene oxide: ECETOC, 1982, 1984; for 2-nitropropane: GSF-Bericht, 1988). Only limited literature on the mutagenicity of vinyl bromide *in vitro* is available (NIOSH/OSHA, 1978; Bartsch et al., 1979). A comparative mutagenicity test in *Salmonella typhimurium* with vinyl chloride and ethylene oxide has been carried out by Victorin and Ståhlberg (1988). Whereas the dose-response curve obtained with vinyl chloride showed saturation kinetics, the corresponding curve for ethylene oxide was strictly linear. In spite of the large amount of mutagenicity data, only limited effort has been made up to now to employ pharmacokinetic principles for a better understanding of the observed dose-effect relationships and for species scaling of the mutagenic risk from animal to man (Rhomberg et al., 1990). In contrast, adequate examples of a pharmacokinetic consideration of the carcinogenic activities of chemicals exist. Therefore, due to the availability of data, we also consider carcinogenicity data as the biological response. Finally, some future trends are discussed.

A. Pharmacokinetic Principles and Models

1. Basic Considerations

A substance N is assumed to be absorbed from the environment or from an organ, distributed via the bloodstream, or eliminated from body, blood, or tissue. The amount of change, dN, per time unit, dt, is related to its actual concentration, c:

$$dN/dt = -k_{(c)} \times Vol_d \times c \tag{1}$$

where Vol_d is the apparent volume of distribution. The minus sign symbolizes the loss of substance. The proportionality factor $k_{(c)}$ (time^{-1}) has a constant value independent of the concentration only if linear pharmacokinetics apply.

METHODS FOR DOSE AND EFFECT ASSESSMENT

However, if saturation kinetics occur, it becomes dependent on the actual concentration and decreases with increasing concentration, since dN/dt reaches a maximum. Vol_d is normally considered constant. Its value can differ significantly from the real tissue volume. For instance, Vol_d can be very high for lipophilic chemicals. Saturation kinetics apply in enzyme-mediated processes as, for instance, in kinetics according to Michaelis-Menten. Here, $k_{(c)}$ can be replaced by

$$k_{(c)} = V_{max} / \left[Vol_d \times \left(K_{mapp} + c \right) \right] \quad (2)$$

with the maximum rate of metabolism, V_{max} and the apparent Michaelis-Menten constant in blood, K_{mapp}. If c becomes much smaller than K_{mapp}, $k_{(c)}$ becomes constant, since:

$$\lim_{c \ll K_{map}} k_{(c)} = V_{max} / \left(Vol_d \times K_{mapp} \right) \quad (3)$$

If c becomes much higher than K_{mapp}, $k_{(c)}$ becomes inversely proportional to c:

$$\lim_{c \gg K_{mapp}} k_{(c)} = V_{max} / \left(Vol_d \times c \right) \quad (4)$$

Under these conditions, dN/dt reaches V_{max} by substituting $k_{(c)}$ from Equation 4 into Equation 1.

Figure 1 gives examples of concentration-time curves characteristic for saturable metabolism. If measured data follow such courses, they can be analyzed for V_{max} and K_{mapp} by means of iterative regression methods such as that of Csanády and Filser (1990). The obtained pharmacokinetic parameters are necessary for the prediction of concentration c for any condition of exposure.

2. Area Under the Curve (AUC) as Dose Surrogate

By analogy to Haber's rule, a specific effect of a substance can be regarded as the product of the concentration at the target place and the time of exposure. This signifies that at the target place, the area under the concentration-time curve (AUC) of a directly acting substance or of the reactive metabolite can be applied as a surrogate of the effective dose (D) (Ehrenberg et al., 1974),

$$D = AUC = \int_0^\infty c_t \times dt \quad (5)$$

with the concentration c_t at time point t.

If, based on pharmacokinetic studies, the concentration-time profile of the ultimately reacting substance in the target tissue can be determined for any

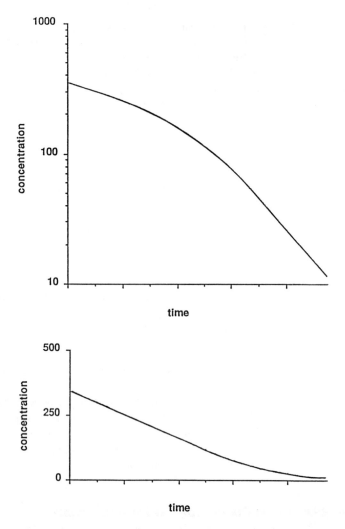

Figure 1. Blood concentration-time curves for saturable metabolism according to Michaelis-Menten kinetics. Both plots represent identical data. Upper curve, logarithmic plot of blood concentration vs. time; lower curve, linear plot of blood concentration vs. time.

condition of exposure, the resulting AUC can be calculated. Various models can be used to estimate the concentration-time profile.

The analysis of experimentally determined concentration-time courses can be done by either physiological pharmacokinetic modeling or compartmental modeling. The former is especially appropriate for species extrapolation if volumes, blood supply, and tissue/blood partition coefficients from the different tissues, and the kinetic parameters of the metabolizing enzymes are known. Compartmental models give less insight into the behavior within the organism, but yield parameters describing the measured curves and provide much more

METHODS FOR DOSE AND EFFECT ASSESSMENT

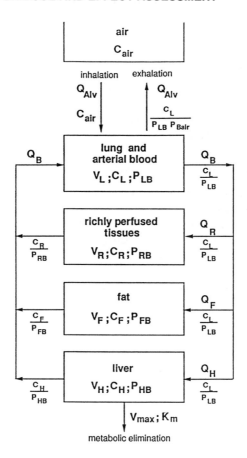

$$V_R \cdot \frac{dC_R}{dt} = +Q_R \cdot \frac{C_L}{P_{LB}} - Q_R \cdot \frac{C_R}{P_{RB}}$$

$$V_F \cdot \frac{dC_F}{dt} = +Q_F \cdot \frac{C_L}{P_{LB}} - Q_F \cdot \frac{C_F}{P_{FB}}$$

$$V_H \cdot \frac{dC_H}{dt} = +Q_H \cdot \frac{C_L}{P_{LB}} - Q_H \cdot \frac{C_H}{P_{HB}} - \frac{V_{max} \cdot \frac{C_H}{P_{HB}}}{K_m + \frac{C_H}{P_{HB}}}$$

$$V_L \cdot \frac{dC_L}{dt} = +Q_F \cdot \frac{C_F}{P_{FB}} + Q_R \cdot \frac{C_R}{P_{RB}} + Q_H \cdot \frac{C_H}{P_{HB}} + Q_{Alv} \cdot C_{air} - Q_{Alv} \cdot \frac{C_L}{P_{LB} \cdot P_{BAir}} - Q_B \cdot \frac{C_L}{P_{LB}}$$

Figure 2. Simple physiological pharmacokinetic model for inhaled substances and corresponding equations. (For detailed information, see, e.g., Johanson and Näslund, 1988; Gargas and Andersen, 1988.) V, volume of tissue; P, partition coefficient; C, average concentration in tissue or organ; Cair, concentration in air; Q, blood flow through tissue or organ; Qalv, alveolar ventilation; L, lung; R, richly perfused tissue group; F, fatty tissue; H, liver; P_{LB}, partition coefficient lung/blood; P_{RB}, partition coefficient richly perfused tissues/blood; P_{FB}, partition coefficient fat/blood; P_{HB}, partition coefficient liver/blood; P_{BAir}, partition coefficient blood/air; V_{max}, maximal rate of metabolism in the liver; K_m, concentration in liver blood at $V_{max}/2$.

precise estimates. Figure 2 is an example of a simple physiological pharmacokinetic model for inhaled gases along with the differential equations describing the model. A detailed discussion of this type of model is given in Balant and Gex-Fabry (1990), Himmelstein and Lutz (1979), and Rowland (1984).

B. Dose Response Curves in Mutagenicity and Carcinogenicity Studies

1. Vinyl Chloride and Vinyl Bromide

Vinyl chloride and vinyl bromide monomers are gases which are used mainly for polymer production. Both chemicals are biotransformed in a first step via cytochrome P 450-dependent monooxygenases in the liver to the corresponding halogenated epoxides, which are the reactive metabolites that alkylate macromolecules such as proteins and DNA (reviewed in Laib, 1982). Pharmacokinetic studies with both vinyl halides *in vivo* have proven the saturability of the first metabolic step, resulting in a maximum internal concentration of the epoxides which cannot be surpassed even at exposure to very high concentrations (Gehring et al., 1978; Filser and Bolt, 1979, 1981). Results of mutagenicity studies with vinyl chloride in *S. typhimurium* TA 100 (Victorin and Ståhlberg, 1988) and in *Drosophila melanogaster* (Verburgt and Vogel, 1977) also showed an increased mutation frequency with concentration, reaching a final plateau consistent with a saturation of metabolism. Corresponding results have been obtained with respect to carcinogenicity in long-term studies with rats exposed to vinyl chloride (Maltoni et al., 1981) and to vinyl bromide (Benya et al., 1982). These carcinogenicity dose-effect curves become linear when the observed tumor incidences are related not to the exposure concentrations but to the amounts of the gaseous monomers metabolized during exposure time. This was done first by Gehring and colleagues on vinyl chloride (Gehring et al., 1978) and extended by Bolt and Filser (1983) and Bolt et al. (1982).

Biotransformation of the metabolic precursor reflects the concentration of the ultimate reactive metabolites only if it represents the rate-limiting step. This appears to be the case with vinyl chloride and vinyl bromide in rats. Therefore, the integrals over biotransformation rates and exposure times paralleled the AUCs of the ultimate reactive metabolites. However, use of these findings for risk extrapolation to man can lead to erroneous results if the ratios of production and detoxification of such metabolites differ in different species. An intelligible extrapolation of results obtained in animal bioassays to man requires for both species knowledge of the AUCs mentioned above, which for vinyl chloride and vinyl bromide have not yet been studied.

2. 2-Nitropropane

2-Nitropropane is a volatile industrial solvent. The pure substance was mutagenic in *S. typhimurium* strains TA 98 and TA 100, resulting in linear dose-response curves. No differences were observed with or without addition of metabolic activation systems derived from liver preparations from rats (Göggelmann et al., 1988). The level of induced DNA repair synthesis was much higher in hepatocytes of male Wistar rats compared to females following exposure to 2-nitropropane, indicating a higher sensitivity of males toward the genotoxic effects of this substance (Andrae et al., 1988). Exposure of Sprague-Dawley rats to gaseous 2-nitropropane at concentrations of 100 or 200 ppm caused hepatotoxicity and liver carcinoma, male rats being more sensitive than females. At exposure levels of about 25 ppm, no adverse effects have been observed in either sex (Lewis et al., 1979; Griffin et al., 1980).

At present, the mechanisms leading to the mutagenicity, hepatotoxicity, and hepatocarcinogenicity of 2-nitropropane are still unclear. To explain the concentration-response relationship and the sex specificity of 2-nitropropane, its inhalation kinetics in Sprague-Dawley rats of both sexes at concentrations between 0 and 250 ppm were studied (Kessler et al., 1989; Denk et al., 1989). In adult rats, 2-nitropropane is metabolized via two different pathways: a saturable process of low capacity and high affinity according to Michaelis-Menten kinetics, and a nonsaturable process following first-order kinetics (Figure 3). Striking sex differences were observed: At concentrations below 60 ppm (males) and 180 ppm (females), more 2-nitropropane was metabolized via the saturable pathway than via the nonsaturable one; above these concentrations, the relationships were reversed. In females, V_{max} of the saturable pathway was about twice as high as in males; whereas first-order metabolism was faster in males. The interaction of both pathways results in an upward concave curve for the metabolic product of the nonsaturable process in relationship to the exposure concentration.

Furthermore, the dose dependence of the carcinogenic potency of 2-nitropropane was evaluated by determining the incidence of preneoplastic enzyme-altered liver foci in mature rats of both sexes by means of the "rat liver foci bioassay" (Denk et al., 1991). This is a short-term carcinogenicity assay described by Oesterle and Deml (1983). The incidence of the foci represents a quantitative measure for hepatocarcinogenicity (Kunz et al., 1978). In the rat liver foci bioassay, 2-nitropropane induced preneoplastic foci, indicating its tumor-initiating potency. In both sexes, the number of foci correlated positively with the exposure concentration (Figure 3). In females, the foci number was significantly higher than that of controls at concentrations above 25 ppm, and in males at all exposure concentrations. The foci incidences reached higher values in males than in females. The concentration dependence of the foci number was found to be upwardly concave in relation to the exposure concentration (Figure 3). These results support the hypothesis that the genotoxic, hepatotoxic, and carcinogenic effects of 2-nitropropane are probably due to the nonsaturable pathway (Denk et al., 1989). This is in accordance with the linear (nonsaturable) dose-response relationships found in the mutagenicity studies (see above) In contrast, the saturable pathway is more likely to lead to less toxic

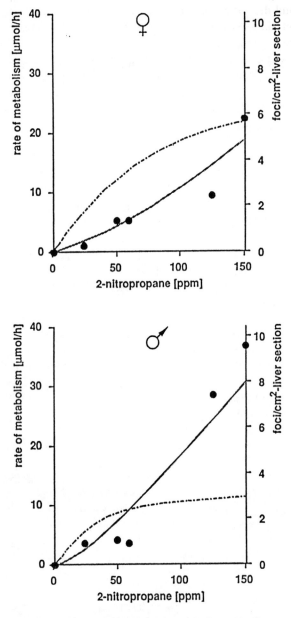

Figure 3. Concentration dependence of rates of metabolism of 2-nitropropane at steady state and induction of preneoplastic liver foci in rats. Dots, number of liver foci/cm² liver section (mean values minus control value); lines, calculated for one rat (250 g). Dashed lines, saturable metabolic pathway; solid lines, nonsaturable metabolic pathway.

or less carcinogenic metabolites. Similar findings of alternative pathways for activation and detoxification have been demonstrated with dichloromethane (Andersen et al., 1987). Although the molecular mechanisms leading to the genotoxic, hepatotoxic, and hepatocarcinogenic effects of 2-nitropropane are not yet known, the aforementioned results allow the correlation of the mutagenicity and carcinogenicity of this compound with its pharmacokinetics.

3. Ethylene and Ethylene Oxide

Ethylene is an ubiquitous gas produced endogenously in mammals, man, and plants (for a review, see NRCC, 1985), and is an important industrial chemical (Kniel et al., 1980). In several studies in rats (Filser and Bolt, 1984) and mice (Ehrenberg et al., 1977; Segerbäck, 1983), ethylene has been shown to be metabolized in a first step to ethylene oxide. This epoxide directly alkylates DNA and proteins (Ehrenberg and Osterman-Golkar, 1980; Segerbäck, 1983; Potter et al., 1989) and was shown to be mutagenic *in vitro* and *in vivo* in mammals and man (Lewis et al., 1986; Generoso et al., 1990; Ehrenberg and Hussain, 1981; Thiess et al., 1981; Garry et al., 1982; Yager et al., 1983; Sarto et al., 1984; Kolman et al., 1986). It was carcinogenic in long-term studies with rats (Dunkelberg, 1982; Snellings et al., 1984; Lynch et al., 1984; Garman et al., 1985) and mice (Dunkelberg, 1981; NTP, 1987), producing tumors in various tissues and organs.

The pharmacokinetics of ethylene oxide in rats (Filser and Bolt, 1984) demonstrated a linear relationship between the concentration of ethylene oxide in the atmosphere and in the organs of the exposed animals below an exposure concentration of 100 ppm ethylene oxide, the highest concentration used in the long-term carcinogenicity study by Snellings et al. (1984). The tumor probability obtained could be described by a linear model (Denk, 1990). Linear dose-response curves have also been found with respect to the alkylation of hemoglobin and liver DNA down to a concentration of 1 ppm (Potter et al., 1989). In contrast, the pharmacokinetics of the metabolic precursor ethylene in rats were saturable. V_{max}, the maximum rate of metabolism, was very low (8.5 µmol/(h × kg) and was reached at about 1000 ppm ethylene (Bolt et al., 1984). The resulting body burden of rats with ethylene oxide was too low for obtaining statistically significant positive tumor results with ethylene, regardless of dose (Bolt and Filser, 1984, 1987). This conclusion, based on the pharmacokinetic data of both substances and on the results of the long-term study with ethylene oxide carried out by Snellings and colleagues, has been confirmed by the negative findings with ethylene in the rat bioassay conducted by Hamm et al. (1984). However, using the pharmacokinetic parameters of both substances, the risk for rats could be quantitated as a function of the exposure concentration (Denk, 1990). Furthermore, after acquiring the respective pharmacokinetic data in man (Shen et al., 1989; Denk, 1990; Filser et al., 1992), its risk was assessed by extrapolating the results of the animal studies with ethylene oxide, considering the internal body burden to be the result of exposure to ethylene and ethylene oxide, respectively (Denk, 1990; Denk and Filser, 1990).

The pharmacokinetic findings *in vivo* are consistent with the mutagenicity data in *S. typhimurium* showing linear dose-response relationships for ethylene oxide between 0 and 200 ppm in the presence and absence of metabolic activating systems. In contrast, in the same study, no mutagenic activity could be detected with ethylene concentrations up to 20%, even after administration of a metabolic activating system obtained as an S9 mix from the livers of Aroclor-induced rats (Victorin and Ståhlberg 1988). As discussed above, the metabolic activation of ethylene to ethylene oxide in the livers of rats is a slow process; as such, the same activation process *in vitro* will also be slow. Therefore, even at the very high substrate concentration used *in vitro,* ethylene oxide production is still limited by the kinetics of the reaction. As the examples have shown, *in vivo* pharmacokinetic data can help to interpret mutagenicity data.

C. Pharmacokinetical Considerations

Knowledge of pharmacokinetics is indispensable for a rational interpretation of dose-response relationships. While the dose-response curves obtained with vinyl chloride and vinyl bromide are the results of saturation of enzymes metabolizing these halogenated ethylenes to active chemical species, the "hockey-stick"-like curve of 2-nitropropane (Figure 3) can be explained by the saturation of inactivating enzymes. The dose at the target tissue is strongly dependent on the rates of uptake, distribution, and elimination, as demonstrated for ethylene and ethylene oxide. Further possible mechanisms influencing kinetic processes, such as the induction of enzyme activities and saturation of DNA repair, are discussed by Hoel et al. (1983) and Lutz (1990). These modifying factors are important when the mutagenic effects of the relatively simple compound ethylene oxide are considered. In germ cells of the mouse, a stage-specific mutagenic activity has been demonstrated. Ethylene oxide was shown to be active in postmeiotic male germ-cell stages (Lewis et al., 1986), whereas results were negative in stem cell spermatogonia (Russell et al., 1984). Although a thorough understanding of mutagenicity dose-response relationships requires knowledge of the pharmacokinetics of xenobiotics, a strong interaction of these two disciplines has been lacking. This may partially result from the fact that pharmacokinetics are traditionally oriented toward more complex chemicals subject to intermediate metabolism, whereas *in vivo* mutagenesis is oriented mainly toward simple, direct-acting mutagens. Further, as will become evident in the later section on *in vivo* mutagenesis, very few dose-response data are available for chemical mutagens. Hopefully, attempts will be made to rectify these shortcomings.

III. MOLECULAR DOSIMETRY

Once a mutagen or its active metabolite has reached the target organ, interaction with DNA, generally by adduct formation, is the next critical step. Characteristic of chemical mutagens is a specificity of mutagen-DNA reaction products leading to a spectrum of DNA adducts. The adducts may be repaired or may eventually lead to mutation fixation via base mispairing, error-prone repair, chromosome breakage, etc. Methods have been developed to identify and measure the levels of DNA adduct formation. These methods may be applied to accurately measure target cell doses. Further, in combination with mutagenicity studies conducted in parallel experiments, the mutagenically relevant DNA adducts may be identified.

A. Methodology

DNA adducts are usually introduced at low frequencies in experimental designs in which animals or cultured cells still have a reasonable survival rate. Examples of such frequencies are given in Table 1. However, it is also clear from measurements in the human population that normally the frequency of DNA adducts may be as low as 5×10^{-10} to 5×10^{-9} adducts per nucleotide (A/N), depending on the type of DNA adduct (Umbenhauer et al., 1985).

The classical method of quantifying DNA adduct formation by chemical mutagens utilizes radioactively labeled forms (Baird, 1979). When the position of the label, usually ^3H or ^{14}C, is properly chosen, several or all DNA adducts caused by the chemical will become labeled. Quantification of the modified bases or nucleosides is carried out by determining the amount of radioactivity in the fraction of the chromatogram containing the modified bases (Beranek et al., 1980, 1983). Labeled mutagens are used because the amount of modified bases is too low to be detected by conventional types of detectors coupled to high-performance liquid chromatography (HPLC) systems. However, the problem in using labeled mutagens is that they can only be applied in experimental systems and not in the human situation. Recently, accelerator mass spectrometry has been applied for the detection of ^{14}C-labeled heterocyclic amine adducts in DNA. This method is extremely sensitive (1×10^{-11} adducts per nucleotide), and the amount of ^{14}C label is so low that radiation exposure, when applied in the human situation, would be insignificant (Turteltaub et al., 1990).

A second technique for measuring DNA adducts at low frequency is the use of monoclonal antibodies specific for certain types of DNA damage. Such antibodies have become available for several alkylated DNA bases (methylated and ethylated), benzo[a]pyrene adducts, cisplatin adducts, acetyl acetoxy amino fluorene adducts, and ultraviolet light-induced pyrimidine dimers and 6-4 photoproducts (Adamkiewitz et al., 1984; Poirier, 1984; Muysken-Schoen et al., 1985).

A very sensitive technique is the ^{32}P-postlabeling method in which DNA adducts are radioactively labeled after the isolation and hydrolysis of DNA

Table 1. Frequency of Various DNA Adducts Under Experimental Conditions

Treatment	Frequency of DNA Adducts
Ultraviolet light (10 J/m^2, 254 nm)	7.9×10^{-5} pyrimidine dimers per nucleotide
Mammalian cells treated with EMS (2 h, 10 mM)	O^6-Ethylguanine: 1×10^{-5} A/N 7-Ethylguanine: 3×10^{-4} A/N
Cells treated with ENU (4 mM)	O^6-Ethylguanine: 1×10^{-5} A/N 7-Ethylguanine: 1.4×10^{-5} A/N
Mouse testis 2 h after i.p. injection with ENU (250 mg/kg)	O^6-Ethylguanine: 2×10^{-6} A/N
Human population background	O^6-Ethylguanine: 4×10^{-9} A/N

A/N, adducts per nucleotide.

(Randerath et al., 1985). This method is especially applicable for measuring DNA damage caused by chemicals which introduce "bulky" adducts in DNA.

Recently, electrochemical detectors have been applied for the measurement of some DNA adducts at low frequency without the use of radioactive labeling (Park et al., 1989). This method is only suitable for electrochemically active adducts such as 7-methylguanine, 7-ethylguanine, and 8-hydroxyguanine. An indication of the sensitivity of the various methods is given in Table 2.

Besides methods for the detection of DNA adducts, monitoring of protein adducts also is used for estimating exposure to genotoxic chemicals. For example, the monitoring of hemoglobin adducts is a very sensitive indicator of exposure to ethylene oxide (Ehrenberg et al., 1974).

B. DNA Adduct Formation

Most genotoxic agents introduce more than one type of DNA modification. A factor which influences the relative proportion of each of these adducts is the chemical reactivity of the chemical or its activated form with the various nucleophilic sites present in DNA. In the case of alkylating agents, the relative reactivity with the different N and O atoms available in DNA can, to a large extent, be deduced from the Swain-Scott s-factor. Compounds with a relatively high s-factor preferentially react with N atoms in DNA. In contrast, chemicals with a low s-factor alkylate N and O atoms in DNA more randomly. Since the total amount of O atoms present in the phosphate backbone of DNA is relatively large, alkylating agents with low s-factors generate to a large extent phosphotriesters (up to 60% with ENU). This does not necessarily mean that the adducts which occur with a high frequency are responsible for the biological effects observed. The distribution of alkylation products in DNA for some alkylating agents is given in Table 3.

Table 2. Sensitivity of Various Detection Methods for DNA Adducts

Method	Sensitivity
Radioactively labeled mutagens	5×10^{-7} A/N
Accelerator mass spectrometry (using ^{14}C)	1×10^{-11} A/N
Immunological detection	1×10^{-8} A/N (depends on antibody)
^{32}P-postlabeling	1×10^{-9} to 1×10^{-10} A/N
HPLC + electrochemical detector	1×10^{-5} to 1×10^{-8} A/N (depends strongly on type of adduct)

A/N, adducts per nucleotide.

Besides a selectivity of adduction to specific bases in DNA, there can also be an unequal distribution of DNA adducts due to primary and secondary DNA structure as well as the DNA-chromatin structure. This uneven distribution can be related to the position of nucleosomes (Smerdon et al., 1990a, b). There are also indications that DNA sequences closely related to the nuclear matrix are more modified by certain genotoxic agents than the genome overall (Arrand and Murray, 1982; Nehls et al., 1984).

C. DNA Adduct Removal by Repair Mechanisms

Elaborate repair processes have evolved to minimize the permanent incorporation of errors into DNA. The main mechanism involved in DNA damage removal is the process of excision repair. It can act on a variety of different types of DNA lesions and involves the removal of a stretch of single-strand DNA with a length of about 13 nucleotides containing the DNA adduct. The resulting gap is then filled in by DNA replication using the complementary DNA strand as a template. This process is called DNA repair replication or unscheduled DNA synthesis (UDS). For some types of DNA damage, an early step in the process is the removal of the modified DNA base by a DNA glycosylase which generates an abasic site. This abasic site is the substrate for the excision repair process. There are indications that in this case the length of the DNA fragment which is replaced by repair replication is shorter than 13 nucleotides.

Recently, it has been shown that removal of cyclobutane pyrimidine dimers by the excision repair process is not the same for all parts of the genome, but that removal is considerably faster from actively transcribed genes compared to nontranscribed genes or to the genome overall (Bohr et al., 1985; Mellon et al., 1986). Moreover, in rodent cells, the fast removal in active genes is primarily due to repair of the transcribed DNA strand, whereas the nontranscribed

Table 3. Distribution of Alkylation Products in DNA, Expressed as Percent of Total DNA Alkylated, Generated by a Number of Monofunctional Alkylating Agents

Chemical	7-Alkyl-guanine	3-Alkyl-adenine	O^6-Alkyl-guanine	O^4-Alkyl-thymidine	Total Phosphotriesters
Methyl methane-sulfonate	82	10	0.3	n.d.	1
Ethyl methane-sulfonate	60	5	2	n.d.	13
Methyl nitrosourea	66	8	6	0.6	12
Ethyl nitrosourea	11	3	7	1	55

Data from Lawley et al. (1975), Singer (1975, 1979), and Beranek et al. (1980).

strand is virtually unrepaired. In human cells, there is a rate difference for removal of pyrimidine dimers from the transcribed strand compared to the nontranscribed strand (Mellon et al., 1987). It is not yet clear whether preferential repair of active genes by excision repair is generally true for most types of DNA damage. The phenomenon of preferential repair is also observed in *Escherichia coli* (Mellon and Hanawalt, 1989) and in yeast (Terleth et al., 1989).

In some organisms and in some mammalian cell types, proteins have been described which can remove methyl groups and, to a lesser extent, ethyl groups from specific positions in alkylated DNA. Such proteins are not enzymes in the strict sense because they participate in the reaction in which the alkyl group is transferred from the DNA to the protein. Each protein molecule can receive one alkyl group and repairs the DNA base adduct. In *E. coli,* the alkyl transferase can remove alkyl groups from the O^6 position of guanine, the O^4 position of thymine, and the oxygens of alkylphosphotriesters (McCarthy et al., 1983, 1984). In mammals, the alkyltransferase only removes alkyl groups from the O^6 position of guanine (Pegg et al., 1983).

Molecular dosimetry offers sensitive methods with which to demonstrate DNA alterations due to chemical exposure. As will become evident in the following section, these methods in combination with comparative mutagenicity data contribute to an understanding of the mechanisms of mutation induction by chemicals and may identify the mutagenically relevant DNA adducts. The methods are sensitive and relatively fast, and may be employed as an initial response to a population exposure to estimate target dose and the resultant genetic risk.

D. Exposure vs. Mutation Fixation

1. Organ Specificity

After exposure to a genotoxic agent, it usually is not known how much of the chemical reaches the target cell or how much of a chemical is transported by the blood to various organs of an animal. DNA adduct frequencies at early times after exposure can be used as indicators of the efficiency of penetration to various organs. Even in the case of direct-acting alkylating agents such as ethyl methanesulfonate (EMS) or *N*-ethyl-*N*-nitrosourea (ENU), DNA adduct frequencies are not the same in various organs of mice after i.p. injection with the chemical (Van Zeeland et al., 1990). Two hours after treatment, DNA adduct frequencies in the liver appeared to be about five times higher than those observed in bone marrow or in testicular DNA. In the case of chemicals activated by enzymes in the liver, i.e., nitrosamines, differences in adduct frequency between the liver and other organs are even larger. The DNA repair capacity of different organs or of various cell types within an organ can account in part for the organ specificity of chemical mutagens. For instance, there are clear differences in the amount of alkyltransferase present in the liver compared to other organs. Furthermore, the capacity of an organ to synthesize more of this protein after exposure to alkylating agents varies between different rodent species (Hall and Montesano, 1990). It is conceivable that differential repair of DNA adducts between different genes, depending on the transcriptional activity of the gene, has an influence on the organ specificity of carcinogens. Mutation fixation is believed to occur upon DNA replication if the DNA damage which has been introduced is not removed. Therefore, it may be expected that fast-growing tissue is more sensitive to exposure to genotoxic agents (Cohen and Ellwein, 1990), since the time interval between DNA adduct formation and replication may not be adequate for repair to take place.

2. Correlation of DNA Adduct Formation with Genetic Effects

An important aspect of quantitative mutagenesis is the efficiency with which an induced DNA adduct is converted to a mutation. Insight into the relative contribution of different classes of DNA adducts caused by monofunctional alkylating agents has been obtained by characterizing for chemicals the ratio of the frequency of chromosomal aberrations to the frequency of gene mutations induced after a certain exposure. This approach, which has been applied successfully to *Drosophila* and cultured mammalian cells, shows that alkylating agents which preferentially react with N atoms in DNA are efficient in inducing chromosome breakage events relative to gene mutations, while alkylating agents, which are more efficient in reacting with O atoms in DNA, generate high frequencies of gene mutations at exposures which do not induce high levels of chromosomal aberrations (Vogel and Natarajan, 1979a, b; Natarajan et al., 1984).

Comparative Mutagenesis of Ethylating Agents on the Basis of Molecular Dosimetry

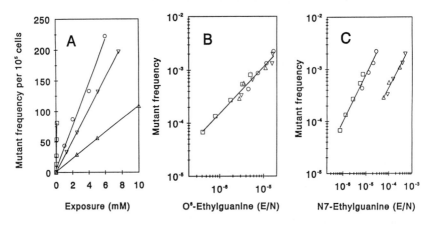

Figure 4. V79 Chinese hamster cells were exposed to ethyl methanesulfonate (△), diethyl sulfate (▽), N-ethyl-N-nitrosourea (○), and N-ethyl-N′-nitro-N-nitrosoguanidine (□). Immediately after exposure, the frequency of O^6-ethylguanine and N7-ethylguanine in DNA was determined in part of the exposed cell population. The remainder of the cells were grown for 7 d, after which the frequency of mutations at the *hprt* locus was determined (Van Zeeland et al., 1985).

The mutagenic potential of chemical agents is primarily determined by the nature and quantity of the lesions introduced in DNA. Most mutagens induce more than one type of lesion, each with different mutagenic properties. An example of a comparative study of the contribution of two types of DNA adducts to the mutagenicity of a series of ethylating agents is shown in Figure 4 (Van Zeeland et al., 1985). When mutation induction by four different ethylating agents was compared on the basis of the exposure concentration, large differences between the chemicals were observed (Figure 4A). However, when the frequency of mutation induction was compared at equal frequencies of O^6-ethylguanine, the response of the four chemicals was very similar (Figure 4B). In contrast, a comparison based on the frequency of 7-ethylguanine showed no such similarity (Figure 4C). These data show that O^6-ethylguanine is the major mutagenic lesion introduced in the DNA of these cells by the four ethylating chemicals.

The situation in germ cells is much more complex. In stem-cell spermatogonia of the mouse, the frequency of mutation induction per locus by ENU was very similar to the frequencies observed in cultured mammalian cells when compared at equal levels of O^6-ethylguanine (van Zeeland et al., 1990). However, when the frequency of alkylations in DNA was compared with mutation induction in postmeiotic cells of mice, no correlation between mutation induction and frequency of O^6-alkylguanine was found (Van Zeeland et al., 1990).

Instead, for these postmeiotic stages, a correlation was observed for mutation induction and adduct formation with the N atoms of DNA (van Zeeland et al., 1990). These differences seem to be due to differences in the repair capacity of the various germ-cell stages and the time period between DNA adduct formation and the next round of DNA replication, which can be very long (up to 16 d). For example, since N alkylations in DNA are chemically unstable, apurinic sites will be generated which in principle are the substrate for excision repair. The more mature stages of the germ cells, in contrast to stem-cell spermatogonia, are deficient in excision repair and therefore, especially in these mature stages, the contribution of N alkylations to mutation induction is expected to be relatively large. Observations in *Drosophila,* where excision repair-deficient mutants have been used to measure the influence of repair in germ cells, strongly support this model (Vogel et al., 1990). Thus, although the ultimate goal of molecular dosimetry is to provide an accurate measure of the target dose for quantification of mutation induction, a deep understanding of the biological processes characteristic of the target cells is of utmost importance.

3. *Molecular Mutation Spectra*

A direct DNA sequence analysis of recovered mutations provides the definitive data for identifying the mutagenically relevant DNA adducts. In order to obtain a clear picture of the various types of mutations that are caused by a mutagen, it is important to use target genes that allow the detection of the majority of induced mutations. The most commonly used systems for mutation studies are cultured mammalian cells, *E. coli* (Richardson et al., 1987), yeast (Kuntz et al., 1989), and *D. melanogaster* (Pastink et al., 1989). Mutation studies in mammalian cells using endogenous genes as the target for mutation induction are usually performed with cultured rodent (Vrieling et al., 1989) or human (Chen et al., 1990) cell lines. Apart from endogenous genes, integrated shuttle vectors containing genes are being used as targets for mutagenesis (Tindall and Stankowski, 1989). The genes which serve as targets for mutation induction in the different systems are listed in Table 4. Of these, molecular mutation spectra are available only for the *hprt, aprt,* and *dhfr* genes (Carothers et al., 1986; De Jong et al., 1988; Vrieling et al., 1989).

Recently, it has become possible to measure mutant frequencies and to determine mutation spectra in the *in vivo* situation (Table 4). *Hprt*-deficient mutants can be selected from lymphocytes obtained from man (Albertini et al., 1982; Henderson et al., 1986; Rossi et al., 1990), mouse (Burkhart-Schultz et al., 1990), and rat (Jansen et al., 1991). In addition, mutants at the autosomal HLA locus can be obtained from populations of human lymphocytes (Morley et al., 1990). Furthermore, transgenic mice are under construction which will make it possible to determine mutation frequencies in different organs (Gossen et al., 1989). This will allow determination of the organ-specific action of a

Table 4. Target Genes for Mutagenesis

Escherichia coli	*lac* I, *gpt*
Yeast	SUP4, URA3
Drosophila melanogaster	White, vermilion, *adh*
Cultured mammalian cells	hprt, aprt, dhtr, tk, ouabain^r, *gpt*
Mouse lymphocytes	hprt
Rat granuloma pouch	hprt
Human lymphocytes	hprt, HLA
Human erythrocytes	Glycophorin A, hemoglobin

genotoxic agent. Differences between species in their response to genotoxic agents are expected to occur as a result of differences in the metabolism of the genotoxic agent. Also, differences in the DNA repair (Hall and Montesano, 1990) or mitotic activity of the target cells (Cohen and Ellwein, 1990) in various organs can vary strongly between different species.

E. Molecular Dosimetry Considerations

Molecular dosimetric methods may provide a sensitive assay for assessing target dose. If strong correlations can be established between adduct formation and mutation induction, these methods may allow a means of mutagenic risk assessment (Ehrenberg et al., 1974; Sobels, 1982). The stage-specific mutagenic action of chemicals must be considered in this context (see also Section V). Of utmost importance for establishing a good comparative database between DNA adduct formation and mutation induction *in vivo* is a method for partitioning the various germ-cell stages for the DNA adduct assay.

IV. SOMATIC CELL MUTAGENESIS

For mutational risk assessments that focus on health risks to the exposed individual (as distinct from effects on subsequent generations), the appropriate potency or dose-response relationships are those involving somatic mutation. Although risk assessments with somatic mutation have not been done to our knowledge, there is a plethora of information on somatic mutation in a wide variety of systems. The so-called short-term tests for genotoxicity generally involve vegetative or somatic cells, and there are, in addition, well-developed somatic mutation assays for gene mutation, chromosome aberration, and aneuploidy that can be applied to either cells *in vitro* or animals and people *in vivo*.

A. Short-Term Assays

The assays used in short-term genotoxicity testing are designed primarily for rapid yes-no answers under conditions of maximum sensitivity for mutation detection. The available testing systems range from free DNA through prokaryotes

METHODS FOR DOSE AND EFFECT ASSESSMENT   85

to eukaryotes, and from isolated cells *in vitro* to intact animals. The protocols are systematized for routine use with unknown chemicals and frequently involve external metabolic activation. Often the biological material has been rendered hypersensitive by inhibiting repair or maximizing access to the agent. Typically, a wide range of doses is tested, not so much for a dose response, but in a search mode for the correct range to adequately determine whether the agent is mutagenic within the tolerance bands dictated by its toxicity. The data are seldom described as response slopes, although the tradition in the field is to publish the raw information at all doses tested so that slopes can be reconstructed.

Several systems have been developed to convert the data for particular short-term tests into a dose response, e.g., Margolin et al. (1981) and Moore and Felton (1983) for the *Salmonella* reverse mutation system, and Margolin (1988) for other assays. However, the combining of such results across tests is fraught with difficulties and has not met with great success or acceptance. Any given set of responses is confounded by the mixture of sensitivity, toxicity, and scale of what is being tested. One must reach across tests with widely different target sizes per cell or with orders of magnitude differences in the numbers of cells being tested. Essentially, single assays must be compared with the typically fivefold structure of the *Salmonella* reverse mutation assay. Each assay has its own confounders, including range of response, vulnerability to secondary effects, and toxicity. And finally, each assay has its own scale of relevance to human risk, including whether the endpoint is limited to DNA damage or is truly mutational, whether the metabolism and repair are analogous to the human, whether chromosomal effects can be expressed, and whether the exposure conditions, including timing, bear some relationship to human exposures. It is not at all surprising that attempts to pool such results for potency have not met with general acceptance (see review and data by McCann et al., 1984).

An alternative approach to comparing short-term tests is based on using the lowest effective dose (LED) and highest ineffective dose (HID) (Waters et al., 1988). For positive outcomes in reasonable protocols, the LED typically lies in the region just above the sensitivity threshold for the method being used. Thus, the LED values for different tests on an agent are roughly normalized by the sensitivity of each assay and are a convenient and efficient way to compare the results. The HID plays a similar role for the negative outcomes and in the Waters approach is plotted in the negative direction, thus keeping the responses separate but neatly summarizing them across the spectrum of tests. This method is now being used by the International Agency for Research on Cancer (IARC) to summarize the mutagenicity data for their carcinogen assessment process.

The International Commission for Protection against Environmental Mutagens and Carcinogens (ICPEMC) Committee 1 has recently completed a variation of the Waters method in which positive (LED) and negative (HID) results are numerically combined and the data, by chemical, organized hierarchically into test scores for replicates with the same test, class scores for similar tests, *in vitro* and *in vivo* family scores for their respective classes, and, finally, a single agent score called Sa for the entire data set (Lohman et al., 1992; Moore et al., 1992; Mendelsohn et al., 1992). In developing this system and applying it to a database containing thousands of outcomes involving 100 or so tests with 117 chemicals, the investigators were able to identify the optimal conditions for merging the positive and negative outcomes and minimizing inherent differences between tests. Thus, the method includes an elaborate scheme for measuring and compensating for the inherent sensitivity to dose of each test system. The final Sa for each chemical ranges from about +50 to –50 and is an overall rank of the mutagenicity of the chemical, including its yield of positive events, the consistency of the response within and across tests, and its toxicity. While not a potency in its present form, this score could be used for comparative ranking of mutagenicity. It could also be easily modified to emphasize somatic vs. heritable effects or to increase the weighting of tests based on relevance to the human.

B. Other Laboratory-Based Mutation Tests

A variety of testing systems have been developed for mechanistic and other research applications in which careful dose-response outcomes can be observed. Much of our knowledge of the inherent shapes and other properties of mutational dose response comes from such systems as applied to prokaryotes, eukaryotic cells *in vitro,* and whole animals. ICPEMC conducted a detailed study of such responses some years ago in the context of whether or not there are thresholds for mutagenicity (Ehling et al., 1983). A similar exercise with an orientation toward absolute or relative potency is feasible and of some interest to the subject at hand. One could also evaluate such data for relevance to human or biotic risk, and conceivably, a coordinated effort could be mounted to use selected systems to generate dose-response relationships for chemicals of regulatory interest.

C. Human Mutation Testing

Over the past decade or two, there have been major advances in our ability to measure somatic mutation frequencies directly in human subjects. At present, there are five test systems that apply to human peripheral blood lymphocytes or erythrocytes and measure locus-specific gene loss or forward mutation. These include the HPRT (Nicklas et al., 1987), HLA (Morley et al., 1990), and TCR (Akiyama et al., 1992) assays in lymphocytes and the glycophorin A (Langlois et al., 1990) and hemoglobin (Bernini et al., 1990) assays in erythrocytes. These assay systems are now being used to define the background behavior and variability of mutant frequencies in human populations. Typically, mutant frequencies increase slowly with age and show considerable interperson variability. They respond dramatically to repair deficiencies, showing one- and two-order-of-magnitude increases in xeroderma pigmentosum, ataxia telangiectasia, and Bloom's syndrome compared to the repair-competent genotype (Bigbee et al., 1989; Cole et al., 1991; Langlois et al., 1989). They also indicate genotoxicity in human exposures to radiation, chemotherapy, and tobacco smoking as well as in preliminary studies on occupational exposures. Preliminary data suggest that the assays have varying capacities for long-term memory of mutagenic events. At present, only the glycophorin mutation assay seems to be able to integrate over the lifetime of a human being.

As this field develops, one can anticipate that we will soon be able to define risk-related semiquantitative responses directly in the human. Formal potency, in the sense of mutation rate change per amount of substance, is unlikely for two reasons: (1) mutant frequency as measured in these systems is not synonymous with mutation rate because of confounding by clonal expansion of mutant cells and uncertainty about the number of cells at risk and (2) the impossibility of deliberately exposing human subjects to graded doses of environmental mutagens. However, it is equally clear that in special situations we will have new and highly relevant mutational data for a number of agents where adventitious or medical exposure to human populations has occurred. The clearest example of this is the evolving data set on atom-bomb survivors in Japan (Akiyama et al., 1992; Langlois et al., 1987). The present measurements on several hundred survivors show linear, nonthresholded responses when plotted against estimated individual radiation dose. The data are highly variable at the individual level, but agglomerate very effectively into well-behaved population estimates. Realizing that this same population is being followed carefully for adverse health effects, one can expect in short order to have information on the correlation of somatic mutation and cancer, heart disease, or even life expectancy. Admittedly this is a very special and, one hopes, singular case, but some approximation to this may be possible for cancer chemotherapy, smoking, and some other widespread or deliberately administered environmental agents. Having information on a subject of direct relevance for human risk estimation is a very powerful opportunity and, if nothing else, should allow validation of the variety of indirect methods of estimation that are currently being practiced.

In the same context, one should not forget that a parallel mutational endpoint involving chromosomal aberration has been available for many years in human subjects (Evans, 1988; Bender et al., 1988). Aberrations show very similar applicability in the atom-bomb survivors or in many other examples of human exposure (Awa et al., 1984). New methods of detection of balanced translocations promise to greatly facilitate the practicality and sensitivity of this approach (Pinkel et al., 1988). Similarly, DNA adducts are detectable in somatic cells of workers receiving occupational exposures of polycyclic aromatic hydrocarbons, and in cancer chemotherapy patients (Santella et al., 1990).

Risk assessments have yet to be done with any of these human-based assays, and are expected to be useful only in very limited situations. However, the opportunity to make the measurements directly in the human provides a very important and unique future opportunity for these methodologies.

V. GERM CELL MUTAGENESIS

The mechanism of mutation induction is a complex process which may involve metabolic activation or detoxification of a chemical mutagen, and the induction of DNA adducts as well as DNA repair and mutation fixation. Further, gametogenesis in mammals is associated with cell differentiation at the genetic, morphologic, and metabolic levels. Germ-cell stages exhibit differences in DNA synthesis activity, and the ability to repair DNA damage as well as the chromosome-associated proteins. The sensitivity to mutation induction may be influenced by such factors due to the accessability of DNA to chemical mutagens, and the interval between DNA damage induction and the next round of DNA replication as well as the repair of DNA damage. Such qualitative and quantitative differences indicate the complexities of mutation induction *in vivo* and emphasize that no single *in vitro* test system can adequately represent the *in vivo* situation. It is generally accepted that germ-cell mutagenesis in humans can most adequately be represented by an *in vivo* mammalian germ-cell test system due to the phylogenetic and physiological similarities. In this section, information regarding some major principles of chemical mutagenesis in the mouse and their implications for an adequate human genetic risk estimation are discussed. We rely on experimental information for gene mutations and do not include data for chromosome structural alterations.

A. Mammalian Germ Cell Mutagenesis

1. *Molecular Characterization of Ethylnitrosourea (ENU)-Induced Mutations*

To date, six independent ENU-induced mutations in germ cells of the mouse have been characterized by direct DNA sequencing or deduced from their amino acid substitutions (Popp et al., 1983; Lewis et al., 1985; Peters et al., 1985; Zdarsky et al., 1990; Lewis, personal communication; Peters, personal communication), and all have been shown to be base substitutions affecting an AT site. The O^6-ethylguanine adduct has been suggested to be the most relevant DNA adduct which leads to mutation (Loveless, 1969; van Zeeland et al., 1985). The mutation resulting from O^6-ethylguanine mispairing is predicted to be a GC-to-AT base substitution. This predicted mechanism of ENU mutagenesis has been experimentally confirmed in *E. coli* (Richardson et al.,

1987) and *Salmonella* (Zielenska et al., 1988), and also predominates in *Drosophila* (Fossett et al., 1990; Pastink et al., 1989, 1990) as well as mammalian cells in culture (Eckert et al., 1988; Vrieling et al., 1988). Thus, the most relevant DNA adduct in germ cells of the mouse differs from that in other test systems. A wide variety of base adducts may be formed following the interaction of ENU with DNA (Montesano, 1981; Singer et al., 1978). Although the O^6-ethylguanine adduct is the most frequent base ethylation which leads to mispairing, its lack of involvement in ENU germ-cell mutagenesis suggests an efficient repair mechanism. In contrast, there is a lack of evidence for an efficient repair mechanism of O-ethyl pyrimidine adducts, which would also lead to base mispairing (Brent et al., 1988). Results therefore emphasize the complexities of germ-cell mutagenesis in mammals and support the assumption that the best experimental system to represent germ-cell mutagenesis in humans is *in vivo* germ-cell test systems in laboratory animals.

2. Test Systems

Table 5 lists a number of mutation test systems developed to screen for transmitted, genetically validated germ-cell mutations in the mouse. A variety of mutational classes may be recovered by the different methods, including recessive and dominant mutations as well as biochemical or immunological mutants. The methods are not mutually exclusive, so experiments may be designed to systematically screen for more than one genetic endpoint in the same experimental population (Ehling et al., 1985).

The specific locus method developed by Russell (1951) is the most efficient method to screen for transmitted germ-cell mutations in the mouse. It has provided virtually all experimental data on factors affecting the mutation process in germ cells of mammals. The advantages of the specific locus test are that the methods to screen for mutations are simple and fast, so that the large numbers of animals required to study an infrequent event such as mutation can be screened. Finally, the animals are alive at examination, and recovered presumed mutations can be subjected to a genetic confirmation test. The specific locus method and results in the mouse have been discussed previously in detail (Ehling and Favor, 1984; Searle, 1974; Selby, 1981). For radiation mutagenesis, dose, dose rate, dose fractionation, dose fractionation interval, radiation quality, and germ-cell stage have all been shown to affect the induced specific-locus mutation rate. For chemical mutagens, germ cell-stage specificity plays an important role. (See the extensive review by Lyon, 1981.) In contrast to radiation, where differences in the relative sensitivity to mutation induction exist among the different spermatogenic stages, qualitative as well as quantitative differences in the sensitivity to mutation induction among the stages may exist for chemicals. Table 6 illustrates this principle for a group of ethylating agents studied in Neuherberg employing the specific locus test. The mutagens diethyl sulfate (DES) and ethyl methanesulfonate (EMS) are effective

Table 5. Tests Developed to Detect Transmitted Mutations in Mouse Germ Cells

Test	Ref.	Methods
Specific locus	Russell, 1951	F_1, external visible traits
Specific locus	Lyon and Morris, 1966	
Recessive lethals	Lüning, 1971	F_2 backcross, embryonic lethals
Dominant visibles	Searle, 1974	F_1, external visible traits
Dominant skeletal	Ehling, 1966; Selby and Selby, 1977	F_1, skeletal preparations
Dominant cataract	Kratochvilova and Ehling, 1979	F_1, ophthalmological examination
Dominant fitness	Green, 1968	F_2, litter size effects
Electrophoretic	Soares, 1979	F_1, variant protein
	Johnson and Lewis, 1981; Pretsch et al., 1982, 1986; Peters et al., 1986	Electrophoretic pattern
Enzyme activity	Charles and Pretsch, 1982, 1986	F_1, specific enzyme activity
Histocompatibility	Bailey and Kohn, 1965	F_1, skin graft rejection by day 80

Table 6. Germ Cell Stage Sensitivity to Mutation Induction in Male Mice

Interval (d)	EMS[a]	DES[b]	ENU[c]
1– 4	0/10,950	0/5,617	5/7.611
5– 8	7/8,276	6/5,271	
9–12	8/7,624	2/6,048	3/4,763
13–16	2/10,816	0/6,375	
17–20	1/11,586	1/6,694	2/3,168
21–42	0/5,138	0/2,609	
>43	2/28,181	0/13,551	173/58,211

[a] Ethyl methanesulfonate; Ehling and Neuhäuser-Klaus, 1989b.
[b] Diethyl sulfate; Ehling and Neuhäuser-Klaus, 1988a.
[c] Ethylnitrosurea; Ehling and Neuhäuser-Klaus, 1984; Favor et al., 1990a.

in the late spermatid and early spermatozoa stages and not effective in spermatogonia. Methyl methanesulfonate (MMS) has the same pattern of germ cell-stage sensitivity to mutation induction in the specific locus test as DES and EMS (Ehling and Neuhäuser-Klaus, 1990). ENU is most effective in stem cell spermatogonia. However, ENU is mutagenically active in postspermatogonial

Table 7. Mutagenic Activity of Chemicals with Conclusive Specific Locus Mutation Test Results in Stem Cell Spermatogonia and Postspermatogonial Stages of the Mouse

Chemical	Postspermatogonia	Spermatogonia
Ethyl nitrosourea	+	+
Methyl nitrosourea	+	+
Procarbazine	+	+
Triethylenemelamine	+	+
Acrylamide	+	+
Chlorambucil	+	−
Cyclophosphamide	+	−
Diethyl sulfate	+	−
Ethyl methanesulfonate	+	−
Mitomycin C	−	+
6-Mercaptopurine	−	−
Adriamycin	−	−
Platinol	−	−
Urethane	−	−

Adapted from Ehling (1978, 1980), Ehling and Neuhäuser-Klaus (1992), Lyon (1981), and Russell (1990).

stages, but mutations mainly occur as mosaics. Table 7 lists the qualitative results for all chemicals with conclusive specific locus mutation results in both postspermatogonial stages and stem cell spermatogonia. All possible patterns of stage sensitivity are apparent. One group of chemicals is active in both postspermatogonial stages and stem cell spermatogonia. A second group of chemicals is active in postspermatogonial stages but not in stem cell spermatogonia. Mitomycin C is not active in postspermatogonia, but is active in stem cell spermatogonia. Finally, a fourth group of chemicals is not active in postspermatogonial stages or in stem cell spermatogonia. It follows that for a complete assessment of the mutagenic activity of a chemical in germ cells of mammals, all germ cell stages must be adequately tested (Ehling, 1989; Ehling and Neuhäuser-Klaus, 1989a, b; Russell, 1989a, b). Further, a risk assessment of a chemical mutagen must take into account the germ cell-stage pattern of mutagenic activity as well as the mode of exposure, chronic or acute.

B. Dose Response

Knowledge of the dose response of a chemical mutagen is required for an adequate estimation of mutagenic effect due to exposure. Otherwise, an assumption of the dose response, usually linearity, must be invoked for a risk estimation. Two chemical mutagens have been extensively studied in stem cell spermatogonia of the mouse with the specific locus test, procarbazine and ENU. Results for procarbazine indicate a humped dose response in which there is a dose-related increase in mutagenic response which reaches a maximum and beyond which the mutagenic effect decreases (Ehling and Neuhäuser, 1979). For ENU, a thresholded or quasithresholded dose response has been demonstrated (Favor et al., 1990b). That both data sets indicate nonlinear dose responses emphasizes the difficulties in an estimation of genetic risk due to exposure, especially if there are large differences between the dose ranges for which experimental data are available upon which an extrapolation is to be based and the actual exposure dose for which a genetic risk is to be estimated. Depending upon (1) the shape of the dose response which exists, (2) the dose interval for which experimental data are available, and (3) the dose for which a risk is to be estimated, either an overestimate or an underestimate of the true risk is possible if the extrapolation is performed assuming linearity.

C. Relevant Genetic Endpoints

In a randomly mating natural population, newly induced recessive mutations will result in a mutant phenotype only when they become homozygous. The yield of homozygotes is a function of the square of the allelic frequency, and in the first generation following mutagen exposure would result in an expected negligible increase in the frequency of mutant individuals. Dominant mutations, by definition, express phenotypic effects as a heterozygote. Thus, newly induced dominant mutations would be detected in the F_1 population following mutagen exposure regardless of the genotype or mating scheme in the parental generation. Table 8 lists the estimated incidence and spontaneous mutational frequency for the various classes of genetic disorders in man. The estimated frequency of spontaneous mutations is based on the assumption of mutation-selection equilibrium. Thus, it is directly proportional to the selection coefficient of a mutant allele in the population and the observed incidence of mutants. It is evident from Table 8 that dominant disorders have a significant incidence in human populations, with a large proportion due to newly occurring spontaneous mutations as compared to recessive or irregularly inherited disorders. Thus, dominant disorders are of major concern in an estimation of the genetic risk due to an increased mutation rate.

D. Genetic Risk Estimation

Two extrapolation procedures have been developed to estimate the genetic risk in man based on experimental data from the mouse. A current status of the methodologies and their shortcomings have been discussed recently (Sankaranarayanan, 1991). The first, termed the doubling dose approach, is based on an estimate in the mouse specific-locus test of that dose which results in an induced mutation rate equal to the per generation spontaneous mutation rate, and on an estimate of the spontaneous mutation rate in humans. An

METHODS FOR DOSE AND EFFECT ASSESSMENT

Table 8. Genetic Disorders in Man

Class	Incidence per 10^6	Spontaneous mutation rate per 10^6
Dominant and X-linked	10,000	1,500
Recessive	2,500	—
Irregularly inherited	90,000	450

Data from UNSCEAR (1982).

indirect method for estimating the spontaneous mutation rate to dominant alleles in humans has been outlined by Childs (1981). Given the population incidence and selection coefficient of a dominant disorder, the spontaneous mutation rate is calculated based on the assumption of mutation-selection equilibrium. For congenital cataract, the population incidence is 4×10^{-5}, the selection coefficient is 0.3, and the spontaneous mutation rate is estimated to be 0.6×10^{-5}. For the entire class of dominant deleterious mutations, approximately 14% are estimated to be due to a newly occurring spontaneous mutation. Employing the doubling dose method and assuming a linear dose response, the number of induced dominant genetic disorders due to mutagenic exposure in humans is calculated as follows:

Induced cases = (exposure dose/doubling dose) × spontaneous mutation rate for dominant deleterious mutations

The second extrapolation procedure for estimating the genetic risk in man is termed the direct approach. It is based on an estimate of the induced mutation rate to dominant alleles in mice and on an estimate of the total number of loci in humans which result in dominant genetic disorders relative to the number of loci in humans controlling the indicator phenotype employed in the mouse experimental studies to estimate the induced mutation rate (to date, skeletal and cataract mutations). Employing the direct method and assuming a linear dose response, the number of induced dominant genetic disorders due to mutagenic exposure in humans is calculated according to Ehling (1984a, 1985) as follows:

$$\text{Induced cases} = \frac{\text{induced mutation rate}}{\text{per gamete per dose}} \times \frac{\text{total dominant loci}}{\text{indicator dominant loci}} \times \text{dose}$$

The direct approach avoids the problem of basing the estimate of induced mutations for one genetic endpoint upon experimental mutagenesis data on a different genetic endpoint. However, the estimate of the total number of loci resulting in dominant genetic disorders and the total number of cataract or skeletal loci in humans is critical. It is based on the tabulations of dominant genetic disorders in humans (McKusick, 1986). To date, 1172 dominant genetic disorders have been identified in man, of which 28 cause cataract. This categorization is based on distinct phenotypes, with no proof that they reflect mutations at distinct loci. The converse is also possible, i.e., similar phenotypes may result from distinct loci. Therefore, there is as yet no way to know if these values are under- or over-estimates.

It should be recalled that the sensitivity to mutation induction by chemical mutagens in the mouse is germ cell stage specific. Thus, the doubling dose and induced mutation rate for dominant alleles determined for a chemical mutagen and employed for an estimation of the genetic risk in man is germ cell stage specific. For an acute exposure, chemicals with a mutagenic effect confined to postspermatogonial stages will have a transitory genetic risk confined to conceptions resulting from gametes which were exposed during the sensitive stages. Conceptions occurring from gametes which were exposed in the nonsensitive stages will have a genetic risk of zero associated with the particular exposure. For chemicals with a mutagenic effect in stem cell spermatogonia, a permanent genetic risk will remain following an acute exposure if the induced mutations are not associated with a reduced survival of the carrier cells, since the population of stem cell spermatogonia constantly cycles to reestablish itself. For a chronic exposure, the genetic risk for a chemical will be the combined genetic risk of the chemical for all stages of spermatogenesis. For chemicals with mutagenic effects confined to postspermatogonial stages, the genetic risk associated with chronic exposure will be equal to the genetic risk in the sensitive postspermatogonial stages regardless of the duration of exposure. Further, upon cessation of exposure, the genetic risk will return to zero. For chemicals with mutagenic effects in stem cell spermatogonia, the genetic risk associated with chronic exposure will constantly increase over the duration of the exposure and will remain at the final level after exposure ceases.

E. Female Germ Cells

The complete assessment of a compound for mammalian germ-cell mutagenesis and the associated genetic risk would require results from female germ-cell stages. In addition to radiation, results for the sensitivity to induction of specific locus mutations in oocytes are only available for four chemicals: procarbazine (Ehling and Neuhäuser-Klaus, 1988b), mitomycin C (Ehling, 1984b), triethylenemelamine (Cattanach, 1982), and ENU (Ehling and Neuhäuser-Klaus, 1988b). Like male germ cells, female germ cells show a germ cell-stage effect on the sensitivity to mutation induction which may vary, depending upon the mutagenic treatment employed. No differences in the sensitivity to mutation induction between sexes were observed when the comparisons were based upon the most sensitive stage identified in each sex for a particular compound. The dynamics of gametogenesis in females (for a review, see Searle, 1974) includes a relatively short period of mitotic proliferation in the early embryonic stage. Meiosis is initiated during embryogenesis and proceeds to the late diplotene stage shortly after birth. Oocytes remain in this stage until shortly before ovulation. This is the meiotic stage of primary importance for mutagenic hazard. Within this stage, differences exist in sensitivity to cell killing and mutation induction, depending upon the stage of development of the associated follicle cells. In comparison to the stem cell spermatogonia of males, the important stage in female gametogenesis may be characterized by the absence of cell division and DNA replication. Like stem cell spermatogonia, oocytes are DNA-repair competent. Thus, DNA damage is not likely to accumulate in oocytes, and there may be a long interval between the induction of DNA damage and the next round of DNA replication, which occurs after fertilization.

F. Germ Cell Mutagenesis Considerations

Extrapolation procedures are only as good as the experimental data upon which they are based. When a particular genotoxic event is to be estimated based upon a comparable endpoint in an experimental system, the extrapolation is relatively straightforward. This situation breaks down when the comparable experimental data do not exist. For example, mammalian *in vivo* germ-cell mutagenicity data are only available for a limited number of chemicals and human data are available for none. Given a specific exposure to a genotoxic agent, two options are available for conducting a risk estimation for genotoxic effects. First, experiments could be undertaken to provide the appropriate mammalian germ-cell mutagenicity data, in which case there would be a long lag time before results became available. This philosophy is advocated by the U.S. EPA guidelines for mutagenicity testing (Dearfield et al., 1991). Second, alternative strategies of risk assessment could be developed, employing available chemical, toxicity, and mutagenicity as well as carcinogenicity data. This second consideration is discussed in the following sections.

VI. COMPARATIVE MUTAGENESIS

A. Experimental Approaches: Qualitative vs. Quantitative Comparisons

1. Qualitative Parameters

The evaluation of the genotoxic potential of a chemical is essentially a two-phase process. First there is the qualitative identification and characterization of the mutagenic activity. This first phase, usually referred to as "mutagenicity testing" and for which a small test battery consisting of *in vitro* and *in vivo* assays has been proposed (Ashby, 1986), is primarily designed to determine whether or not the chemical under test is a mutagen. The information provided by mutagen-screening programs has a purely qualitative dimension, i.e., mutational response-mutagenic effectiveness, in the terminology developed by Ehrenberg (Hussein and Ehrenberg, 1975; Osterman-Golkar et al., 1970), is related to the exposure dose or exposure concentration:

$$\text{Effectiveness} = \frac{\text{mutation frequency}}{\text{unit exposure dose}}$$

In the widely used *Salmonella* assay, for instance, mutagenic effectiveness is usually measured by relating the number of revertants per plate to the concentration (micromolar or microgram). This parameter, however, is not suitable for quantitative comparisons because even closely related genotoxins may differ vastly in their mutagenic effectiveness. This is a result of the large number of factors having an impact on the amount of chemical reacting with the DNA: uptake, penetration, and migration of the chemical to the target cells, rate of hydrolysis, and metabolic fate. For instance, in the *Salmonella* plate assay, N-methyl-N-nitrosourea (MNU) was 116-fold more active than N-methyl-N'-nitro-N-nitrosoguanidine (MNNG), despite their close similarity with respect to type (methylation) and distribution (initial ratio of O^6-:N7-methylguanine) of DNA adduction (Bartsch et al., 1983).

2. Quantitative Comparisons

In the second phase, mutagenic agents are assessed quantitatively, the ultimate goal of the second phase being to understand the potential risk to human populations. A focal point in discussions on feasible approaches for assessing potential risks due to genotoxic agents has been (1) the question of how "mutagenic potency" could be quantified and (2) whether cause-effect relationships for adverse effects of genotoxins can be generalized beyond the experimental or environmental conditions under which the correlations are established. The two experimental approaches used to investigate the "potency" of genotoxic agents and to identify correlations between chemical property, type of DNA modification, and genetic damage are determinations of *efficiency,* and *relative efficiency,* respectively (Bartsch et al., 1983; Barbin and Bartsch, 1989; Ehrenberg et al., 1966; Osterman-Golkar et al., 1970; Vogel and Ashby, 1993).

2.1 Efficiency

In the ideal situation, both the absolute number of adducts per nucleotide and the relative distribution of distinct types of DNA adducts resulting from interaction with a genotoxic agent are measured. This method is called molecular dosimetry. The molecular dosimetry approach enables one to quantify the genotoxic effects per unit of DNA modification:

$$\text{Effectiveness} = \frac{\text{mutation frequency}}{\text{number of adducts per nucleotide}}$$

Determination of genotoxic efficiency by molecular dosimetry seems to be the method of choice for quantitative comparisons and extrapolations across species. Unfortunately, such data mainly exist for only a few cellular systems and a limited number of chemical classes, i.e., for small alkylating agents (alkylnitrosamides, dialkylnitrosamines, alkylalkanesulfonates, dialkylsulfates), aromatic amines (2-acetylaminofluorene), and polycyclic hydrocarbons (benzo[a]pyrene). For instance, when mutation induction by ethylating agents in mammalian cells in culture and in bacteria was determined and compared with the frequency of various ethylation products in DNA, a quantitative relationship (Figure 4) was found between the frequency of O^6-ethylguanine and the frequency of gene mutations determined (Heflich et al., 1982; Mohn et al., 1984; Natarajan et al., 1984; Van Zeeland et al., 1985; Van Zeeland, 1988).

2.2 Relative Efficiency

Clearly, the most desirable procedure for risk assessment would be to relate chronic toxic effects, mutation or cancer, to DNA adduction after exposure to genotoxic agents, i.e., the dosimetry approach. Where such data are not available, as is usually the case, two or more genetic endpoints, measured in the same tissue or cell population, may be compared to determine the *relative efficiency* of an agent to produce one type of genetic damage in relation to another one, e.g.:

$$\text{Relative clastogenic efficiency} = \frac{\text{Freq. chromosome aberrations}}{\text{Freq. forward mutations}}$$

Although the actual dose at the DNA target molecule is unknown when using this procedure, it enables one to rank genotoxins on a relative potency scale and has proved its usefulness for studying cause-effect relationships between mutagenesis and relative carcinogenic potency (see Section VI.B).

B. Carcinogenic Potency Ranking: Interspecies Comparisons in Rodents

Extensive cancer bioassays on 770 compounds are available and have provided the database to compare cancer incidence across animal species for a wide variety of compounds (Gold et al., 1984, 1986, 1987), with major emphasis on the large group of alkylating agents. In terms of the underlying mechanisms of this group, analysis of reactions of small AAs (alkylating agents) with DNA in animals showed that neither the absolute amount of reaction with DNA (number of alkylations per nucleotide) nor the reaction at the N7 atom of guanine, a major DNA modification after alkylation, could be correlated with carcinogenic potency (Schoental, 1967; Swann and Magee, 1968, 1971). On the other hand, the relative ability of methylating and ethylating agents to react with the O^6 atom of guanine correlated well with carcinogenic action (O'Connor et al., 1979; Lawley, 1976; Frei et al., 1978), confirming the prediction of Loveless (1969) that the induction of mutations and cancer is correlated with O^6-G alkylation.

In more recent attempts to examine quantitative relationships between modes of DNA modifications and tumor incidence, Bartsch and co-workers (Barbin and Bartsch, 1986, 1989; Bartsch et al., 1983) used median TD_{50} estimates for rodents (mouse, rat, and hamster) and compared these to the nucleophilic selectivity of mono- and bifunctional direct-acting alkylating agents. (TD_{50} is defined as the total dose of carcinogen in milligrams per kilogram of body weight required to reduce by 50% the probability of the animal being tumor-free throughout a standard lifetime; Barbin and Bartsch, 1989.) The nucleophilic selectivity of AAs was expressed by their Swain-Scott (Swain and Scott, 1953) constants s and/or the ratios of N7-/O^6-alkyl-guanine in DNA. A positive linear relationship between the log of TD_{50} estimates and s-values was established for 17, mostly monofunctional, AAs:

$$\log(TD_{50}) = 4.54\,s - 0.36\,(r = 0.88;\ p < 0.005)$$

(Barbin and Bartsch, 1989; Vogel et al., 1990). The tumorigenic potencies (median TD_{50} in milligrams per kilogram of body weight) of alkylating carcinogens varied over a 10,000-fold range in dose, and the empirical relationship enabled monofunctional AAs to be ranked on a relative carcinogenic potency scale (Figure 5, Table 9). By comparison, the variations between TD_{50} estimates for any of these compounds were only 20-fold (variations within routes of administration) or 40-fold (variations between routes) (Barbin and Bartsch, 1989; Ehrenberg et al., 1992). Although further validation and refinement of the correlation is necessary, e.g., determination of actual tissue doses, detoxification reactions, and analysis of repair processes, the above equation may be used to *predict* relative carcinogenic potency for members of this large class of agents.

Similar comparisons were made for agents capable of cross-linking DNA, which consisted of a set of bifunctional alkylating agents (all antineoplastic drugs) and of chloroethylene oxide (CEO). In contrast to the linear relationships between TD_{50} and the s-value established for monofunctional chemicals, no such relationship was found for cross-linking agents. They showed a higher carcinogenic potency than would have been predicted on the basis of their nucleophilic selectivity, and no simple correlation was apparent between the TD_{50} of these AAs and their s-value. Strikingly, their TD_{50}

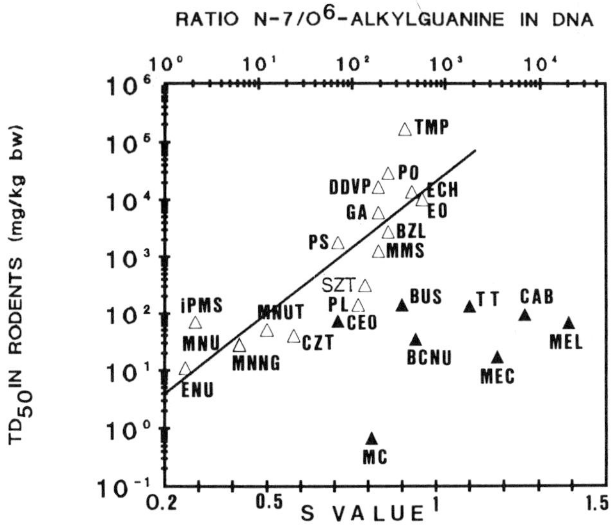

METHODS FOR DOSE AND EFFECT ASSESSMENT

[Figure: Scatter plot with x-axis "RATIO N-7/O⁶-ALKYLGUANINE IN DNA" (top, log scale 10^0 to 10^4) and "S VALUE" (bottom, 0.2 to 1.5), y-axis "TD$_{50}$ IN RODENTS (mg/kg bw)" (10^{-1} to 10^6). Data points include TMP, PO, DDVP, ECH, GA, EO, BZL, PS, MMS, SZT, iPMS, PL, BUS, TT, CAB, MNU, MNUT, CEO, CZT, MNNG, BCNU, MEC, MEL, ENU, MC.]

Figure 5. Relationship between the carcinogenic potency in rodents (median TD$_{50}$ estimate, expressed as lifetime dose in milligrams per kilogram of body weight) of direct alkylating agents and their nucleophilic selectivity (Swain-Scott constant s or initial ratio of 7-alkylguanine/O^6-alkylguanine in DNA), $r = 0.88$, $p <0.001$, $n = 17$. The 7-/O^6-alkylguanine ratio refers solely to compounds that have been shown to react with oxygen atoms in nucleic acid bases. Open symbols, monofunctional agents; closed symbols, cross-linking agents. BCNU, 1,3-*bis*(2-chloroethyl)-2-nitrosoureas; BUS, busulfan; BZL, benzyl chloride; CAB, chlorambucil; CEO, chloroethylene oxide; CZT, chlorozotocin; DDVP, dichlorvos; ECH, epichlorohydrin; ENU, N-ethyl-N-nitrosourea; EO, ethylene oxide; GA, glycidaldehyde; IPMS, isopropyl methanesulfonate; MEC, mechlorethamine; MEL, melphalan; MC, mitomycin C; MMS, methyl methanesulfonate; MNNG, *N*-methyl-*N*'-nitro-*N*-nitrosoguanidine; MNU, *N*-methyl-*N*-nitrosourea; MNUT, *N*-methyl-*N*-nitrosourethane; PL, β-propiolactone; PO, propylene oxide; PS, 1,3-propane sultone: SZT, streptozotocin; TMP, trimethyl phosphate; TT, Thio-TEPA. (Redrawn from Barbin, A. and H. Bartsch (1989), *Mutat. Res.*, 215, 95–107; Vogel E. W., A. Barbin, M. J. M. Nivard, and H. Bartsch (1990), *Carcinogenesis*, 11, 2211–2217.)

values remained within narrow limits, despite the s-values ranging from 0.90 (busulphan) to 1.39 (melphalan, MEL) (Table 9). The data suggested that their ability for cross-linking DNA, but not a definite nucleophilic selectivity, is the predominant cause of their high carcinogenic potency (Vogel et al., 1990).

Comparisons of the relative carcinogenic potencies of chemicals in rodent species and humans also revealed strong correlations for different groups of carcinogens and a great similarity in sensitivity between species (in general, within one order of magnitude) (Ehrenberg et al., 1993; Kaldor et al., 1988). Allen et al. (1988) compared carcinogenic potency estimates (TD$_{25}$s, 25% incidence of exposure-related cancer) from epidemiological data with those from animal cancer bioassays for 23 chemicals and found a highly statistically

Table 9. Comparison of Median TD_{50} (mg/kg b.w.) in Rodents and s-Values for Monofunctional and Cross-Linking Agents[a]

Compound	O^6-/7-Alk guanine[b]	s	TD_{50}[c] (mg/kg bw)
Cross-Linking Agents			
Mitomycin C (MitC)		0.81	0.68
Busulphan (BUS)		0.90	<140
Bis(chloroethyl)nitrosourea (BCNU)		0.94	34
Thio-tepa (t-TEPA)		1.10	131
Mechlorethamine (MEC)		1.18	17
Chlorambucil (CAB)		1.26	92
Melphalan (MEL)		1.39	67
Monofunctional Methylating Agents			
N-Methyl-N-nitrosourea (MNU)	0.11	0.42	28
N-Methyl-N'-nitro-N-nitrosoguanidine (MNNG)	0.11	0.42	103
Dimethylnitrosamine (DMN)	0.10/0.13	—	70.2[d]
Methyl methanesulfonate (MMS)	0.004	0.86	12,200
Trimethyl phosphate (TMP)	—	0.91	167,000

[a] Vogel et al. (1990).
[b] For reference, see Vogel and Natarajan (1982).
[c] From Barbin and Bartsch (1989); bw, body weight.
[d] Procarcinogen.

significant correlation (Figure 6). It was concluded that the animal TD_{25}s and human TD_{25}s are within the same order of magnitude (Allen et al., 1988).

C. Relative Carcinogenic Potency: Relationship to Genetic Activity Profiles

It is now well established that tumorigenesis is a multistage process consisting of a sequential series of both genetic and nongenetic events. Nevertheless, the fact that correlations between relative carcinogenic potency and DNA interaction patterns could be established for groups of structurally related genotoxins suggests a strong impact of the initial events on cancer incidence: the chemical modification of DNA, i.e., the types of DNA adducts formed in relation to their repair or persistence. The question as to the major determinants of these correlations initiated a number of studies. In the absence of sufficient data on molecular dosimetry, the concept followed was to first characterize, by chemical reaction parameters, structurally related genotoxins and then estimate their ability to produce a genetic effect in relation to another, using the "relative efficiency approach". The relative ranking scale established in this way for individual members of groups of chemicals was then compared with their positions on the relative carcinogenic potency scale.

METHODS FOR DOSE AND EFFECT ASSESSMENT

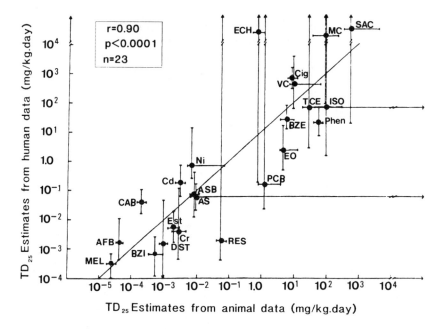

Figure 6. Relationship between carcinogenic potencies of chemicals in humans and rodents. (Ehrenberg, et al., in press). AFB, aflatoxin B_1; As, arsenic; ASB, asbestos; BZE, benzene; BZI, benzidine; CAB, chlorambucil; Cd, cadmium; Cig, cigarette smoke; Cr, chromium; DST, diethylstilbestrol; ECH, epichlorohydrin; EO, ethylene oxide; Est, estrogens (conjugated); ISO, isoniazid; MC, methylene chloride; MEL, melphalan; Ni, nickel; PCB, polychlorinated biphenyl; Phen, phenacetin (analgesics containing phenacetin); RES, reserpine; SAC, saccharin; TCE, trichloroethylene; VC, vinyl chloride.

In one study of this type, Roldan-Arjona et al. (1990) compared the relative efficiencies of 11 direct-acting monofunctional alkylating agents in the Ara test of *Salmonella* with their TD_{50}s in rodents. The relative mutagenic efficiency of the alkylating agents was estimated by relating the maximum yield of forward mutations (Ym) to both the dose (Dm) and the number of lethal hits (Zm) at which the maximum yield occurred:

$$\text{Relative mutagenic efficiency} = \frac{Ym}{Zm(-\ln S) \times Dm}$$

where Ym was the number of induced mutants at the maximally effective dose (Dm in micromoles per plate) and Zm ($-\ln$ survival) the number of lethal hits giving the maximum number of induced mutants (Ym). Only when making this three-parameter comparison was a highly significant correlation ($r_{10} = 0.86$) found between the relative mutagenic efficiency of the compounds in the Ara test and their relative carcinogenic potency in rodents, expressed as TD_{50} values (Figure 7).

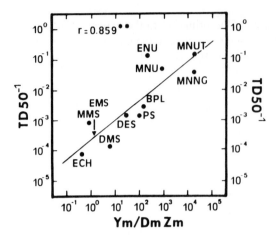

Figure 7. Comparison of carcinogenic potency in rodents vs. mutagenic activity in the Ara test of *S. typhimurium* of 11 alkylating agents, on a double logarithmic scale. The reciprocals of the TD_{50} values were plotted against mutagenic activity expressed as $Y_m/(D_m \times Z_m)$ values. The mutagenic activity of EMS is indicated by an arrow in the absence of a precise TD_{50} value. The solid line is the regression line for ten compounds (EMS excluded): $\ln(TD_{50}^{-1}) = -8.416 + 0.642 \ln[Y_m/(D_m \times Z_m)]$. DES, diethyl sulfate; DMS, dimethyl sulfate; ECH, epichlorohydrin; EMS, ethyl methanesulfonate; ENU, *N*-ethyl-*N*-nitrosourea; MMS, methyl methanesulfonate; MNNG, *N*-methyl-*N*'-nitro-*N*-nitrosoguanidine; MNU, *N*-methyl-*N*-nitrosourea; MNUT, *N*-methyl-*N*-nitrosourethane; BPL, β-propiolactone; PS, 1,3-propane sultone. (From Roldan-Arjona, T., F. L. Luque-Romero, M. Ruiz-Rubio, and C. Pueyo (1990), *Carcinogenesis,* 11, 975–980. With permission; courtesy of Dr. C. Pueyo.)

In a similar study on *E. coli,* the mutagenic (M), recombinogenic (R), and SOS repair induction (I) potencies of nine monofunctional alkylating agents and six bifunctional directly acting alkylating agents were measured as the integral under the yield-dose curve obtained for each event (Quinto and Radman, 1987; Quinto et al., 1990).

Among several combinations tested, a positive correlation between relative carcinogenic potencies (TD_{50} in milligrams per kilogram of body weight) and the product of mutagenic and recombinogenic potencies was found for the monofunctional agents (Quinto et al., 1990). With the cross-linking agents, however, forward mutation activity was detected for only two (mitomycin C and thiotepa) of the six agents. Except for bis(2-chloroethyl)ether, all agents were recombinogenic and induced SOS repair. When the three parameters M, R, and I for these bifunctional alkylating agents were compared, either separately or in combination with the respective carcinogenic potencies in rodents, a highly significant correlation was obtained with both the recombinogenic and SOS-inducing potentials, while there appeared to be a lack of correlation between the mutagenic and carcinogenic potencies (Quinto and Radman, 1987). However, the low activity of cross-linking agents in the forward mutation assay

METHODS FOR DOSE AND EFFECT ASSESSMENT 103

with *E. coli* may be due primarily to the experimental design, since growth prior to mutation selection increased the mutagenic effectiveness of several of these agents (Zijlstra, 1989).

The criterion used in a *Drosophila* repair assay was to determine, again for groups of structurally related chemicals, compound-specific hypermutability indices. Hypermutability was defined as the enhancement seen when measuring the induction of recessive lethal mutations in excision repair-defective (exr⁻) cells and in repair-proficient (exr⁺) genotypes:

$$\text{Hypermutability index} = \frac{\text{mutation induction}_{exr^-}(F-)}{\text{mutation induction}_{exr^+}(F+)}$$

This analysis revealed a positive linear relationship between the hypermutability index (enhancement ratio) and an increased ability for N-alkylation in relation to O-alkylation in DNA for a series of 18 chemicals, representing either monofunctional or cyclic alkylating agents (Figure 8):

$$f_{exr^-}/f_{exr^+} = 12.4s - 1.9 \qquad (r = 0.79;\ p < 0.01)$$

where f_{exr^-}/f_{exr^+} denotes the enhancement ratio for mutation induction and s the Swain-Scott constant (Vogel, 1989). This hypermutability effect, seen in the exr⁻ condition, increases with the preference of agents for alkylating nitrogens in DNA and has therefore been attributed to the level and persistence of N-alkylated bases or of secondary lesions derived from these adducts.

The striking parallelism between the relative potency ranking of alkylating carcinogens in rodents (Figure 5) and the ranking obtained for the same group of carcinogens in the *Drosophila* repair assay (Figure 8) suggests a strong function of DNA repair in the removal of N-alkylation damage. In rodents, efficient, error-free repair of alkylated ring nitrogens appears to be the key mechanism responsible for the high exposure doses needed for tumor incidence by strong N-alkylators (S_N2 mechanism). In support of this, the up to 20-fold increases in mutation rates found in *Drosophila* in the absence of excision repair shows that products of N-alkylation damage, such as apurinic sites (Laval et al., 1990), are efficient promutagenic lesions if not repaired.

At this stage it is interesting to consider the test performance of alkylating carcinogens in the specific-locus test against the background of their nucleophilic selectivity (s-values). Interestingly, efficient, error-free repair seems to be the reason why most monofunctional alkylating agents of high nucleophilic selectivity failed to produce significant numbers of specific-locus mutations in early germ cells in the mouse. This point is exemplified in Table 10 for diethyl sulfate (DES), ethyl methanesulfonate (EMS), methyl methanesulfonate (MMS) and ethylene oxide (EO). Thus, for genotoxins with preference for N-alkylation

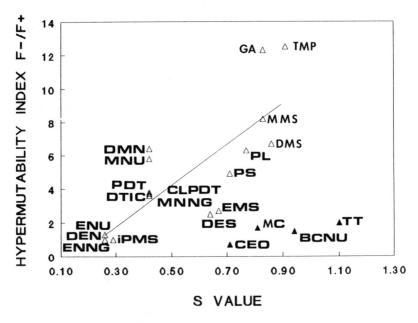

Figure 8. Relationship between the hypermutability index F−/F+ (f_{exr-}/f_{exr+}) in *Drosophila* of 18 alkylating agents and their nucleophilic selectivity. ($r = 0.79$; $p < 0.01$; $n = 18$. The F−/F+ index is the ratio of mutation frequencies induced in an excision repair-deficient strain over that induced in the wild-type (Vogel, 1989). Open symbols, monofunctional; closed symbols, cross-linking agents. BCNU, 1,3-bis(2-chloroethyl)-2-nitrosoureas; CEO, chloroethylene oxide; DEN, *N*-nitrosodiethylamine; DES, diethyl sulfate; DMN, *N*-nitrosodimethylamine; DMS, dimethyl sulfate; DTIC, 5-(3,3-dimethyl-1-triazeno)imidazole-4-carboxamide; EMS, ethyl methanesulfonate; ENNG, *N*-ethyl-*N*′-nitro-*N*-nitrosoguanidine; ENU, *N*-ethyl-*N*-nitrosourea; GA, glycidaldehyde; IPMS, isopropyl methanesulfonate; MC, mitomycin C; MMS, methyl methanesulfonate; MNNG, *N*-methyl-*N*′-nitro-*N*-nitrosoguanidine; MNU, *N*-methyl-*N*-nitrosourea; PDT, 1-phenyl-3,3-dimethyltriazene; CLPDT, 1-(2,4,6-trichlorophenyl)-3,3-dimethyltriazene; PL, β-propiolactone; PS, 1,3-propane sultone; TMP, trimethyl phosphate; TT, Thio-TEPA.

in DNA, there seems to exist a parallelism in the high doses required for tumor incidence in the somatic tissue of rodents (Table 9) and their low genotoxic effectiveness in immature male germ cells of the mouse (Table 10).

D. Comparative Mutagenesis Considerations

Comparisons of data on the relative tumorigenic potency in rodents and relative genetic efficiencies in three nonmammalian systems reveal correlations between the DNA interaction patterns and genetic activity profiles of monofunctional and cross-linking alkylating agents. On the relative ranking scale, carcinogenic potency varies over a 10,000-fold range in dose (Barbin and Bartsch, 1989), with the extremes having the following characteristics:

Table 10. Induction of Specific Locus Mutations in (102/E1 × C3H/E1) F_1 Male Mice[a] Vs. Carcinogenicity Data

Compound	s-Value	TD_{50} (mg/kg bw)	Germ cell stage				Ref.
			Postspermatogonial[b]		Spermatogonia		
			Activity	DD (mg/kg)	Activity	DD (mg/kg)	
Monofunctional							
Ethyl nitrosourea	0.26	11	−/+[c]		+	4	5,6
Procarbazine, HCl[d]	—	262	+		+	110	9,10
Diethyl sulfate	0.64	[100]	+	17	−		3
Ethyl methanesulfonate	0.67	NA	+	9	−		4
Methyl methanesulfonate	0.83	12200	+	2	−		7,8
Ethylene oxide	0.96	10100	+		−		
Cross-Linking							
Chlorambucil	1.26	92	+		−		
Mitomycin C	0.81	0.68	−		+	1	8
Mechlorethamine	1.18	17	+	0.1	−		1
Cyclophosphamide[e]	—	2080	+	6	−		2
Platinol, cisplatin	20		−		−		
Miscellaneous							
6-Mercaptopurine			−		−		
Adriamycin			−		−		
Urethane			−		−		

DD, doubling dose in mg/kg body weight (bw); NA, not available.

[a] From Neuhäuser-Klaus, personal communication.
[b] For specification of different postspermatogonial germ cell stages, see cited references.
[c] Among 15,542 offspring, only one complete specific locus mutation, but nine mosaic mutations were obtained. References (specific-locus test): (1) Ehling and Neuhäuser-Klaus, 1989a; (2) Ehling and Neuhäuser-Klaus, 1988a; (3) Ehling and Neuhäuser-Klaus, 1988b; (4) Ehling and Neuhäuser-Klaus, 1989b; (5) Favor et al., 1990a; (6) Ehling and Neuhäuser-Klaus, 1984; (7) Ehling and Neuhäuser-Klaus, 1990; (8) Ehling, 1978; (9) Kratochvilova et al., 1988; (10) Ehling and Neuhäuser, 1979; (11) Ehling et al., 1988.
[d] Since procarbazine is a promutagen, an s-value cannot be determined. However, in in vivo studies, the ratio O^6 : N-7 alkylguanine has been determined and an "s-value" of the methylating species was determined to be similar to MNU, i.e., 0.42 (Barbin and Bartsch, 1989; Vogel et al., 1990).
[e] Cyclophosphamide requires metabolic activation in vivo. Studies comparing O^6 to N7 DNA adduct formation are not available, but a relatively high yield of N7 alkyl guanine is reported with an associated high nucleophilic selectivity.

1. Chemicals of a relatively high carcinogenic potency, as indicated by a low TD_{50} in rodents, either have low nucleophilic selectivity (and therefore mainly act through O-alkylation in DNA) or are capable of cross-linking DNA. The monofunctional members of this group, typified by *N*-ethyl-*N*-nitrosourea (ENU), are active in both the spermatogonia and postspermatogonial stages in the mouse (Table 10) and in *Drosophila*. Cross-linking agents in this group also have a low TD_{50} value in rodents, but generally are not expected to display genetic action in premeiotic stages (exception: TEM, Table 10).
2. A relatively low carcinogenic potential is associated with monofunctional chemicals of high Swain-Scott *s*-values, typified by ethylene oxide (EO), trimethyl phosphate (TMP), glycid aldehyde (GA), or methyl methanesulfonate (MMS). Efficient, error-free repair (by excision repair) of N-alkylation damage appears to be the mechanism responsible for their high TD_{50} in rodents and why they tend to be inactive in repair-competent (spermatogonia) germ cells of the mouse.
3. When the functionality (monofunctional vs. cross-linking) and nucleophilic selectivity (*s*-value and/or the O^6-/N7-alkylguanine ratio) of a chemical are known, two general predictions regarding its relative genotoxic potency should be possible. (1) Tumorigenic potency could be calculated according to the equation $\log (TD_{50}) = 4.54\ s - 0.36$ (Barbin and Bartsch, 1989; Vogel et al., 1990). (2) Monofunctional AAs with *s*-values >0.8 and cross-linking agents are not expected to produce significant numbers of mutations in repair-active germ cells (spermatogonia) in the mouse (see Table 10) and in *Drosophila*.
4. The principal objective of relative potency ranking is the identification of chemical action principles associated with high potency for adverse effects. As a second step, this may lead to prioritization in terms of hazard evaluation and detailed mechanistic studies. The mechanistic studies may concentrate on the types and distribution of DNA adducts, their persistence in relation to repair, and the elaboration of molecular mutation spectra. Of most interest are those chemicals showing an extremely low TD_{50} across species (e.g., mitomycin C).
5. This multi-endpoint analysis and relative potency ranking procedure should be extended to other classes of genotoxins, i.e., those forming "bulky" adducts (aromatic compounds) and those capable of forming cyclic DNA adducts (vinyl chloride, etc.). Correlations similar to those established for small alkylating agents are also expected for other groups of genotoxic carcinogens. However, systematic studies with, e.g., aromatic amines, heterocyclic amines, aflatoxins, and substituted and nonsubstituted polycyclic hydrocarbons have not been conducted yet.
6. In discussing whether certain types of genetic damage might constitute higher genetic risks than others, i.e., point mutations as compared to clastogenic damage or vice versa, it may be a misconception to keep this discussion at the genetic endpoint level. One reason is that DNA modifications may cause both base-pair substitutions and DNA-strand breakage, as was shown for N-alkylation damage (Brookes, 1990; Laval et al., 1990; Osterman-Golkar et al., 1970). Rather, a more relevant question to ask

METHODS FOR DOSE AND EFFECT ASSESSMENT

might be whether promutagenic lesions exist which are repaired only slowly or not at all. Lindahl (1990) considered minor adducts such as 1-methyladenine or 3-methylthymine, lesions which do not seem to be a substrate for DNA repair, as candidates for slowly accumulating DNA damage in long-lived nonreplicating cells.

VII. OVERALL CONCLUSIONS

As illustrated in the previous sections, the complexities of mutagenesis *in vivo* may be approached by different but not mutually exclusive methods. Pharmacokinetic principles may be employed with the goal of determining the concentration × time burden of the target organ to the active chemical species associated with an exposure to a chemical. Identifying the relevant target organ dose, the time integral of concentration, is of utmost importance in predicting the biological response in man associated with the exposure. This information is essential in the case of nonlinear associations of effect to exposure due to saturable kinetics. Such findings allow a more accurate extrapolation to a low dose range as well as estimating genotoxic effects from exposure data. The consequences of a saturable repair or detoxification process are obvious, i.e., a thresholded response. Similarly, a saturable metabolic activation system would result in an exposure dose range for which the relevant target organ dose, the time integral of concentration, is linearly or approximately linearly related to the exposure concentration. At the upper end of this exposure concentration range, the relevant target organ dose reaches a maximum, and beyond this exposure concentration the relevant target organ dose remains constant. Knowledge of the pharmacokinetics of a particular chemical is essential in the interpretation of animal studies or human epidemiological studies of, e.g., positive but not dose-related effects. Furthermore, pharmacokinetics are the basis for species scaling, necessary for the extrapolation of dose-effect relationships observed in animal assays to the human situation.

Molecular dosimetry procedures bypass pharmacokinetics and allow a direct measurement of DNA adduct formation in the relevant target. Given that molecular dosimetry procedures may be developed, pharmacokinetics could then serve as a method to determine the mechanisms required to explain the molecular dosimetry observations. A requirement for molecular dosimetry procedures is the ability to isolate the DNA from the target cell population and to anticipate and measure the relevant DNA adduct. These requirements are not trivial. For example, there are at present no methods available to subdivide the various spermatogenic germ-cell stages in order that DNA may be isolated from the different cell populations. This is of utmost importance, as evidenced by the highly specific germ cell stage-specific action of chemical mutagens. Of equal importance is the ability to measure the various DNA adducts which may ultimately result in mutation. Given that these requirements may be satisfied, the molecular dosimetry methods may allow an accurate measurement of the relevant dose in the target cells, provide insights into the mechanisms of mutation induction, and may substitute for *in vivo* mutagenicity studies.

In vivo mutagenicity data in mammals are the most comparable to the human situation when pharmacokinetic data for species scaling are available. The measured mutagenic effect is dependent on all of the complexities outlined in Section II as well as the processes of mutation fixation, mutation survival, and mutation transmission. The stage-specific action of chemical mutagens has important implications. It defines the upper and lower time limits in determining the relevant dose, the time integral of the concentration of the active chemical species in the target organ. Finally, as indicated above, it identifies the target cells upon which molecular dosimetry procedures may be applied.

Although most comparable to the human situation, *in vivo* mammalian mutagenicity procedures are slow, resource intensive, and could never keep pace with the rapidly expanding group of chemicals with potential mutagenic hazard. Two options are available. Upon identification of a chemical exposure with a potential genotoxic risk and for which the appropriate *in vivo* data are not available, experiments could be undertaken to provide the relevant *in vivo* mutagenicity data. Alternatively, an assessment of the risk associated with exposure to a chemical may be based on an alternative set of data. However, a genetic risk estimation based on such procedures only provides an indication of the relative mutagenic potential, and important information as to mutagenic activity in mammalian germ cells may not be available for a complete assessment of risk *in vivo*.

VIII. PERSPECTIVES

The previous sections have illustrated the complexities of *in vivo* germ-cell mutagenicity and have argued that the best experimental basis for a genetic risk estimation in humans would be *in vivo* mammalian germ-cell mutagenicity data as well as the pharmakokinetic evaluation of the compound in question required for cross-species extrapolation. The paucity of *in vivo* mammalian germ-cell mutagenicity experimental data available has forced a compromise in which genetic risk assessment may be based on the more extensive short-term mutagenicity data available in an evaluation of chemicals. Recently, two independent *in vivo* mutagenicity assays based on transgenic mice have been developed (Gossen et al., 1989; Kohler et al., 1990). Both test systems are based on a λ-shuttle vector containing an easily scorable target gene which is incorporated into the mouse genome. Thus, the system promises to combine the speed and precision of microbial mutagenicity assays with the complexities of the mammalian *in vivo* situation. If these methods prove to accurately reflect the *in vivo* situation, alternative genetic risk estimation procedures may be considered based upon such experimental results. More importantly, due to the speed with which such results may be obtained, one could reasonably envisage a strategy in which the required *in vivo* experiments would be undertaken after identification of a human exposure to a potential chemical mutagen.

REFERENCES

Adamkiewitz, J., O. Ahrens, N. Huh, P. Nehls, and M. F. Rajewsky (1984), High affinity monoclonal antibodies for the specific recognition and quantification of deoxynucleosides structurally modified by N-nitroso compounds, in *N-Nitroso Compounds: Occurrence, Biological Effects, and Relevance to Human Cancer*, O'Neill, I. K., R. C. von Borstel, C. T. Miller, J. Long, and H. Bartsch, Eds., IARC Sci. Publ. No. 57, International Agency for Research on Cancer, Lyon, 581–587.

Akiyama, M., Y. Kusunoki, S. Umeki, N. Nakamura, and S. Kyoizumi (1992), Evaluation of four somatic mutation assays as biological dosimeter in humans, Int. Congr. Radiation Research, Toronto.

Albertini, R. J., K. S. Castle, and W. R. Borcherding (1982), T cell cloning to detect the mutant 6-thioguanine resistant lymphocytes present in human peripheral blood, *Proc. Natl. Acad. Sci. U.S.A.*, 79, 6617–6621.

Allen, B. C., K. S. Crump, and A. M. Shipp (1988), Correlation between carcinogenic potency of chemicals in animals and humans, *Risk Anal.*, 8, 531–544.

Andersen, M. E., H. J. Clewell, III, M. L. Gargas, F. A. Smith, and R. H. Reitz (1987), Physiologically based pharmacokinetics and the risk assessment process for methylene chloride, *Toxicol. Appl. Pharmacol.*, 87, 185–205.

Andrae, U., H. Homfeld, L. Vogl, J. Lichtmannegger, and K.-H. Summer (1988), 2-Nitropropane induces DNA repair synthesis in rat hepatocytes in vitro and in vivo, *Carcinogenesis*, 9, 811–815.

Arrand, J. E. and A. M. Murray (1982), Benzpyrene groups bind preferentially to the DNA of active chromatin in human lung cells, *Nucleic Acids Res.*, 10, 1547–1555.

Ashby, J. (1986), The prospects for a simplified and internationally harmonized approach to the detection of possible human carcinogens and mutagens, *Mutagenesis*, 1, 3–16.

Awa, A. A., T. Sofuni, T. Honda, H. B. Hamilton, and S. Fujita (1984), Preliminary reanalysis of radiation-induced chromosome aberrations in relation to past and newly revised estimates for Hiroshima and Nagasaki A-bomb survivors, in *Biological Dosimetry: Cytometric Approaches to Mammalian Systems*, Springer-Verlag, New York, 77–82.

Bailey, D. W. and H. I. Kohn (1965), Inherited histocompatibility changes in progeny of irradiated and unirradiated inbred mice, *Genet. Res.*, 6, 330–340.

Baird, W. K. (1979), The use of radioactive carcinogens to detect DNA modifications, in *Chemical Carcinogens and DNA*, Grover, D. L., Ed., CRC Press, Boca Raton, FL, 59–83.

Balant, L. P. and M. Gex-Fabry (1990), Review. Physiological pharmacokinetic modelling, *Xenobiotica*, 20(11), 1241–1257.

Barbin, A. and H. Bartsch (1986), Mutagenic and promutagenic properties of DNA adducts formed by vinyl chloride metabolites, in *The Role of Cyclic Nucleic Acid Adducts in Carcinogenesis and Mutagenesis,* Singer, B. and H. Bartsch, Eds., IARC Sci. Publ. No. 70, International Agency for Research on Cancer, Lyon.

Barbin, A. and H. Bartsch (1989), Nucleophilic selectivity as a determinant of carcinogenic potency (TD_{50}) in rodents: a comparison of mono- and bi-functional alkylating agents and vinyl chloride metabolites, *Mutat. Res.,* 215, 95–107.

Bartsch, H., C. Malaveille, A. Barbin, and G. Planche (1979), Mutagenic and alkylating metabolites of haloethylenes, chlorobutadienes and dichlorobutenes produced by rodent or human liver tissues, *Arch. Toxicol.,* 41, 249–277.

Bartsch, H., B. Terracini, C. Malaveille, L. Tomatis, J. Wahrendorf, G. Brun, and B. Dodet (1983), Quantitative comparisons of carcinogenicity, mutagenicity and electrophilicity of 10 direct-acting alkylating agents and of the initial O^6:N7-alkylguanine ratio in DNA with carcinogenic potency in rodents, *Mutat. Res.,* 110, 181–219.

Bender, M. A., A. A. Awa, A. L. Brooks, H. J. Evans, L. P. G. Groer, L. G. Littlefield, C. Pereira, R. J. Preston, and B. W. Wacholz (1988), Current status of cytogenetic procedures to detect and quantify previous exposures to radiation, *Mutat. Res.,* 196, 103–159.

Benya, T. J., W. M. Busey, M. A. Dorato, and P. E. Berteau (1982), Inhalation carcinogenicity bioassay of vinyl bromide in rats, *Toxicol. Appl. Pharmacol.,* 64, 367–379.

Beranek, D. T., C. C. Weis, and D. H. Swenson (1980), A comprehensive quantitative analysis of methylated and ethylated DNA using high pressure liquid chromatography, *Carcinogenesis,* 1, 595–606.

Beranek, D. T., R. H. Heflich, R. L. Kodell, S. M. Morris, and D. A. Casciano (1983), Correlation between specific DNA-methylation products and mutation induction at the HGPRT locus in Chinese hamster ovary cells, *Mutat. Res.,* 11, 171–180.

Bernini, L. F., A. T. Natarajan, A. H. M. Schreuder-Rotteveel, P. C. Giordano, J. S. Ploem, and A. Tates (1990), Assay of somatic mutation of human hemoglobins, in *Mutation and the Environment, Part C: Somatic and Heritable Mutation, Adduction and Epidemiology,* Mendelsohn, M. L. and R. J. Albertini, Eds., Wiley-Liss, New York, 57–67.

Bigbee, W. L., R. G. Langlois, M. Swift, and R. H. Jensen (1989), Evidence for an elevated frequency of in vivo somatic cell mutations in ataxia telangectasia, *Am. J. Hum. Genet.,* 44, 402–408.

Bohr, V. A., C. A. Smith, D. S. Okumoto, and P. C. Hanawalt (1985), DNA repair in an active gene: removal of pyrimidine dimers from the dhfr gene of CHO cells is much more efficient than in the genome overall, *Cell,* 40, 359–369.

Bolt, H. M. and J. G. Filser (1983), Quantitative Aspekte der Karzinogenität von Vinylbromid, in *Verhandlungen der Deutschen Gesellschaft für Arbeitsmedizin, Bericht über die 23. Jahrestagung,* Gentner-Verlag, Stuttgart, 433–437.

Bolt, H. M. and J. G. Filser (1984), Olefinic hydrocarbons: a first risk estimate for ethene, *Toxicol Pathol.,* 12, 101–105.

Bolt, H. M. and J. G. Filser (1987), Kinetics and disposition in toxicology. Example: carcinogenic risk estimate for ethylene, *Arch. Toxicol.,* 60, 73–76.

Bolt, H. M., J. G. Filser, and F. Störmer (1984), Inhalation pharmacokinetics based on gas uptake studies. V. Comparative pharmacokinetics of ethylene and 1,3-butadiene in rats, *Arch. Toxicol.,* 55, 213–218.

Bolt, H. M., R. J. Laib, and J. G. Filser (1982), Reactive metabolites and carcinogenicity of halogenated ethylenes, *Biochem. Pharmacol.,* 31, 1–4.

Brent, T. P., M. E. Dolan, H. Fraenkel-Conrat, J. Hall, P. Karran, F. Laval, G. P. Margison, R. Montesano, A. E. Pegg, P. M. Potter, B. Singer, J. A. Swenberg, and D. B. Yarosh (1988), Repair of O-alkylpyrimidines in mammalian cells: a present consensus, *Proc. Natl. Acad. Sci. U.S.A.*, 85, 1759–1762.

Brookes, P. (1990), The early history of the biological alkylating agents, 1918–1968, *Mutat. Res.*, 233, 3–14.

Burkhart-Schultz, K., C. J. Strout, and I. M. Jones (1990), Mouse model for somatic mutation at the HPRT gene — molecular and cellular analyses, *Mutat. Environ. Part C*, 340, 5–14.

Carothers, A. M., G. Urlaub, R. W. Steigerwalt, L. A. Chasin, and D. Grunberger (1986), Characterization of mutations induced by 2-(N-acetoxy-N-acetyl)aminofluorene in the dihydrofolate reductase gene of cultured hamster cells, *Proc. Natl. Acad. Sci. U.S.A.*, 83, 6519–6523.

Cattanach, B. M. (1982), Induction of specific-locus mutations in female mice by triethylenemelamine (TEM), *Mutat. Res.*, 104, 173–176.

Charles, D. J. and W. Pretsch (1982), Activity measurements of erythrocyte enzymes in mice. Detection of a new class of gene mutations, *Mutat. Res.*, 97, 177–178.

Charles, D. J. and W. Pretsch (1986), Enzyme activity mutations detected in mice after paternal fractionated irradiation, *Mutat. Res.*, 160, 243–248.

Chen, R. H., V. M. Maher, and J. J. Mccormick (1990), Effect of excision repair by diploid human fibroblasts on the kinds and locations of mutations induced by (+/–) 7-Beta,8-alpha-dihydroxy-9-alpha,10-alpha-epoxy-7,8,9,10-tetrahydrobenzo[a]pyrene in the coding region of the HPRT gene, *Proc. Natl. Acad. Sci. U.S.A.*, 87, 8680–8684.

Childs, J. D. (1981), The effect of a change in mutation rate on the incidence of dominant and X-linked recessive disorders in man, *Mutat. Res.*, 83, 145–158.

Cohen, S. M. and L. B. Ellwein (1990), Cell proliferation in carcinogenesis, *Science*, 249, 1007–1011.

Cole, J., C. F. Arlett, M. H. L. Green, P. G. Norris, and M. L. Price (1991), An assessment of the circulating T-lymphocyte 6-thioguanine resistance system for population monitoring, in *Human Carcinogen Exposure: Biomonitoring and Risk Assessment*, Oxford University Press, Cambridge, 233–240.

Csanády, G. A. and J. G. Filser (1990), Solvekin: a new program for solving pharmaco- and toxicokinetic problems (abstr.), Int. Workshop on Pharmacokinetic Modelling in Occupational Health, Leysin, Switzerland, March 4–8, 35.

Dearfield, K. L., A. E. Auletta, M. C. Cimino, and M. M. Moore (1991), Considerations in the U.S. Environmental Protection Agency's testing approach for mutagenicity, *Mutat. Res.*, 258, 259–283.

De Jong, P. J., A. J. Grosovsky, and B. W. Glickman (1988), Spectrum of spontaneous mutation at the APRT locus of Chinese hamster ovary cells: an analysis at the DNA level, *Proc. Natl. Acad. Sci. U.S.A.*, 85, 3499–3503.

Denk, B. (1990), Abschätzung des kanzerogenen Risikos von Ethylen und Ethylenoxid für den Menschen durch Speziesextrapolation von der Ratte unter Berücksichtigung der Pharmakokinetik, GSF-Bericht 20/90, GSF-Forschungszentrum für Umwelt und Gesundheit, GmbH, Munich.

Denk, B. and J. G. Filser (1990), Abschätzung des durch Ethylen und Ethylenoxid bedingten kanzerogenen Risikos für den Menschen — Vergleich mit dem Risiko durch endogenes Ethylen, in *Verhandlungen der Deutschen Gesellschaft für Arbeitsmedizin, Bericht über die 30. Jahrestagung*, Gentner-Verlag, Stuttgart, 397–401.

Denk, B., M. Baumann, and J. G. Filser (1989), Pharmacokinetics and hepatotoxicity of 2-nitropropane in rats, *Arch. Toxicol. Suppl.,* 13, 330–332.

Denk, B., W. Kessler, E. Deml, D. Oesterle, and J. G. Filser (1991), 2-Nitropropane: Beziehung zwischen Induktion präneoplastischer Inseln in der Rattenleber und Metabolismus, in *Verhandlungen der Deutschen Gesellschaft für Arbeitsmedizin, Bericht über die 31. Jahrestagung,* Gentner-Verlag, Stuttgart, 563–567.

Dunkelberg, H. (1981), Kanzerogene Aktivität von Ethylenoxid und seinen Reaktionsprodukten 2-Chlorethanol, 2-Bromethanol, Ethylenglykol und Diethylenglykol. I. Kanzerogenität von Ethylenoxid im Vergleich zu 1,2-Propylenoxid bei subkutaner Applikation an Mäusen, *Zentralbl. Bakterid. Parasitenk. Infektionskr. Hyg. Abt. Orig. Reihe B,* 174, 383.

Dunkelberg, H. (1982), Carcinogenicity of ethylene oxide and 1,2-propylene oxide upon intragastric administration to rats, *Br. J. Cancer,* 46, 924.

ECETOC (1982), *Toxicity of Ethylene Oxide and its Relevance to Man,* Tech. Rep. No. 5, European Chemical Industry Ecology and Toxicology Centre, Brussels.

ECETOC (1984), *Ethylene Oxide Toxicology and its Relevance to Man: An Updating of ECETOC Technical Report No 5,* Tech. Rep. No. 11, European Chemical Industry Ecology and Toxicology Centre, Brussels.

ECETOC (1988), *The Mutagenicity and Carcinogenicity of Vinyl Chloride: A Historical Review and Assessment,* Tech. Rep. No. 31, European Chemical Industry Ecology and Toxicology Centre, Brussels.

Eckert, K. A., C. A. Ingle, D. K. Klinedinst, and N. R. Drinkwater (1988), Molecular analysis of mutations in human cells by N-ethyl-N-nitrosourea, *Mol. Carcinogen.,* 1, 50–56.

Ehling, U. H. (1966), Dominant mutations affecting the skeleton in offspring of X-irradiated male mice, *Genetics,* 54, 1381–1389.

Ehling, U. H. (1978), Specific locus mutations in mice, in *Chemical Mutagens,* Vol. 5, Hollaender, A. and F. J. de Serres, Eds., Plenum Press, New York, 233–256.

Ehling, U. H. (1980), Comparison of the mutagenic effect of chemicals and ionizing radiation in germ cells of the mouse, in *Progress in Environmental Mutagenesis,* Alacevic, M., Ed., Elsevier, Amsterdam, 47–58.

Ehling, U. H. (1984a), Schätzung des strahlengenetischen Risikos, in *Tagungsbericht: Strahlung und Radionuklide in der Umwelt, Tagung der Arbeitsgemeinschaft der Großforschungseinrichtungen,* Bonn-Bad Godesberg, 70–72.

Ehling, U. H. (1984b), Methods to estimate the genetic risk, in *Mutations in Man,* Obe, G., Ed., Springer-Verlag, Berlin, 292–318.

Ehling, U. H. (1985), Induction and manifestation of hereditary cataracts, in *Assessment of Risk from Low-Level Exposure to Radiation and Chemicals,* Woodhead, A. D., C. J. Shellabarger, V. Pond, and A. Hollaender, Eds., Plenum Press, New York, 345–367.

Ehling, U. H. (1989), Induction of specific-locus mutations in male mice by diethyl sulfate (DES), *Mutat. Res.,* 214, 329.

Ehling, U. H. and J. Favor (1984), Recessive and dominant mutations in mice, in *Mutation, Cancer, and Malformation,* Chu, E. H. Y. and W. M. Generoso, Eds., Plenum Press, New York, 389–428.

Ehling, U. H. and A. Neuhäuser (1979), Procarbazine-induced specific-locus mutations in male mice, *Mutat. Res.,* 59, 245–256.

Ehling, U. H. and A. Neuhäuser-Klaus (1984), Dose-effect relationships of germ-cell mutations in mice, in *Problems of Threshold in Chemical Mutagenesis,* Tazima, Y., S. Kondo and Y. Kuroda, Eds., Kokusai-bunken, Tokyo, 15–25.

Ehling, U. H. and A. Neuhäuser-Klaus (1988a), Induction of specific-locus and dominant-lethal mutations in male mice by diethyl sulfate (DES), *Mutat. Res.*, 199, 191–198.

Ehling, U. H. and A. Neuhäuser-Klaus (1988b), Induction of specific-locus mutations in female mice by 1-ethyl-1-nitrosourea and procarbazine, *Mutat. Res.*, 202, 139–146.

Ehling, U. H. and A. Neuhäuser-Klaus (1989a), Induction of specific-locus and dominmant lethal mutations in male mice by chlormethine, *Mutat. Res.*, 227, 81–89.

Ehling, U. H. and A. Neuhäuser-Klaus (1989b), Induction of specific-locus mutations in male mice by ethyl methanesulfonate (EMS), *Mutat. Res.*, 227, 91–95.

Ehling, U. H. and A. Neuhäuser-Klaus (1990), Induction of specific-locus and dominant lethal mutations in male mice in the low dose range by methyl methanesulfonate (MMS), *Mutat. Res.*, 230, 61–70.

Ehling, V. H. and A. Neuhäuser-Klaus (1992), Reevaluation of the induction of specific-locus mutations in spermatogonia of the mouse acrylamide, *Mutat. Res.*, 283, 185–191.

Ehling, U. H., D. J. Charles, J. Favor, J. Graw, J. Kratochvilova, A. Neuhäuser-Klaus, and W. Pretsch (1985), Induction of gene mutations in mice: the multiple endpoint approach, *Mutat. Res.*, 150, 393–401.

Ehling, U. H., J. Kratochvilova, W. Lehmacher, and A. Neuhäuser-Klaus (1988), Mutagenicity testing of vincristine in germ cells of male mice, *Mutat. Res.*, 209, 107–113.

Ehling, U. H., D. Averbeck, P. A. Cerutti, J. Friedman, H. Greim, A. C. Kolbye, Jr., and M. L. Mendelsohn (1983), Review of the evidence for the presence or absence of thresholds in the induction of genetic effects of genotoxic chemicals, *Mutat. Res.*, 123, 281–341.

Ehrenberg, L. and S. Hussain (1981), Genetic toxicity of some important epoxides, *Mutat. Res.*, 86, 1–113.

Ehrenberg, L. and S. Osterman-Golkar (1980), Alkylation of macromolecules for detecting mutagenic agents, *Teratogen. Carcinogen., Mutagen.*, 1, 105–127.

Ehrenberg, L., D. Segerbäck, H. Bartsch, and A. Barbin (in press), Quantitative cross-species extrapolation using macromolecular adducts, in *SCOPE 52, Methods to Assess DNA Damage and Repair: Interspecies Comparisons,* R. G. Tardiff, P. H. M. Lohman, and G. N. Wogan, Eds., John Wiley & Sons, Ltd.

Ehrenberg, L., K. D. Hiesche, S. Osterman-Golkar, and I. Wennberg (1974), Evaluation of genetic risks of alkylating agents: tissue dose in the mouse from air contaminated with ethylene oxide, *Mutat. Res.*, 24, 83–103.

Ehrenberg, L., U. Lundquist, S. Osterman, and B. Sparrman (1966), On the mutagenic action of alkanesulfonic esters in barley, *Hereditas,* 56, 277–305.

Ehrenberg, L., S. Osterman-Golkar, D. Segerbäck, K. Svensson, and C. J. Calleman (1977), Evaluation of genetic risks of alkylating agents. III. Alkylation of hemoglobin after metabolic conversion of ethene to ethylene oxide in vivo, *Mutat. Res.*, 45, 175–184.

Evans, H. J. (1988), Mutation cytogenetics: past, present and future, *Mutat. Res.*, 204, 355–363.

Favor, J. (1983), A comparison of the dominant cataract and recessive specific-locus mutation rates induced by treatment of male mice with ethylnitrosourea, *Mutat. Res.*, 110, 367–382.

Favor, J., A. Neuhäuser-Klaus, and U. H. Ehling (1990a), The frequency of dominant cataract and recessive specific-locus mutations and mutation mosaics in F_1 mice derived from post-spermatogonial treatment with ethylnitrosourea, *Mutat. Res.*, 229, 105–114.

Favor, J., M. Sund, A. Neuhäuser-Klaus, and U. H. Ehling (1990b), A dose-response analysis of ethylnitrosourea-induced recessive specific-locus mutations in treated spermatogonia of the mouse, *Mutat. Res.*, 231, 47–54.

Filser, J. G. and H. M. Bolt (1979), Pharmacokinetics of halogenated ethylenes in rats, *Arch. Toxicol.*, 42, 123–136.

Filser, J. G. and H. M. Bolt (1981), Inhalation pharmacokinetics based on gas uptake studies. I. Improvement of kinetic models, *Arch. Toxicol.*, 47, 279–292.

Filser, J. G. and H. M. Bolt (1984), Inhalation pharmacokinetics based on gas uptake studies. VI. Comparative evaluation of ethylene oxide and butadiene monoxide as exhaled reactive metabolites of ethylene and 1,3-butadiene in rats, *Arch. Toxicol.*, 55, 219–223.

Filser, J. G., B. Denk, M. Törnqvist, W. Kessler, and L. Ehrenberg (1992), Pharmacokinetics of ethylene in man; body burden with ethylene oxide and hydroxyethylation of hemoglobin due to endogenous and environmental ethylene, *Arch. Toxicol.*, 66, 157–163.

Fossett, N. G., P. Arbour-Reily, G. Kilroy, M. McDaniel, J. Mahmoud, A. B. Tucker, S. H. Chang, and W. R. Lee (1990), Analysis of ENU-induced mutations at the *Adh* locus in *Drosophila melanogaster*, *Mutat. Res.*, 231, 73–85.

Frei, J. V., D. H. Swenson, W. Warren, and P. D. Lawley (1978), Alkylation of deoxyribonucleic acid in various organs of C57BL mice by the carcinogens N-methyl-N-nitrosourea, N-ethyl-N-nitrosourea and ethyl methanesulphonate in relation to induction of thymic lymphoma, *Biochem. J.*, 174, 1031–1044.

Gargas, M. L. and M. E. Andersen (1988), Physiologically based approaches for examining the pharmacokinetics of inhaled vapors, in *Toxicology of the Lung*, Gardner, D. E., J. D. Crapo, and E. J. Massaro, Eds., Raven Press, New York, 449–476.

Garman, R. H., W. M. Snellings, and R. R. Maronpot (1985), Brain tumors in F344 rats associated with chronic inhalation exposure to ethylene oxide, *Neurotoxicology*, 6, 117–138.

Garry, V. F., C. W. Opp, J. K. Wiencke, and D. Lakatua (1982), Ethylene oxide induced sister chromatid exchange in human lymphocytes using a membrane dosimetry system, *Pharmacology*, 25, 214–221.

Gehring, P. J., P. G. Watanabe, and C. N. Park (1978), Resolution of dose-response toxicity data for chemicals requiring metabolic activation: example — vinyl chloride, *Toxicol. Appl. Pharmacol.*, 44, 581–591.

Generoso, W. M., K. T. Cain, C. V. Cornett, N. L. A. Cacheiro, and L. A. Hughes (1990), Concentration-response curves for ethylene-oxide-induced heritable translocations and dominant lethal mutations, *Environ. Mol. Mutagen.*, 16, 126–131.

Göggelmann, W., M. Bauchinger, U. Kulka, and E. Schmid (1988), Genotoxicity of 2-nitropropane and 1-nitropropane in Salmonella typhimurium and human lymphocytes, *Mutagenesis*, 3, 137–140.

Gold, L. S., C. B. Sawyer, R. Magaw, M. Backman, R. de Veciana, N. K. Hooper, W. R. Havender, J. Bernstein, R. Peto, M. C. Pike, and B. N. Ames (1984), A carcinogenic potency database of the standardized results of animal bioassays, *Environ. Health Pespect.*, 58, 9–319.

Gold, L. S., M. de Veciana, G. M. Backman, R. Magaw, P. Lopipero, M. Smith, M. Blumenthal, R. Levinson, L. Bernstein, and B. N. Ames (1986), Chronological supplement to the carcinogenic potency database, standardized results of animal bioassays published through December 1982, *Environ. Health Perspect.*, 67, 161–200.

Gold, L. S., T. H. Slone, G. M. Backman, R. Magaw, M. Da Costa, P. Lopipero, M. Blumenthal, and B. N. Ames (1987), Second chronological supplement to the carcinogenic potency database: standardized results of animal bioassays published through December 1984 and by the National Toxicology Program through May 1986, *Environ. Health Perspect.*, 74, 237–329.

Gossen, J. A., W. J. F. De Leeuw, C. H. T. Tan, E. C. Zwarthoff, F. Berends, P. H. M. Lohman, D. L. Knook, and J. Vijg (1989), Efficient rescue of integrated shuttle vectors from transgenic mice — a model for studying mutations in vivo, *Proc. Natl. Acad. Sci. U.S.A.*, 86, 7971–7975.

Green, E. L. (1968), Genetic effects of radiation on mammalian populations, *Annu. Rev. Genet.*, 2, 87–120.

Griffin, T. B., F. Coulston, and A. A. Stein (1980), Chronic inhalation exposure of rats to vapors of 2-nitropropane at 25 ppm, *Ecotoxicol. Environ. Saf.*, 4, 267–281.

GSF-Bericht (1988), Gentoxische Wirkungen in vitro und in vivo von 2- und 1-Nitropropan. Bericht über eine gemeinschaftlich durchgeführte experimentelle Studie in der GSF, Wolff, T., Ed., GSF-Forschungszentrum für Umwelt und Gesundheit, 8042 Neuherberg, Germany.

Hall, J. and R. Montesano (1990), DNA alkylation damage — consequences and relevance to tumour production, *Mutat. Res.*, 233, 247–252.

Hamm, T., D. Guest, and J. Dent (1984), Chronic toxicity and oncogenicity bioassay of inhaled ethylene in Fisher-344 rats, *Fundam. Appl. Toxicol.*, 4, 473–478.

Heflich, R. H., D. T. Beranek, R. J. Kodell, and S. M. Morris (1982), Induction of mutation and sister-chromatid exchanges in Chinese hamster ovary cells by ethylating agents. Relationship to specific DNA adducts, *Mutat. Res.*, 106, 147–161.

Henderson, L., H. Cole, J. Cole, S. E. James, and M. Green (1986), Detection of somatic mutations in man: evaluation of the microtiter cloning assay for T-lymphocytes, *Mutagenesis*, 1, 195–200.

Himmelstein, K. J. and R. J. Lutz (1979), A review of the applications of physiologically based pharmacokinetic modelling, *J. Pharmacokinet. Biopharm.*, 7, 127–145.

Hoel, D. G., N. L. Kaplan, and M. W. Anderson (1983), Implication of nonlinear kinetics on risk estimation in carcinogenesis, *Science*, 219, 1032–1037.

Hussein, S. and L. Ehrenberg (1975), Prophage inductive efficiency of alkylating agents and radiations, *Int. J. Radiat. Biol.*, 27, 355–362.

Jansen, J., H. Vrieling, G. R. Mohn, and A. A. van Zeeland (1992), The gene coding for hypoxanthine-guanine phosphoribosyl transferase as target for mutational analysis: PCR-cloning and sequencing of the cDNA from the rat, *Mutat. Res.*, 266, 105–116.

Johanson, G. and P. H. Näslund (1988), Spreadsheet programming — a new approach in physiologically based modeling of solvent toxicokinetics, *Toxicol. Lett.*, 41, 115–127.

Johnson, F. M. and S. E. Lewis (1981), Electrophoretically detected germinal mutations induced in the mouse by ethylnitrosourea, *Proc. Natl. Acad. Sci. U.S.A.,* 78, 3138–3141.

Kaldor, M. J., N. E. Day, and K. Hemminki (1988), Quantifying the carcinogenicity of antineoplastic drugs, *Eur. J. Cancer Clin. Oncol.,* 24, 703–711.

Kessler, W., B. Denk, and J. G. Filser (1989), Species-specific inhalation pharmacokinetics of 2-nitropropane, methyl ethyl ketone, and n-hexane, in *Biologically Based Methods for Cancer Risk Assessment,* Vol. 159, Travis, C. C., Ed., NATO ASI Ser. A: Life Sciences, Plenum Press, New York, 123–139.

Kniel, L., O. Winter, and C. H. Tsai (1980), Ethylene, in *Kirk-Othmer, Encyclopedia of Chemical Technology,* Vol. 9, 3rd ed., Grayson, M., Ed., John Wiley & Sons, New York.

Kohler, S. W., G. S. Provost, P. L. Kretz, M. J. Dycaico, J. A. Sorge, and J. M. Short (1990), Development of a short-term, in vivo mutagenesis assay: the effects of methylation on the recovery of a lambda phage shuttle vector from transgenic mice, *Nucleic Acids Res.,* 18(10), 3007–3013.

Kolman, A., M. Näslund, and C. J. Calleman (1986), Genotoxic effects of ethylene oxide and their relevance to human cancer, *Carcinogenesis,* 7(8), 1245–1250.

Kratochvilova, J. and U. H. Ehling (1979), Dominant cataract mutations induced by gamma-irradiation of male mice, *Mutat. Res.,* 63, 221–223.

Kratochvilova, J., J. Favor, and A. Neuhäuser-Klaus (1988), Dominant cataract and recessive specific-locus mutations detected in offspring of procarbazine-treated male mice, *Mutat. Res.,* 198, 295–301.

Kunz, B. A., B. G. Ayre, A. M. T. Downes, S. E. Kohalmi, C. R. McMaster, and M. G. Peters (1989), Base-pair substitutions alter the site-specific mutagenicity of UV and MNNG in the SUP4-o gene of yeast, *Mutat. Res.,* 226, 273–278.

Kunz, W., K. E. Appel, R. Rickart, M. Schwarz, and G. Stöckle (1978), Enhancement and inhibition of carcinogenic effectiveness of nitrosamines, in *Primary Liver Tumors,* Remmer, H., H. M. Bolt, P. Bannasch, and H. Popper, Eds., MTP Press, Lancaster, 261–284.

Laib, R. J. (1982), Specific covalent binding and toxicity of aliphatic halogenated xenobiotics, *Rev. Drug Metab. Drug Interact.,* 4(1), 1–48.

Langlois, R. G., W. L. Bigbee, R. H. Jensen, and J. German (1987), Evidence for increased somatic cell mutations at the glycophorin A locus in atomic bomb survivors, *Science,* 236, 445–448.

Langlois, R. G., B. A. Nisbet, W. L. Bigbee, D. N. Ridinger, and R. H. Jensen (1990), An improved flow cytometric assay for somatic mutations at the glycophorin A locus in humans, *Cytometry,* 11, 513–521.

Langlois, R. G., W. L. Bigbee, S. Kyoizumi, N. Nakamura, M. A. Bean, M. Akiyama, and R. H. Jensen (1989), Evidence for increased in vivo mutation and somatic recombination in Bloom's syndrome, *Proc. Natl. Acad. Sci. U.S.A.,* 86, 670–674.

Laval, J., S. Boiteux, and T. R. O'Connor (1990), Physiological properties and repair of apurinic/apyrimidinic sites and imidazole ring-opened guanines in DNA, *Mutat. Res.,* 233, 73–80.

Lawley, P. D. (1976), Methylation of DNA by carcinogens: some applications of chemical analytical methods, in *Screening Tests in Chemical Carcinogenesis,* Montesano, R., H. Bartsch, and L. Tomatis, Eds., IARC Publ. No. 12, International Agency for Research on Cancer, Lyon, 181–208.

Lawley, P. D., D. J. Orr, and M. Jarman (1975), Isolation and identification of products from alkylation of nucleic acids: ethyl and isopropyl-purines, *Biochem. J.,* 145, 73–84.

Lewis, S. E., F. M. Johnson, L. C. Skow, D. Popp, L. B. Barnett, and R. A. Popp (1985), A mutation in the β-globin gene detected in the progeny of a female mouse treated with ethylnitrosourea, *Proc. Natl. Acad. Sci. U.S.A.,* 82, 5829–5831.

Lewis, S. E., L. B. Barnett, C. Felton, F. M. Johnson, L. C. Skow, N. Cacheiro, and M. D. Shelby (1986), Dominant visible and electrophoretically expressed mutations induced in male mice exposed to ethylene oxide by inhalation, *Environ. Mutagen.,* 8, 867–872.

Lewis, T. R., C. E. Ulrich, and W. M. Busey (1979), Subchronic inhalation toxicity of nitromethane and 2-nitropropane, *J. Environ. Pathol. Toxicol.,* 2, 233–249.

Lindahl, T. (1990), Repair of intrinsic DNA lesions, *Mutat. Res.,* 238, 305–311.

Lohman, P. H. M., M. L. Mendelsohn, D. H. Moore, II, M. D. Waters, D. J. Brusick, J. Ashby, and W. J. A. Lohman (1992), A method for comparing and combining short-term genotoxicity test data: the basic system, *Mutat. Res.,* 266, 7–25.

Loveless, A. (1969), Possible relevance of O-6 alkylation of deoxyguanosine to the mutagenicity and carcinogenicity of nitrosamines and nitrosamides, *Nature (London),* 223, 206–207.

Lüning, K. G. (1971), Testing for recessive lethals in mice, *Mutat. Res.,* 11, 125–132.

Lutz, W. K. (1990), Dose-response relationship and low dose extrapolation in chemical carcinogenesis, *Carcinogenesis,* 11(8), 1243–1247.

Lynch, D. W., T. R. Lewis, W. J. Moorman, J. R. Burg, D. H. Groth, A. Khan, L. J. Ackerman, and B. Y. Cockrell (1984), Carcinogenic and toxicologic effects of inhaled ethylene oxide and propylene oxide in F344 rats, *Toxicol. Appl. Pharmacol.,* 76, 85–95.

Lyon, M. F. (1981), Sensitivity of various germ-cell stages to environmental mutagens, *Mutat. Res.,* 87, 323–345.

Lyon, M. F. and T. Morris (1966), Mutation rates at a new set of specific loci in the mouse, *Genet. Res.,* 7, 12–17.

Maltoni, C., G. Lefemine, A. Ciliberti, G. Cotti, and D. Carretti (1981), Carcinogenicity bioassays of vinyl chloride monomer: a model of risk assessment on an experimental basis, *Environ. Health Perspect.,* 41, 3–29.

Margolin, G. (1988), Statistical aspects of using biologic markers, *Stat. Sci.,* 3, 351–357.

Margolin, B. H., N. Kaplan, and E. Zeiger (1981), Statistical analysis of the Ames Salmonella/microsome test, *Proc. Natl. Acad. Sci. U.S.A.,* 78, 3779–3783.

McCann, J., L. Horn, and J. Kaldor (1984), An evaluation of Salmonella (Ames) test data in the published literature: application of statistical procedures and analysis of mutagenic potency, *Mutat. Res.,* 134, 1–47.

McCarthy, J. G., B. V. Edington, and P. F. Schendel (1983), Inducible repair of phosphotriesters in Escherichia coli, *Proc. Natl. Acad. Sci. U.S.A.,* 80, 7380–7384.

McCarthy, T. V., P. Karran, and T. Lindahl (1984), Inducible repair of O-alkylated DNA pyrimidines in Escherichia coli, *EMBO J.,* 3, 545–550.

McKusick, V. A. (1986), *Mendelian Inheritance in Man,* 7th ed., Johns Hopkins University Press, Baltimore.

Mellon, I. and P. C. Hanawalt (1989), Induction of the Escherichia coli lactose operon selectively increases repair of its transcribed DNA strand, *Nature (London),* 342, 95–98.

Mellon, I., V. A. Bohr, C. A. Smith, and P. C. Hanawalt (1986), Preferential DNA repair of an active gene in human cells, *Proc. Natl. Acad. Sci. U.S.A.,* 83, 8878–8882.

Mellon, I., G. Spivak, and P. C. Hanawalt (1987), Selective removal of transcription-blocking DNA damage from the transcribed strand of the mammalian DHFR gene, *Cell,* 51, 241–249.

Mendelsohn, M. L., D. H. Moore, II, and P. H. M. Lohman (1992), A method for comparing and combining short-term genotoxicity test data: results and interpretation, *Mutat. Res.,* in press.

Mohn, G. R., P. R. M. Kerklaan, A. A. van Zeeland, R. A. Baan, P. H. M. Lohman, and F. W. Pons (1984), Methodologies for the determination of various genetic effects in permeabilized strains of E. coli K12 differing in DNA repair capacity. Quantitation of DNA adduct formation, experiments with organ homogenates and hepatocytes and animal-mediated assays, *Mutat. Res.,* 125, 153–184.

Montesano, R. (1981), Alkylation of DNA and tissue specificity in nitrosoamine carcinogenesis, *J. Supramol. Struct. Cell. Biochem.,* 17, 259–273.

Moore, D. and J. S. Felton (1983), A microcomputer program for analyzing Ames test data, *Mutat. Res.,* 119, 95–102.

Moore, D. H., II, M. L. Mendelsohn, and P. H. M. Lohman (1992), A method for comparing and combining short-term genotoxicity test data: the optimal use of dose information, *Mutat. Res.,* 266, 27–42.

Morley, A. A., S. A. Grist, D. R. Turner, A. Kutlaca, and G. Bennet (1990), Molecular nature of in vivo mutation in human cells at the autosomal HLA-A locus, *Cancer Res.,* 50, 4584–4587.

Muysken-Schoen, M. A., R. A. Baan, and P. H. M. Lohman (1985), Detection of DNA adducts in N-acetoxy-2-acetylaminofluorene-treated human fibroblasts by means of immunofluorescence microscopy and quantitative immuno autoradiography, *Carcinogenesis,* 6, 999–1004.

Natarajan, A. T., J. W. I. M. Simons, E. W. Vogel, and A. A. van Zeeland (1984), Relationship between killing, chromosomal aberrations, sister chromatid exchanges and point mutations induced by monofunctional alkylating agents in Chinese hamster cells. A correlation with different ethylation products in DNA, *Mutat. Res.,* 128, 31–40.

Nehls, P., M. F. Rajewsky, E. Spiess, and D. Werner (1984), Highly sensitive sites for guanine-O^6 ethylation in rat brain DNA exposed to N-ethyl-N-nitrosourea in vivo, *EMBO J.,* 3, 327–332.

Nicklas, J. A., T. C. Hunter, L. M. Sullivan, J. K. Berman, J. P. O'Neill, and R. J. Albertini (1987), Molecular analyses of in vivo hprt mutations in human T-lymphocytes. I. Studies of low frequency 'spontaneous' mutants by Southern blots, *Mutagenesis,* 2, 341–347.

NIOSH/OSHA (1978), Vinyl halides carcinogenicity, *Curr. Intelligence Bull.,* 28, 1–10.

NRCC (1985), Ethylene in the Environment: Scientific Criteria for Assessing its Effects on Environmental Quality, NRCC/CNRC No. 22496, National Research Council of Canada, NRCC Associate Committee on Scientific Criteria for Environmental Quality, Subcommittee on Air, Ottawa.

NTP (1987), Toxicology and Carcinogenesis Studies of Ethylene Oxide in B6C3F1 Mice (Inhalation Studies), National Toxicology Program, Tech. Rep. 326, NIH-88-2582, U.S. Department of Health & Human Services.

O'Connor, P. J., R. Saffhill, and G. P. Margison (1979), N-nitroso compounds: biochemical mechanisms of action, in *Environmental Carcinogenesis: Occurrence, Risk Evaluation and Mechanisms,* Emmelot, P. and E. Kriek, Eds., Elsevier/North-Holland, New York, 73–96.

Oesterle, D. and E. Deml (1983), Promoting effect of polychlorinated biphenyls on development of enzyme-altered islands in livers of weanling and adult rats, *J. Cancer Res. Clin. Oncol.,* 105, 141–147.

Osterman-Golkar, S., L. Ehrenberg, and C. A. Wachtmeister (1970), Reaction kinetics and biological action in barley of monofunctional methanesulphonic esters, *Radiat. Bot.,* 10, 303–327.

Park, J.-W., K. C. Cundy, and B. N. Ames (1989), Detection of DNA adducts by high performance liquid chromatograophy with electrochemical detector, *Carcinogenesis,* 10, 827–832.

Pastink, A., C. Vreeken, E. W. Vogel, and J. C. J. Ecken (1990), Mutations induced at the *white* and *vermillion* loci in *Drosophila melanogaster, Mutat. Res.,* 231, 63–71.

Pastink, A., C. Vreeken, M. J. M. Nivard, L. L. Searles, and E. W. Vogel (1989), Sequence analysis of N-ethyl-N-nitrosourea-induced vermillion mutations in *Drosophila melanogaster, Genetics,* 123, 123–129.

Pegg, A. E., L. Wiest, R. S. Foote, S. Mitra, and W. Perry (1983), Purification and properties of O^6-methylguanine-DNA transmethylase from rat liver, *J. Biol. Chem.,* 258, 2327–2333.

Peters, J., S. J. Andrews, J. F. Loutit, and J. B. Clegg (1985), A mouse β-globin mutant that is an exact model of hemoglobin Rainier in man, *Genetics,* 110, 709–721.

Peters, J., S. T. Ball, and S. J. Andrews (1986), The detection of gene mutations by electrophoresis, and their analysis, *Prog. Clin. Biol. Res.,* 209B, 367–374.

Pinkel, D., J. Landegent, C. Collins, J. Fuscoe, R. Segraves, J. Lucas, and J. Gray (1988), Fluorescence in situ hybridization with human chromosome-specific libraries: detection of trisomy 21 and translocations of chromsome 4, *Proc. Natl. Acad. Sci. U.S.A.,* 85, 9138–9142.

Poirier, M. C. (1984), The use of carcinogen-DNA adduct antisera for quantitation and localization of genomic damage in animal models and the human population, *Environ. Mutagen.,* 6, 879–888.

Popp, R. A., E. G. Bailiff, L. C. Skow, F. M. Johnson, and S. E. Lewis (1983), Analysis of a mouse α-globin gene mutation induced by ethylnitrosourea, *Genetics,* 105, 157–167.

Potter, D., D. Blair, R. Davies, W. P. Watson, and A. S. Wright (1989), The relationship between alkylation of hemoglobin and DNA in Fischer 334 rats exposed to [^{14}C]ethylene oxide, *Arch. Toxicol. Suppl.,* 13, 254–257.

Pretsch, W. (1986), Protein-charge mutations in mice, *Prog. Clin. Biol. Res.,* 209B, 383–388.

Pretsch, W., D. J. Charles, and K. R. Narayanan (1982), The agar contact replica technique after isoelectric focusing as a screening method for the detection of enzyme variants, *Electrophoresis,* 3, 142–145.

Quinto, I. and M. Radman (1987), Carcinogenic potency in rodents versus genotoxic potency in E. coli: a correlation analysis of bifunctional alkylating agents, *Mutat. Res.,* 181, 235–242.

Quinto, I., L. Tennenbaum, and M. Radman (1990), Genotoxic potency of monofunctional alkylating agents in E. coli: comparison with carcinogenic potency in rodents, *Mutat. Res.,* 228, 177–185.

Randerath, E., H. P. Agrawal, J. A. Weaver, C. B. Bordelon, and K. Randerath (1985), ^{32}P-postlabeling analysis of DNA adducts persisting for up to 42 weeks in the skin, epidermis and dermis of mice treated topically with 7,12-dimethyl benz(a)anthracene, *Carcinogenesis*, 6, 1117–1126.

Rhomberg, L., V. L. Dellarco, C. Siegel-Scott, K. L. Dearfield, and D. Jackobson-Kram (1990), Quantitative estimation of the genetic risk associated with the induction of heritable translocations at low-dose exposure: ethylene oxide as an example, *Environ. Mol. Mutagen.*, 16, 104–125.

Richardson, K. K., F. C. Richardson, R. M. Crosby, J. A. Swenberg, and T. R. Skopek (1987), DNA base changes and alkylation following *in vivo* exposure of *Escherichia coli* to N-methyl-N-nitrosourea or N-ethyl-N-nitrosourea, *Proc. Natl. Acad. Sci. U.S.A.*, 84, 344–348.

Roldan-Arjona, T., F. L. Luque-Romero, M. Ruiz-Rubio, and C. Pueyo (1990), Quantitative relationship between mutagenic potency in the Ara test of Salmonella typhimurium and carcinogenic potency in rodents, *Carcinogenesis*, 11, 975–980.

Rossi, A. M., J. C. P. Thijssen, A. D. Tates, H. Vrieling, A. T. Natarajan, P. H. M. Lohman, and A. A. van Zeeland (1990), Mutations affecting RNA splicing in man are detected more frequently in somatic than in germ cells, *Mutat. Res.*, 244, 351–357.

Rowland, M. (1984), Physiologic pharmacokinetic models: relevance, experience and future trends, *Drug Metab. Rev.*, 15, 55–74.

Russell, L. B. (1990), Patterns of mutational sensitivity to chemicals in poststem-cell stages of mouse spermatogenesis, *Prog. Clin. Biol. Res.*, 340C, 101–113.

Russell, L. B., R. B. Cumming, and P. R. Hunsicker (1984), Specific-locus mutation rates in the mouse following inhalation of ethylene oxide, and application of the results to estimation of human genetic risk, *Mutat. Res.*, 129, 381–388.

Russell, W. L. (1951), X-ray-induced mutations in mice, *Cold Spring Harbor Symp. Quant. Biol.*, 16, 327–356.

Russell, W. L. (1989a), Comment on mutagenicity of diethyl sulfate in mice and on germ-cell mutagenicity testing, *Mutat. Res.*, 225, 127–129.

Russell, W. L. (1989b), Reply to U. H. Ehling, *Mutat. Res.*, 214, 331–332.

Sankaranarayanan, K. (1991), Ionizing radiation and genetic risks. IV. Current methods, estimates of risk of Mendelian disease, human data and lessons from biochemical and molecular studies of mutations, *Mutat. Res.*, 258, 99–122.

Santella, R. M., X. Y. Yang, L. L. Hsieh, and T. L. Young (1990), Immunologic methods for the detection of carcinogen adducts in humans, in *Mutation in the Environment, Part C: Somatic and Heritable Mutation, Adduction and Epidemiology,* Mendelsohn, M. L. and R. J. Albertini, Eds., Wiley-Liss, New York, 247–257.

Sarto, F., I. Cominato, A. M. Pinton, P. G. Brovedani, C. M. Faccioli, V. Bianchi, and A. G. Levis (1984), Cytogenetic damage in workers exposed to ethylene oxide, *Mutat. Res.*, 138, 185–195.

Schoental, R. (1967), Methylation of nucleic acids by N[^{14}C]-methyl-N-nitrosourethane in vitro and in vivo, *Biochem. J.*, 102, 5c-7c.

Searle, A. G. (1974), Mutation induction in mice, in *Advances in Radiation Biology*, Vol. 4, Lett, J. T., H. I. Adler, and M. Zelle, Eds., Academic Press, New York, 131–207.

Segerbäck, D. (1983), Alkylation of DNA and hemoglobin in the mouse following exposure to ethene and ethene oxide, *Chem. Biol. Interact.*, 45, 139–151.

Selby, P. B. (1981), Radiation genetics, in *The Mouse in Biomedical Research,* Vol. 1, Foster, H. L., J. D. Small, and J. G. Fox, Eds., Academic Press, New York, 263–283.

Selby, P. B. and P. R. Selby (1977), Gamma-ray-induced dominant mutations that cause skeletal abnormalities in mice. I. Plan, summary of results and discussion, *Mutat. Res.,* 43, 357–375.

Shen, J., W. Kessler, B. Denk, and J. G. Filser (1989), Metabolism and endogenous production of ethylene in rat and man, *Arch. Toxicol. Suppl.,* 13, 237–239.

Singer, B. (1975), The chemical effect of nucleic acid alkylation and its relation to mutagenesis and carcinogenesis, *Prog. Nucleic Acid Res. Mol. Biol.,* 15, 219–284.

Singer, B. (1979), N-nitroso alkylating agents: formation and persistance of alkyl derivatives in mammalian nucleic acids as contributing factors in carcinogenesis, *J. Natl. Cancer. Inst.,* 62, 1329–1339.

Singer, B. (1986), O-alkyl pyrimidines in mutagenesis and carcinogenesis: occurrence and significance, *Cancer Res.,* 46, 4879–4885.

Singer, B., H. Fraenkel-Conrat, and J. T. Kusimerek (1978), Preparation and template activities of polynucleotides containing O^2- and O^4-alkyluridine, *Proc. Natl. Acad. Sci. U.S.A.,* 75, 1722–1726.

Smerdon, M. J. and F. Thoma (1990b), Site-specific DNA repair at the nucleosome level in a yeast minichromosome, *Cell,* 61, 675–684.

Smerdon, M. J., J. Bedoyan, and F. Thoma (1990a), DNA repair in a small yeast plasmid folded into chromatin, *Nucleic Acids Res.,* 18, 2045–2051.

Snellings, W. M., C. S. Weil, and R. R. Maronpot (1984), A two-year inhalation study of the carcinogenic potential of ethylene oxide in Fischer 344 rats, *Toxicol. Appl. Pharmacol.,* 75, 105–117.

Soares, E. R. (1979), TEM-induced gene mutations at enzyme loci in the mouse, *Environ. Mutat.,* 1, 19–25.

Sobels, F. H. (1982), The parallelogram: an indirect approach for the assessment of genetic risk from chemical mutagens, in *Progress in Mutation Reseach,* Vol. 3, Bora, K. C., G. R. Douglas, and E. R. Nestman, Eds., Elsevier, Amsterdam, 323–327.

Swain, C. G. and C. B. Scott (1953), Quantitative correlation of relative rates: comparison of hydroxide ion with other nucleophilic reagents towards alkyl halides, esters, epoxides and acyl halides, *J. Am. Chem. Soc.,* 75, 141–147.

Swann, P. F. and P. N. Magee (1968), Nitrosamine-induced carcinogenesis: the alkylation of nucleic acids of the rat by N-methyl-N-nitrosourea, dimethylnitrosamine, dimethyl sulphate and methyl methanesulphonate, *Biochem. J.,* 110, 39–47.

Swann, P. F. and P. N. Magee (1971), Nitrosamine-induced carcinogenesis: the alkylation of N-7 of guanine of nucleic acids of the rat by diethylnitrosamine, N-ethyl-N-nitrosourea and ethyl methanesulphonate, *Biochem. J.,* 125, 841–847.

Terleth, C., C. A. van Sluis, and P. van de Putte (1989), Differential repair of UV damage in Saccharomyces cerevisiae, *Nucleic Acids Res.,* 17, 4433–4439.

Thiess, A. M., H. Schwegler, I. Fleig, and W. G. Stocker (1981), Mutagenicity study of workers exposed to alkylene oxides (ethylene oxide/propylene oxide) and derivatives, *J. Occup. Med.,* 23, 343–347.

Tindall, K. R. and L. F. Stankowski (1989), Molecular analysis of spontaneous mutations at the gpt locus in Chinese hamster ovary (AS52) cells, *Mutat. Res.,* 220, 241–253.

Turteltaub, K. W., J. S. Felton, B. L. Gledhill, J. S. Vogel, J. R. Southon, M. W. Caffee, R. C. Finkel, D. E. Nelson, I. D. Proctor, and J. C. Davis (1990), Accelerator mass spectrometry in biomedical dosimetry: relationship between low-level exposure and covalent binding of heterocyclic amine carcinogens to DNA, *Proc. Natl. Acad. Sci. U.S.A.,* 87, 5288–5292.

Umbenhauer, D., C. P. Wild, R. Montesano, R. Saffhill, J. M. Boyle, N. Huh, U. Kirstein, J. Thomale, M. F. Rajewsky, and S. H. Lu (1985), O^6-methyldeoxyguanosine in oesophageal DNA among individuals at high risk of oesophageal cancer, *Int. J. Cancer,* 36, 661–665.

UNSCEAR (1982), Ionizing Radiation: Sources and Biological Effects, United Nations Scientific Committee on the Effects of Atomic Radiation, New York.

Van Zeeland, A. A. (1988), Molecular dosimetry of alkylating agents: quantitative comparison of genetic effects on the basis of DNA adduct formation, *Mutagenesis,* 3, 179–191.

Van Zeeland, A. A., A. de Groot, and A. Neuhäuser-Klaus (1990), DNA adduct formation in mouse testis by ethylating agents. A comparison with germ cell mutagenesis, *Mutat. Res.,* 231, 55–62.

Van Zeeland, A. A., G. R. Mohn, A. Neuhäuser-Klaus, and U. H. Ehling (1985), Quantitative comparison of genetic effects of ethylating agents on the basis of DNA-adduct formation. Use of O6-ethylguanine as molecular dosimeter for extrapolation from cells in culture to the mouse, *Environ. Health Perspect.,* 62, 163–169.

Verburgt, F. G. and E. Vogel (1977), Vinyl chloride mutagenesis in Drosophila melanogaster, *Mutat. Res.,* 48, 327–346.

Victorin, K. and M. Ståhlberg (1988), A method for studying the mutagenicity of some gaseous compounds in Salmonella typhimurium, *Environ. Mol. Mutagen.,* 11, 65–77.

Vogel, E. W. (1989), Nucleophilic selectivity of carcinogens as a determinant of enhanced mutational response in excision repair-defective strains in Drosophila: effects of 30 carcinogens, *Carcinogenesis,* 10, 2093–2106.

Vogel, E. W. and J. Ashby (in press), Structure-Activity Relationships: Experimental Approaches, in *Methods to Assess DNA Damage and Repair: Interspecies Comparisons,* R. G. Tardiff, P. H. M. Lohman, and G. N. Wogan, Eds., John Wiley & Sons, Ltd.

Vogel, E. W. and A. T. Natarajan (1979a), The relationship between reaction kinetics and mutagenic action of monofunctional alkylating agents in higher eukaryotic systems. I. Recessive lethal mutations and translocations in Drosophila, *Mutat. Res.,* 62, 51–100.

Vogel, E. W. and A. T. Natarajan (1979b), The relationship between reaction kinetics and mutagenic action of monofunctional alkylating agents in higher eukaryotic systems. II. Total and partial sex chromosome loss in Drosophila, *Mutat. Res.,* 62, 101–123.

Vogel, E. W. and A. T. Natarajan (1982), The relation between reaction kinetics and mutagenic action of mono-functional alkylating agents in higher eukaryotic systems. III. Interspecies comparisons, in *Chemical Mutagens,* Hollaender, A. and F. J. de Serres, Eds., Plenum Press, New York, 295–336.

Vogel, E. W., A. Barbin, M. J. M. Nivard, and H. Bartsch (1990), Nucleophilic selectivity of alkylating agents and their hypermutability in Drosophila as predictors of carcinogenic potency in rodents, *Carcinogenesis,* 11, 2211–2217.

Vrieling, H., M. J. Niericker, J. W. I. M. Simons, and A. A. van Zeeland (1988), Molecular analysis of mutations induced by N-ethyl-N-nitrosourea at the HPRT locus in mouse lymphoma cells, *Mutat. Res.,* 198, 99–106.

Vrieling, H., M. L. van Rooijen, N. A. Groen, M. Z. Zdzienicka, J. W. I. M. Simons, P. H. M. Lohman, and A. A. van Zeeland (1989), DNA strand specificity for UV-induced mutations in mammalian cells, *Mol. Cell. Biol.,* 9, 1277–1283.

Waters, M. D., H. F. Stack, A. L. Brady, P. H. M. Lohman, L. Haroun, and H. Vaino (1988), Use of computerized data listings and activity profiles of genetic and related effects in the review of 195 compounds, *Mutat. Res.,* 205, 295–312.

Yager, J. W., C. J. Hines, and R. C. Spear (1983), Exposure to ethylene oxide at work increases sister chromatid exchanges in human peripheral lymphocytes, *Science,* 219, 1221–1223.

Zdarsky, E., J. Favor, and I. J. Jackson (1990), The molecular basis of brown, an old mouse mutation, and of an induced revertant to wild-type, *Genetics,* 126, 443–449.

Zielenska, M., D. Beranek, and J. B. Guttenplan (1988), Different mutational profiles induced by N-nitroso-N-ethylurea: effects of dose and error-prone DNA repair and correlations with DNA adducts, *Environ. Mol. Mutagen.,* 11, 473–485.

Zijlstra, J. A. (1989), Liquid holding increases mutation induction by formaldehyde and some other cross-linking agents in Escherichia coli, *Mutat. Res.,* 210, 255–262.

CHAPTER 4

Risk Characterization Strategies for Genotoxic Environmental Agents

J. Lewtas, D. DeMarini, J. Favor, D. Layton, J. MacGregor, J. Ashby, P. Lohman, R. Haynes, and M. Mendelsohn

TABLE OF CONTENTS

I. Introduction 126

II. Risk Assessment 128
    A. Hazard Identification 131
    B. Dose-Response Assessment 132
    C. Exposure Assessment 133
    D. Environmental and Human Monitoring 134
    E. Risk Characterization of Genotoxic Agents 136

III. Risk Characterization for Nonhuman Biota 136

IV. Risk Characterization for Human Exposure to
Genotoxic Agents 138
    A. Qualitative Risk-Characterization Strategies 139
        1. Classification of Genotoxicity 139
            1.1 Organism and Relevance to Humans 139
            1.2 Nature of the Genetic Effect 140
            1.3 Weight-of-Evidence Classification 142
        2. Qualitative Exposure Assessment 143
        3. Risk Categorization 143
    B. Quantitative Risk-Characterization Strategies 144
        1. Dose-Response Extrapolation 145
            1.1 Potency Estimation Methods 145
                1.1.1 Slope 146
                1.1.2 Doubling Dose and Relative Mutagenic
                      Effect (RME) 146

        1.1.3 Lowest Effective Dose (LED) 148
        1.1.4 Genotoxicity Score 149
        1.1.5 Radiation Dose Equivalence 149
    1.2 Low-Dose Extrapolation Methods 150
    1.3 Human Risk Extrapolation Methods 150
        1.3.1 Probabilistic Methods 151
        1.3.2 Comparative Methods 152
            1.3.2.1 Radiation Equivalence Method 152
            1.3.2.2 Comparative Potency Method 152
            1.3.2.3 Relative Ranking 153
    2. Quantitative Exposure Assessment 153
    3. Risk Quantitation 156
        3.1. Probabilistic Risk Assessment 157
        3.2   Comparative Risk Assessment 158

V. Factors That Affect Risk 158
    A. Endogenous Factors 159
    B. Background Mutation Rate 159
    C. Cell Replication 160
    D. Age 160
    E. Diet 160
    F. Economic and Social Factors 161
    G. Duration of Exposure 161

VI. Significance of Comparative Genotoxic Risk Assessment 162

References 163

I. INTRODUCTION

Deoxyribonucleic acid (DNA) is the master molecule of life. Its presence and integrity in all cellular organisms on the planet are essential for life as we know it. Genes and chromosomes are composed of DNA, which codes for the myriad of enzymes, structural proteins, regulatory factors, and other molecules that constitute the structure of organisms and enable them to function. Because of its pivotal role, the integrity of the DNA's structure is critical. Thus, DNA damage and changes in the informational sequences in DNA, defined as genotoxic events, including mutations, are considered deleterious in themselves and should be avoided to the extent that is reasonably possible.

Life, however, exists within a milieu of agents, both physical (e.g., radiation) and chemical, that damage and mutate DNA. No component of the environment is free of mutagens, including most air, water, and soil samples (Lewtas, 1991; MacGregor et al., 1992). Some mutagens occur naturally, whereas others are introduced into the environment by human activities. Perhaps because organisms have evolved in the presence of mutagens, most organisms have developed ways by which they repair most, but not all, of the damage caused to their DNA by mutagens (Setlow, 1979).

Nonetheless, the potential harm posed to organisms by environmental mutagens demands that exposure to such agents should be kept to a reasonable minimum. This chapter describes both qualitative and quantitative strategies to characterize the human risk associated with exposure to genotoxic agents and addresses the problems associated with assessing the risk to nonhuman biota. When adequate data are available to link exposure and genetic changes to human disease, the increased probability of a specific disease may be estimated. In most cases, however, the data are inadequate to estimate disease outcome. It is then practical to use comparative methods to estimate which genotoxic agents or which sources of these agents pose the greatest risk. These methods of estimating "genotoxic risk" without reference to any specific disease outcome are presented here to facilitate efforts in many countries to set priorities for action to reduce the risk of exposure to various environmental pollutants. By prioritizing risk, limited resources may be concentrated on decreasing exposures to the environmental pollutants which present major rather than minor risk.

Assessment of the risks that result from exposure to genotoxic agents has in the past focused on either characterization of the risk of heritable disease via the production and transmission of germ-cell mutations (ICPEMC, 1983; U.S. EPA, 1986a; Ehling, 1991) or on the risk of cancer via induction of somatic-cell mutations (Albert et al., 1977; IRLG, 1979; U.S. EPA, 1986b; Moolgavkar and Knudson, 1981). Risk characterization has, therefore, meant estimating the risk of increased disease incidence or mortality in humans. However, a wide diversity of adverse health effects may result from genetic damage (e.g., gene mutations, DNA damage, and chromosomal aberrations) induced by genotoxic agents as a result of somatic- or germ-cell mutations (Figure 1).

Genetic damage to somatic cells can result in cancer (Barbacid, 1985; Bishop, 1987; Aust, 1991), and there is evidence suggesting that such damage may also contribute to aging (Hartman, 1983; Rattan, 1991) and cardiovascular disease (Bridges et al., 1990). Cancer and cardiovascular disease may, in some cases, develop from at least two mutations, a germ-cell mutation, which alters the individual's susceptibility to disease, followed by a second (somatic-cell) mutation. The hereditary form of retinoblastoma, a cancer of the retina, has been hypothesized to arise from such a two-step mechanism (Aust, 1991), as well as other cancers (e.g., FAP, Wilms tumor of the kidney, and neurofibromatosis type-1).

Genetic damage to germ cells may affect exposed individuals (e.g., effects on fertility), but the primary risk is to the exposed individual's offspring who are at increased risk of suffering from genetic disorders. The range of genetic disorders affecting human health is vast; all organ systems and biological processes are subject to genetic disorders, and humans may be affected at any time from conception to old age. Examples of genetic diseases determined by single genes include cystic fibrosis (Dean, 1988) and sickle cell anemia. In humans there are at least 5000 known genetic loci, many of which are associated with human disorders (McKusick, 1990) with a relatively simple pattern of inheritance, suggesting that they result from mutation of a single gene. In contrast, there are also many diseases with a genetic component but an irregular pattern of inheritance (Mohrenweiser, 1991), such as certain forms of diabetes, psychoses, and cardiovascular disease. These diseases are often referred to as multifactorial diseases and are thought to result from interactions between multiple genes and environmental factors in ways that are not yet fully understood. Nonetheless, it is clear that genetic disease has a major impact on human health, and efforts to minimize exposure to environmental agents that might lead to these diseases should be encouraged.

The chemical and physical properties of genotoxic agents (e.g., electrophilicity) that confer the ability to react with DNA (Miller and Miller, 1966, 1977; Brooks, 1966) also give these agents the ability to react with RNA, proteins, and other nucleophilic molecules present in cells (Ehrenberg, 1985). Therefore these agents often also induce toxic effects in addition to DNA damage, mutations, and chromosomal alterations. Because of this wide diversity of possible effects, we focus on comparative or relative risk, in contrast to the probabilistic characterization of risk of a specific disease outcome.

II. RISK ASSESSMENT

Risk assessment attempts to define the nature and magnitude of the risk of a process, situation, or environment. The assessment process is usually divided into four stages (Figure 2; NRC, 1983): (1) hazard identification, in which the harmfulness of an agent and its presence in the environment is determined, (2) dose-response assessment, in which the dose of the agent and the response in laboratory or human studies are used to predict the relationship between dose and disease or other adverse outcome in humans through extrapolation methods, (3) exposure assessment, in which the agent is located in the environment and its transport and access to the target species are estimated, and (4) risk characterization, in which exposure assessment and dose-response (potency) assessment are combined to predict risk or disease outcome. Research and risk assessment provide the scientific basis for risk management.

The diversity of possible adverse effects that may result from exposure to genotoxic environmental pollutants presents a special problem for anyone attempting to adapt the current cancer risk assessment methods to "genotoxic risk". The basic elements of the process of risk assessment, however, remain the same regardless of the biological effect in question. In response to the complexity of the problem, we introduce several strategies for either qualitatively

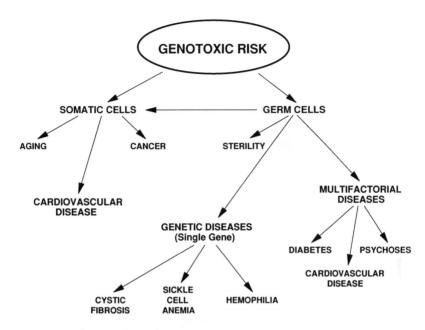

Figure 1. Scheme illustrating adverse health effects resulting from genetic damage induced by genotoxic agents.

or quantitatively characterizing the genotoxic risk from environmental exposure within the risk-assessment framework.

Three approaches to risk characterization, one qualitative and two quantitative, are presented. The selection of a particular approach or combination of approaches to use will depend on the objectives of the assessment process and the quality and quantity of information that is available. When information is limited or when the objectives of the assessment require only a general classification of risk, a qualitative assessment of risk is recommended. In exceptional cases, when available information includes either human or mammalian data of direct significance to human genotoxic risk, risk characterization can be quantitative and expressed as a probabilistic risk of mutation occurring in humans. Examples of such cases are radiation (Neel and Lewis, 1990; Ehling, 1991), several cytotoxic drugs (e.g., cyclophosphamide, mitomycin C, and natulan), which have been evaluated by the International Commission for Protection against Environmental Mutagens and Carcinogens (ICPEMC, 1983), and ethylene oxide (Dellarco, 1990).

Of the thousands of chemicals and complex mixtures identified in the environment, only about 10% have been studied for genotoxicity (Graedel et al., 1986). The data available are often limited to one or two short-term genetic bioassays. Here, we present several approaches for selecting and expressing such data in a comparative risk characterization that includes a quantitative ranking of relative risk but does not express the risk in probabilistic units. Each

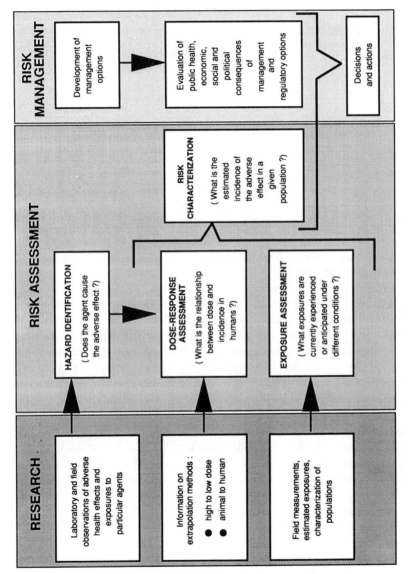

Figure 2. Elements of risk assessment and risk management resulting from research.

of these approaches to risk characterization is useful: (1) qualitative assessment guides further study of an agent and serves as a potential alert, (2) comparative assessment prioritizes genotoxic risks, and (3) probabilistic risk characterization of a single agent or exposure generates risk units that can be compared to other types of risks (e.g., disease or death).

Today, quantitative risk-assessment procedures for environmental chemicals are commonly applied to assess risks of cancer from human population exposure to these agents (U.S. EPA, 1986). Risk-assessment procedures for heritable mutations, reproductive effects, and other noncancer effects have also been developed more recently (U.S. EPA, 1986a, b), and procedures for the assessment of risk to the ecosystem are being developed.

The purpose of this study requested by the United Nations Environment Programme (UNEP) was to develop recommendations and guidelines for the comparative risk assessment of genotoxic agents. The ICPEMC conducted the study through working groups established for each of the steps of risk assessment. Each of these components of risk assessment is discussed in detail in accompanying papers on hazard identification (Chapter 1, this volume), dose-response assessment (Chapter 3, this volume), and exposure assessment (Chapter 2, this volume). An additional working group on human and environmental biomonitoring developed guidelines for using genetic biomonitoring methods to provide information for all aspects of risk assessment (Chapter 5, this volume). Because this chapter presents risk characterization within the framework of risk assessment, a brief summary of each of the components of risk assessment of genotoxic agents is summarized here.

A. Hazard Identification

Hazard identification often begins with the listing of potentially genotoxic agents. These are identified using a combination of the following elements: chemical structure-activity relationships (SARs), short-term genotoxicity tests, human and environmental monitoring, and gross estimates of exposure. The universality of DNA as the basic genetic material is the basis on which any agent detected as a mutagen in one organism should be considered to be a possible human mutagen. The hazard identification step can also include a rough prioritization of the agents according to their prevalence in the environment. These approaches are reviewed by DeMarini et al. in Chapter 1.

Hazard identification is an important first step that determines whether there is sufficient evidence to proceed to dose-response and exposure assessment or whether more research is needed. Although new industrial and chemical agents are now screened for genotoxicity before being introduced into commerce in the U.S., Japan, and the European community, many existing chemicals have not been evaluated. In addition to synthetic chemicals, the environment contains many thousands of chemicals produced from combustion, industrial processes, other human activities, natural processes, and natural products from organisms such as fungi and plants. Unfortunately, fewer than 200 environmental chemicals, and very few complex environmental mixtures, have been well characterized with respect to their genotoxicity. The most thoroughly studied agents are radiation, laboratory chemicals (e.g., methyl methanesulfonate, MMS), pharmaceutical agents (e.g., cyclophosphamide), pesticides, and several other industrial chemicals that have been regulated.

B. Dose-Response Assessment

Dose-response assessment involves quantitation of the potency with which an agent produces genetic damage. Where possible, these studies focus on mutational endpoints that are directly relevant to the species of interest. Thus, for human risk assessment, the optimal data would be from humans or animals (e.g., rodents) and the endpoint might be somatic- or germ-cell mutations. As discussed below, the level of quantitation and its proximity to the human can vary enormously.

We define genotoxic dose-response assessment as an evaluation of the ability of an agent to alter or mutate DNA as a function of dose. The genotoxic response per unit dose is expressed as a unit or index of potency and is discussed in more detail later. The measure of dose is, for most practical purposes, the dose applied to the test system, although increasing efforts are being made to determine the concentration of the active agent and the time it remains at a target tissue or molecule (e.g., DNA; see Chapter 3). The issues surrounding dose-effect assessment for genotoxic agents based on *in vivo* and *in vitro* studies are discussed in detail by Favor et al. in Chapter 3. Ideally, human or *in vivo* mammalian mutagenicity dose-response data are needed to permit extrapolation from high dose to low dose and from animal data to human risk. Such data are not often available. The alternative use of *in vitro* and *in vivo* data allows an estimation of the relative genotoxic or mutagenic potency.

Measures of mutagenic potency include the slope of the dose-response curve (e.g., genetic alterations per unit dose applied) and the dose at which a specific response is obtained, such as the lowest effective dose (LED) or the doubling dose. The advantage of using a method (e.g., LED) that expresses potency in units of dose is that data from several different genotoxicity assays may be combined more readily than measures that include the response unit (e.g., mutation frequency). Unfortunately, dose-response data and even reliable measures of the doubling dose from relevant test systems are frequently not available. Therefore, the LED based on *in vitro* and short term test systems is often used as a rough potency index (Waters et al. 1988, 1991). ICPEMC working groups have recently developed a method for comparing and combining genotoxicity data from different organisms and endpoints by using LED as the measure of potency and combining the potency from several test systems into a single genotoxicity score (Lohman et al., 1992; Mendelsohn et al., 1992; Moore et al., 1992).

C. Exposure Assessment

Measurement and modeling methods are used to quantitatively assess exposures to genotoxic agents. Exposure refers generally to the contact with genotoxic agents at the exchange boundaries of an organism (e.g., lung, GI tract) prior to uptake by cells and the systemic circulation. Exposures may occur through multiple routes (e.g., oral, inhalation, and dermal) and media (e.g., air, water, and food). If all of these pathways cannot be assessed, it is important to identify those that are the dominant routes of exposure. Assessment of exposure to genotoxic agents is reviewed in more detail by Layton et al. in Chapter 2.

Exposure assessment should ideally be tailored to the specific agent or chemical and the site of exposure. In addition, it is also useful to have some prior knowledge of the nature of the dose-response function of the agent being assessed so that the most meaningful determination of the exposure can be made. If the exposure assessment will be used in the assessment of chronic genotoxic effects, an average annual exposure may be adequate. If the dose-response assessment indicates that the response is confined to a critical exposure period, e.g., during one stage of reproduction, then either the peak exposure or total exposure during that period of time would be needed to link the exposure assessment with the dose-response data.

The initial efforts in an exposure assessment focus on the identification of the sources of an agent and estimation of emission or release rates to environmental media. Modeling and measurement methods are then used to determine the concentrations of the agent in air, water, soil, or food. Assessments of total human exposures to genotoxic agents are rarely possible, and it is inevitable that the use of modeling and monitoring techniques will result in exposure assessments with some uncertainties. Therefore, it is important to identify and, if possible, quantitate the uncertainties. Exposure assessments result in estimates of both the concentrations of an agent in various media (e.g., micrograms per cubic meter of air or liter of drinking water) and exposure rates (e.g., milligrams per kilogram of body weight per day) for specific routes of exposure.

If environmental concentrations are constant in time, the average population exposure to an agent can be determined by the product of the concentration and a pathway exposure factor (PEF). The PEF relates the concentration of a pollutant in the environment to the chronic daily intake during the exposure period. This factor translates a chemical concentration in each of the primary environmental media (air, water, and soil) into exposure rates (e.g., milligrams per kilogram per day) for specific routes of exposure. These factors include information on human physiology (e.g., breathing rate) and lifestyle (e.g., time spent indoors, volume of water consumed), as well as data describing pollutant behavior in food chains or in microenvironments, such as indoor air (McKone and Daniels, 1991). Individual or population exposures may then be used to determine the chronic daily intake of a genotoxic agent during a specified time period.

D. Environmental and Human Monitoring

Biomonitoring genotoxicants or genotoxic events in the environment or human populations encompasses a broad range of tools that can be used at several stages of risk assessment (Figure 3), e.g., (1) identification of genotoxic hazards in populations or environments of interest, (2) direct measurement of population exposures using protein or DNA adducts, (3) determination of dose-response relationships by evaluation of the relationship between exposure (e.g., protein adducts), dose (e.g., DNA adducts), and genetic effects (e.g., mutations), and (4) after characterizing the risk of a population using biomonitoring methods to evaluate intervention or pollution control efforts. This integral and supportive role of monitoring for genetic effects as a part of the risk-assessment process is discussed in more detail by MacGregor et al. in Chapter 5.

Monitoring for genotoxic activity in an environment is useful for hazard identification because it does not require knowledge about the specific chemicals, which often number in the thousands in complex environmental mixtures of interest. Once genotoxic activity is discovered in an environment, specific agents responsible for the activity can then be isolated using chemical fractionation and separation techniques along with bioassays as tools to track and identify the responsible agents.

After hazard identification, measurement of specific types of exposure and damage in populations at risk allows a direct determination of the risk of those types of genetic damage under actual environmental conditions and in specific organisms of interest. Examples include measuring covalent binding of genotoxic agents to protein and DNA as well as cytogenetic effects in the population at risk. Once the relationship between exposure and a specific biological response is known, monitoring for that response can be used as a dosimetric method for estimating the risk of that population.

The quantitative impact of suspected environmental risk factors on relevant genetic endpoints in human populations can be determined by "biochemical or molecular epidemiological" studies in which statistical associations between suspected risk factors and observed genetic damage are determined, or in which genetic damage is monitored, while suspected risk factors are specifically modified in intervention trials. In addition, population studies provide the information necessary to establish the critical linkage between exposure and biological outcome in the populations studied.

Tools for conducting biological environmental monitoring studies have developed rapidly during the last decade. Currently available assays allow environmental samples, such as air, soil, or water, to be assayed in the laboratory for contaminants that are genotoxic to various organisms, and they also permit *in situ* monitoring of organisms in the environment and direct measurement of genetic damage in humans. Laboratory assays using *in vitro* techniques for mutagenicity in bacteria or cultured cells, assays for mutations or chromosomal

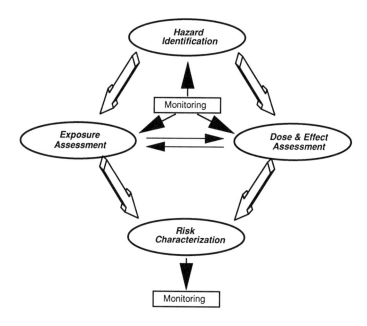

Figure 3. Scheme illustrating the components of risk assessment.

damage in aquatic and terrestrial plants and animals, including humans, and assays for various genetic endpoints in mammalian laboratory animals permit rapid and inexpensive screening for the presence of genotoxic activity in environmental samples, mechanistic studies of the molecular lesions induced, and determination of dose-response relationships.

In situ bioassays can be performed using either "sentinel" organisms placed into an environment of interest or by monitoring indigenous species for genotoxic damage. A variety of terrestrial and aquatic species, including terrestrial plants, aquatic invertebrates, fish, and terrestrial mammals, can be used. This method is important because adverse effects induced by environmental chemicals often result from complex interactions among various chemical and physical components of a given environment. Not only may these effects be due to degradation products formed in the environment, but additive, synergistic, and antagonistic interactions often occur among the many components present in complex environmental mixtures. These *in situ* assays allow specific environments to be monitored for particular types of genotoxic activity in the real-world situation that includes these interactions.

Methods of monitoring genotoxic damage in humans are still limited, but they are expanding. Cytogenetic methods that permit the monitoring of chromosomal aberrations in peripheral blood lymphocytes of humans have been available for many years and have been used to demonstrate chromosomal damage in individuals exposed to many genotoxic agents, including ionizing radiation, industrial chemicals, chemotherapeutic agents, and cigarette smoke. Methods that permit the measurement of gene mutations and heritable (stable) chromosomal aberrations in somatic cells have become available only recently. It is now possible to screen for cytogenetic damage, to monitor heritable cytogenetic damage with increased efficiency, and to measure mutations in somatic cells directly in humans. It is anticipated that recent advances in molecular biology will soon make it possible to monitor mutations at many more loci and in additional tissues.

E. Risk Characterization of Genotoxic Agents

Risk characterization takes the results of exposure assessment and dose-response assessment and integrates them to generate an estimate of the risk or probability of adverse outcome. The result should be interpreted in terms of the uncertainties and caveats that inevitably accompany such a complex analysis, and which is simplified by appropriate assumptions. Evaluations of the public health, societal, and political consequences of the different options for managing risk are conceptually distinct from risk characterization and are considered components of risk management (NRC, 1983). Because mutation is not itself a disease, the outcome of genotoxicity risk assessments is, necessarily, an estimation of risk of adverse genetic damage that could result in disease or death. We describe both qualitative and quantitative strategies to characterize the human risk associated with exposure to genotoxic agents, and we address the problems associated with assessing the risk to nonhuman biota.

III. RISK CHARACTERIZATION FOR NONHUMAN BIOTA

Homo sapiens is an integral part of the biogeosphere, i.e., the global ecosystem of Earth. The human race is dependent for its sustenance primarily on the domesticated plants and animals of agriculture and on certain wild organisms, such as fish and crustaceans, that are harvested in large quantities. However, all organisms are mutually interdependent through complex food chains and the biogeochemical cycles of the local and global ecosystems in which they live. Thus, it is important to assess the nature and significance of any chronic increase in mutation rates on nonhuman biota and ecosystems, inasmuch as significant alterations in ecosystem structure, function, and stability could have adverse effects on people. Relatively little empirical research has been performed in the area of "ecogenotoxicology". Thus, our discussion must be based primarily on various theoretical generalizations (Wügler and Kramers, 1992).

In the context of ecological health and human food production, primary concern is focused generally on *populations* rather than individuals. Our degree of concern usually depends on the apparent roles these populations play in ecosystem stability and in agricultural and marine productivity. In some situations, of course, individual organisms can become centers of attention, e.g., prize breeding stock and personal pets. The loss of endangered species may upset the balance of nature or decrease biological diversity in the ecosystem. Historically, the primary catastrophic ecological disasters have resulted from the release of acutely toxic chemicals that may not be genotoxic.

Because our concern with nonhuman biota focuses on *populations* rather than individuals, somatic mutations in such environmental organisms are of much less concern than their occurrence in humans. Somatic mutations will most frequently disappear with the death of the individual. Large populations, with large reproductive rates, will not be affected by the premature loss of a few individuals to diseases or malformations associated with such mutations. Only in extremely small populations are somatic mutations likely to contribute to extinction (Wügler and Kramers, 1992). Although a heavy mutational burden could produce a general decrease in "fitness" or effects such as immunosuppression, the consequences of such effects are not practical to assess with current technology. It should be noted, however, that the observation of genotoxic events from exposure to environmental mutagens in nonhuman biota can be important as practical biological monitors of such pollution.

Enhanced frequencies of new germline mutations of significant effect might be expected to reduce the fitness of a population, and this could affect ecosystem structure. However, the strong selective pressures normally operating against deviant individuals within well-adapted populations could reasonably be expected to eliminate such mutations well before they become fixed in the gene pool. Again, practical quantitative methods of monitoring such effects are currently not available.

Thus, a long-term increase in population exposure to environmental mutagens could lead ultimately to an increase in the rate of occurrence and frequency of neutral or nearly neutral mutant alleles in the gene pool. This would have no significant effect on a population evolutionarily adapted to a stable environment. Should the environment change, the fitness attributable to this expanded genetic variability could be either positive or negative, but in any case, its character would be unpredictable and likely to have little consequence to the overall balance and integrity of an ecosystem.

The most interesting and important situation arises in cases where new heritable mutations may prove advantageous for the species in which they occur but disadvantageous to humans (Wügler and Kramers, 1992). Examples include pesticide, herbicide, or antibiotic resistance, and changes in virulence and host range of pathogens. In some cases, the genes controlling these phenotypes may be propagated "horizontally" from organism to organism and even across "species boundaries". This is a case where selection is more important than mutation in fixing the advantageous allele.

In summary, our knowledge of the ecological consequences of an increase in mutation rates in wild populations of nonhuman biota, or of the appearance of new mutant organisms in an ecosystem, is extremely limited. General predictions of the significance of such events are, for all practical purposes, impossible to make. Ecology and evolutionary theory remain much more descriptive than predictive sciences. Ecological risk assessment can, at best, be performed only on a case-by-case basis. Studies of genetically engineered genes inserted into organisms released into the environment illustrate that such genes can be spread (Lenski, 1987). Clearly, a conservative approach to these issues is the wisest. Most new mutant alleles that may persist in wild populations are likely to be selectively neutral, and those rare persisting mutations with significant positive effects may do so only under specific conditions of selection. However, precisely because of our lack of knowledge in this area, it would be wise to control against significant exposure to mutagenic substances in the environment.

IV. RISK CHARACTERIZATION FOR HUMAN EXPOSURE TO GENOTOXIC AGENTS

Available information on the genotoxicity of and exposure to an agent is the first of several factors that are necessary to determine which risk characterization strategy may be used. The availability of information will divide agents (e.g., chemicals, radiation, mixtures, and pollution sources) into five subsets:

1. Those agents with no information on exposure or genotoxicity and, hence, cannot be subjected to risk assessment
2. Those agents with so little dose-response or exposure information that risk assessment can only be approached qualitatively
3. Those agents with sufficient information on genotoxicity and exposure to permit comparative risk assessment
4. Those agents that have sufficient information (e.g., *in vivo* mammalian data with endpoints relevant to or in humans) to permit probabilistic risk assessment
5. Those agents that have been studied extensively in humans and may serve as a comparative standard (e.g., radiation, cyclophosphamide, chlorambucil, etc.)

The largest number of agents are in the first subset, with the numbers reducing dramatically to the very small number that will fit into the fourth and fifth. Thus, we anticipate that the emphasis in actual practice will be on developing strategies for risk characterization of those agents that fall into either the second or third subset described above.

The purpose of the final risk assessment will influence the selection of a risk-characterization strategy. If the purpose of the assessment is purely scientific, e.g., as an exercise to improve risk assessment methodologies, then the selection of the strategy will be dictated by the scientific issues being addressed. Risk assessment organizes and presents scientific information about

RISK CHARACTERIZATION STRATEGIES 139

public health hazards in a clear and informative manner so that policy decisions (risk management) can be based on the best possible science. Therefore, an understanding of the risk-management objectives should provide input into setting the risk-assessment requirements. Certain risk-management objectives can be achieved with qualitative rankings of risk, whereas others require quantitative information to establish safe levels of exposure to an agent. Examples of risk-management objectives, and the matching risk-assessment strategy, are shown in Table 1 and illustrate the range of needs for information. Although scientific attention has been given primarily to improving the methodology and reducing the uncertainties in probabilistic risk assessments, there would appear to be an even greater need for developing methods for qualitative and comparative risk assessments.

A. Qualitative Risk-Characterization Strategies

For agents where there is so little dose-response and/or exposure data that only a qualitative risk characterization can be considered, a simple graded classification of both genotoxic potency and environmental prevalence is sufficient to carry out a rough, qualitative risk assessment. As few as two or three categories of genotoxicity and environmental availability can provide three qualitative levels of risk (high, medium, and low), as discussed below. Obviously, the more refined the classification, the greater the resolution will be among the qualitative risks. Such assessments can be thought of as a preliminary way to prioritize which agents receive attention, either for further measurements or for definitive action.

1. Classification of Genotoxicity

The weight-of-evidence classification of genotoxicity (Figure 4) takes into account the organism in which genotoxicity was detected and its relevance to humans. The nature of the genetic effect, although not providing information about the degree of hazard, does confer information about the nature of the hazard and, more importantly, the nature of the evidence, as discussed below.

1.1 Organism and Relevance to Humans

Humans, the most relevant organism, are exposed to a multitude of complex mixtures that include genotoxic agents. Therefore, it is rare that definitive human data are available to evaluate the genotoxicity of a particular exposure or agent. The most commonly employed genotoxicity assays are those for gene mutation in bacteria. Although the least relevant to humans, the bacterial gene-mutation assays have been essential to understanding the genotoxicity of complex environmental mixtures and continue to be the most widely used tool for initial genotoxicity screening. The use of eukaryotes such as yeast, plants, insects, and rodents make it possible to detect all of the genetic effects that may

Table 1. Examples of Risk Assessment Strategies Meeting Different Risk Management Objectives

Risk Management Objectives	Assessment Strategy
Ranking the genotoxic risk of various environmental media (e.g., air, water, soil/waste) for the purpose of setting research funding priorities	Qualitative or comparative
Prioritizing the risk of environmental pollution sources (e.g., energy production, chemical industry, and agricultural runoff) in a geographical region for the purpose of deciding on the expenditure of funds for pollution-control devices	Comparative
Assessing the risks of all air pollution sources in a country to determine which sources make the greatest contribution to genotoxic risk in order to initiate and/or evaluate air pollution control	Comparative
Assessing the genotoxic risks posed by various water pollution control technologies (e.g., chlorination vs. ozonation) so that specific emission standards can be set for each technology to ensure public safety	Probabilistic
Assessing the genotoxic risk posed by allowing a food additive to be used in baby cereal, soft drinks, and other food products	Probabilistic
Assessing various waste chemicals to determine whether they should be classified as potentially hazardous and, therefore, should not be mixed with normal municipal waste	Qualitative
Assessing new chemicals prior to their introduction into commerce to determine if they should be allowed without regulation, regulated, or prohibited	Qualitative followed by probabilistic if regulation is required

impact on human risk from exposure to genotoxic agents. One of the largest differences between laboratory studies and human studies, however, is the tremendous interindividual variation observed in humans in exposure, uptake, metabolism, DNA adduct dosimetry, and disease outcome.

1.2 Nature of the Genetic Effect

The structural nature of mutations that are the basis for human genetic disease are discussed in Chapter 1. The three major categories of genetic damage (referred to here as endpoints) that are associated with human disease are gene or point mutations, chromosomal mutations, and aneuploidy. These

RISK CHARACTERIZATION STRATEGIES

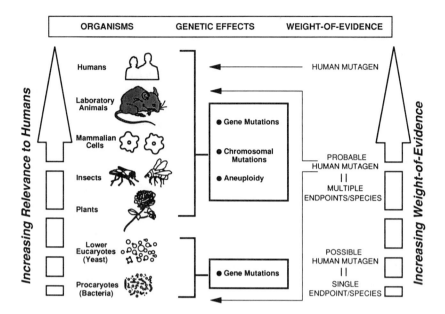

Figure 4. Classification of genotoxicity.

three categories of genetic damage have been definitively associated with a variety of human genetic diseases and have also been linked with carcinogenesis.

The evidence for a role by other types of genetic or related events in the etiology of genetic disease and cancer is less well established. These events include premutational lesions (e.g., DNA adducts), damage to DNA (e.g., unscheduled DNA synthesis), markers of chromosomal breaks (e.g., micronucleus assay), gene conversion and recombination (resulting, e.g., in loss of heterozygosity, sister chromatid exchanges, and gene amplification), and modification of processes related to an error-free replication of DNA (e.g., effects on nucleotide pools, modified nucleotides, DNA repair, and cell cycle control). However, information from these assays can also play an important role in detecting genotoxicity and elucidating mutagenic mechanisms.

In multicellular organisms, such as humans, the biological consequences of mutations depend on whether the mutations occur in somatic or germ cells. Mutations in germ cells can be transmitted to, and potentially affect, the next generation of organisms, whereas mutations in somatic cells have biological consequences only to the individual in which they arise. Thus, when considering genetic consequences of mutagen exposure, it is useful to categorize mutagens as somatic-cell mutagens (those demonstrated to cause mutations in somatic cells under a defined exposure condition) or as germ-cell mutagens (those demonstrated to induce mutations in germ cells and that may be transmitted to the next generation).

Germ-cell mutagens can only be classified definitively as such by the use of *in vivo* assays. The best experimental system for prediction of germ-cell mutations in humans are mammalian germ-cell assays, as described in Chapter 3. In the absence of germ-cell mutagenicity data, evidence that an agent is a somatic-cell mutagen, together with evidence that the agent reaches the germinal tissue, provides presumptive evidence that the agent may be a germ-cell mutagen. Conversely, an agent that has been shown to be a germ-cell mutagen may be presumed to also be a somatic-cell mutagen.

1.3 Weight-of-Evidence Classification

The longest-standing approach to risk assessment is the qualitative assessment of risk using experts to evaluate the weight of evidence that an agent presents a human health hazard and to define the exposure level at which an effect might occur. Classification schemes have been used extensively for the categorization of chemical carcinogens, based on weight- or strength-of-evidence evaluations (U.S. EPA, 1986b; IARC, 1987). In this context, mutagenicity data have always played a supportive role based on the genetic theory of cancer induction. Recently, genotoxicity data have become even more crucial in the IARC process of classifying carcinogens by providing evidence that a carcinogen may act via a genotoxic mechanism (IARC, 1991). Classification schemes have also been developed for the categorization of genotoxic agents that may induce heritable genetic damage based on a weight-of-evidence determination (U.S. EPA, 1986a, b, c).

We suggest the following classification categories to define, in general terms, the application of the weight-of-evidence procedure for assessing the genotoxic hazard of an agent to humans. Potency does not play an explicit role in this classification. However, whenever genotoxic effects are detected at unusually low exposures, the level of concern is always heightened.

Possible Human Mutagen — Because of the universality of the genetic material and the mechanism of DNA damage in all cells, an agent that has been clearly demonstrated to be mutagenic in one organism should be viewed as a possible mutagen in other organisms.

Probable Human Mutagen — Due to the difficulty of determining induced mutations in humans, it is necessary to establish indirect criteria for probable human mutagens. These criteria will vary from country to country and over time as new technologies are developed. The evidence would be derived from either mammalian mutagenicity data or, as a minimum, multispecies and/or multi-endpoint data. The guiding principle for defining the criteria for a probable human mutagen is that the heaviest weight of evidence should be given to agents that are mutagenic in organisms most closely related to humans and to agents that demonstrate genotoxic effects in multiple organisms.

Human Mutagen — An agent that has been demonstrated to induce mutations in humans would be classified as a human mutagen with the highest weight of evidence. Although no human heritable mutagens have been identified through human genetic epidemiological studies (Neel and Lewis, 1990), agents have been identified as inducing somatic-cell mutations (chromosomal and/or gene mutations) in human lymphocytes (Albertini et al., 1990; MacGregor et al., Chapter 5). Chromosomal changes have also been detected in human sperm cells (Brewen et al., 1975; Martin et al., 1986; Wyrobeck, 1983, 1990).

2. Qualitative Exposure Assessment

A qualitative exposure assessment relies on surrogate measures of exposure to estimate the potential magnitude of human contact with a given genotoxic substance. The basic surrogates involve the amounts of a substance released into the environment and the nature and extent of potential human contact. The simplest approach for ranking or categorizing the release of a substance is to determine the quantities produced in or imported into a country. This information, when combined with knowledge of the uses of the chemicals (and residuals associated with their consumption or processing), can be used to rank order or categorize substances according to their release potential.

These release categories can be refined further with information on the potential extent of human contact. Human contact can be divided into potential exposures that are widespread across a population (e.g., a food additive) or those that are localized to a smaller, defined group (occupational settings or areas around industrial facilities). Exposure categories that are based on combinations of the exposure surrogates, releases, and contact can then be used with information on the genotoxicity of the chemical to prepare qualitative assessments of the relative risks of genotoxic chemicals.

3. Risk Categorization

Qualitative categorization of genotoxic risk into several categories such as low, moderate, and high is illustrated in Figure 5. Both the qualitative assessment of exposure and classification of genetic activity must be considered in categorizing the risk. A rare synthetic laboratory chemical that is not found in the environment, but is a mutagen in multiple test systems, would be categorized as a low risk; whereas a ubiquitous chemical, to which humans are widely exposed and which had only been tested with positive mutagenic results in microbes, might be given a moderately high priority, especially for further genotoxicity evaluation.

This approach to risk characterization is accomplished most appropriately through a committee of experts reviewing the evidence and reaching consensus on the category of risk. This general approach has been used widely in many

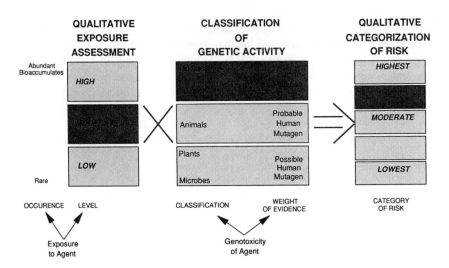

Figure 5. Qualitative risk characterization.

countries and is illustrated by the International Agency for Research on Cancer (IARC) working groups that are convened to classify the weight of evidence for carcinogenicity (IARC, 1987).

B. Quantitative Risk-Characterization Strategies

Quantitative estimation of genetic risks and the possible increased incidence of genetic disease due to human exposure to mutagens and the issues and problems associated with these estimates have been addressed earlier by various committees (Committee 17, 1975; Biological Effects of Ionizing Radiations (BEIR), 1980; United Nations Scientific Committee on the Effects of Atomic Radiation (UNSCEAR), 1977, 1982; National Academy of Sciences (NAS), 1982; ICPEMC, 1983; U.S. EPA, 1986a). These groups have addressed the issue of genetic risk assessment by focusing primarily on the production and hereditary transmission of germ-cell mutations. A review by Neel and Lewis (1990) describes the current status of determining germ-cell mutations induced in mice and humans by ionizing radiation and provides estimates of risk calculated from these data. However, assessment of risk for chemically induced germ-cell mutations has not been pursued to the same extent as has assessment of the risk of genetic diseases resulting from ionizing radiation-induced germ-cell mutations. A recent example of quantitative characterization of the human risk of heritable germ-cell mutations from a genotoxic chemical agent is the assessment of ethylene oxide by the U.S. EPA (Dellarco, 1990). Although this group predicted the excess risk of human heritable translocations per parts per million per hour, they did not extrapolate to a human disease frequency.

It has been suggested that genotoxic chemicals be compared to ionizing radiation and that the genetic risk of chemicals be expressed as an equivalent dose of radiation (Committee 17, 1975; Ehrenberg, 1979). ICPEMC Committee 4 reviewed these and other approaches to estimating the heritable genetic risk of chemicals. Although Committee 4 recommended the use of either the doubling dose or related relative mutagenic effect (RME) as discussed below, Ehling (1991) recommends a direct method for estimation of first-generation heritable effects. NCRP (1989) has more recently updated and extended the use of ionizing radiation data in a comparative approach to assessing the heritable genetic risk of chemicals. These approaches to genetic risk assessment have all addressed only heritable genetic risk and have not been extended to the genetic risk from somatic-cell mutations.

Carcinogenic risk assessments have been made for many years based on human epidemiological data and rodent carcinogenicity data (Doll and Peto, 1981; Crump, 1981). Such assessments may be viewed as a type of somatic-cell genotoxic risk assessment. In this regard, assessments of risk for incurring leukemia and other cancers from exposure to ionizing radiation have been made using data from the atomic-bomb survivors in Japan (Neel and Lewis, 1990). Also, the risk of death from developing lung cancer and cardiovascular diseases from exposure to cigarette smoke has also been estimated (Health & Human Services [HHS], 1986; NRC, 1986; U.S. EPA, 1992). These are multifactorial diseases in which the induction of somatic-cell mutations have been proposed to be an initiating or contributing event (Crump et al., 1979; Moolgavkar and Knudson, 1981).

1. Dose-Response Extrapolation

The problems associated with dose-response extrapolation from high dose to low dose and from rodent to humans have been addressed in the past (NRC, 1983; ICPEMC, 1983) and still remain the two major controversial scientific issues for quantitative risk assessment regardless of the nature of the biological effect or disease outcome. In the context of genotoxic risk assessment, we separate these issues into the issues surrounding the estimation of potency, which incorporate the high-to-low-dose issues and the issue of risk extrapolation from laboratory data to human risk.

1.1 Potency Estimation Methods

Genotoxic potency relates the magnitude of genotoxic or mutagenic effect to either exposure or dose. Several methods have been used to transform experimental dose-response data into potency estimates (Figure 6). The most commonly used methods are the slope of the dose response, the doubling dose or relative mutagenic effect, the lowest effective dose, and the radiation dose equivalence method. In attempting to assign a potency ranking to a chemical based on any of these values, there is often an inherent assumption of a linear dose response. This assumption of linearity has been considered conservative, because a violation of the linearity assumption was considered to be due to a sublinear or threshold response. However, the other alternative to linearity, supralinearity, should not be excluded, for it could result in an underestimation of the relative potency of a compound if dose-response data in the low-dose range are missing. As the dose range tested approximates the relevant dose range of human exposure, the magnitude of the error due to the false assumption of linearity decreases.

1.1.1 Slope

Slope is defined as the increase in the number of genotoxic or mutagenic events per unit exposure or per unit dose. It may be calculated from a zero dose control and a single experimental dose assuming linearity, or preferably via a dose-response analysis of a more extensive data set. Although the simple linear model is most commonly used, many different statistical models may be used to describe the dose response and estimate the initial linear slope when toxicity, saturation, or other biological mechanisms result in nonlinear dose-response curves. The most common of these models exhibit a decrease in slope at higher doses (Stead et al., 1981; Bernstein et al., 1982; Drinkwater and Klotz, 1981).

The advantage of this method of potency analysis is that it relies directly on human or laboratory dose-response or exposure-response relationships. Using slope as an expression of potency for a particular agent is consistent with traditional risk-evaluation procedures, where the risk is determined by multiplying the potency slope times the estimated dose or exposure to humans (i.e., risk = slope × exposure).

To combine or compare potencies from animal or other laboratory bioassays where the response being measured may differ (e.g., mutations, DNA adducts, chromosomal aberrations, and tumors) requires the conversion to a relative scale. This is one of the disadvantages of the slope method as compared to one of the methods described below that expresses potency in terms of dose (e.g., doubling dose or lowest effective dose). The disadvantage of these methods is that they may not take advantage of the entire dose response, but may rely on only one dose tested.

1.1.2 Doubling Dose and Relative Mutagenic Effect (RME)

Doubling dose is defined as the dose at which the observed genotoxic effect or mutation rate is twice the spontaneous or background genotoxic effect or mutation rate. This definition is equivalent to stating that the doubling dose is the dose at which the induced genotoxic effect is equal to the spontaneous mutation rate. The relative mutagenic effect (RME) of an agent is the ratio of the induced mutation rate per unit dose to the spontaneous mutation rate, which is the reciprocal of the doubling dose (Figure 6). The RME was proposed by ICPEMC Committee 4 (1983) to serve as the measure of potency for the estimation of heritable genetic risks and increased incidence of genetic disease.

RISK CHARACTERIZATION STRATEGIES

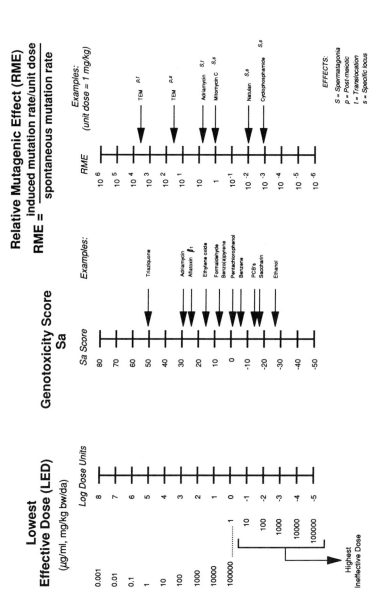

Figure 6. Genotoxic potency scales.

These two interchangeable units have been used extensively in heritable mutagenesis studies in rodents, where a minimum number of doses are tested.

The doubling dose or RME may also be calculated from experimental data consisting of a control and a single dose group assuming linearity or from a dose-response analysis. Algebraically, the doubling dose may be expressed as the spontaneous mutation rate divided by the slope of the dose response. Because the spontaneous mutation rate is usually not known with precision, including it in the calculation of a value to be used in expressing the potency of chemicals results in an increased variability of the estimate as compared to simply using the slope. A further consideration is that the doubling dose is dependent directly upon the magnitude of the spontaneous mutation rate. Thus, an observed difference in the doubling dose as determined for different mutational endpoints could be due to the spontaneous mutation rates, not the genotoxic potency of the compounds.

These disadvantages are offset by the formal simplicity of risk estimation employing the doubling dose approach, which was developed initially for radiation risk assessment (UNSCEAR, 1977, 1982; BEIR, 1980). In this approach, the spontaneous mutation rate for different classes of mutation is estimated from human epidemiological data, the doubling dose is estimated in experimental studies with laboratory animals (e.g., the mouse specific-locus test), and it is assumed that the doubling dose may be extrapolated across species. As is true for slope, the doubling dose of a chemical is also specific for a particular mutagenic test.

1.1.3 Lowest Effective Dose (LED)

The LED is defined as the lowest dose tested that resulted in a significant increase in the genotoxic effect. Waters et al. (1988) developed a computerized approach graphically presenting genetic activity profiles (GAPs), which utilized the log-transformed LED and the highest ineffective dose (HID) tested (Figure 6). GAPs have been used by IARC (Waters et al., 1988, 1991) in classifying the genotoxicity of agents. The LED value can only be meaningful if a range of dose points, including a no-observed-effect dose, is tested. Further, the use of the LED for a particular mutagenicity test for the purpose of ranking a group of chemicals for genotoxic potency is only valid if all chemicals have been tested similarly over a comparable dose range. The LED may in some cases depend upon the spontaneous mutation rate if the criteria for a positive result are that the observed mutation rate is greater than some multiple of the spontaneous mutation rate. The LED has not been considered in traditional quantitative risk-estimation methods, but has recently been used to combine data from the GAP database into a genotoxicity score, as discussed below.

In a single test series, the LED provides a rough, estimate of the dose corresponding to the sensitivity limit (or the boundary between positive and negative outcomes) for a test-agent combination. When comparing the results for two agents using the same test, the one with the steeper dose response

RISK CHARACTERIZATION STRATEGIES

should have the smaller LED, and the two slopes should correspond roughly to the two reciprocal LEDs. The same applies for an agent compared across multiple tests. Each test will provide an LED scaled to its own sensitivity limit and efficiency in the use of agent, and the reciprocal LEDs could serve as a surrogate for the slopes of the dose responses. Even in the absence of such normalization, averaging reciprocal LEDs across tests would be roughly analogous to averaging the slopes of dose responses. Even though there is no explicit event-related numerator in such an estimate, the numerator can be thought of as the sensitivity limit of each test. In practice, the broad range one typically finds for LED values (Figure 6) has led to the convention of using the log LED.

1.1.4 Genotoxicity Score

A general approach for creating a computer-based composite-scoring system for genotoxicity test data, called Genotoxicity Score, has recently been developed by Committee 1 of the ICPEMC (Lohman et al., 1992; Moore et al., 1992; Mendelsohn et al., 1992). This system is based on the LED methodology developed by Waters et al. (1988) and discussed above. Both positive and negative data are included by combining the LED and highest ineffective dose (HID) into one response for estimating the overall genotoxicity of an agent. The LED and HID scores are normalized between assays by their respective sensitivities and efficiencies (Moore et al., 1992).

The practical significance of this approach is its inclusion of evidence from negative tests. Thus, it weighs the consistency of the genotoxic response. Using this system, 113 agents have been ranked, with Sa scores ranging from a low of −27.70 (ethanol) to a high of 49.67 (triaziquone) (Figure 6). Although this approach to combining data has used the log LED, it could be applied to the doubling dose or slope by generalizing methods for each test. This may be more difficult than using the LED, however, because of the varying levels of quantitation in the data and the dramatically different ranges of response.

1.1.5 Radiation Dose Equivalence

Ionizing radiation was the first environmental agent for which risk-assessment methods were developed and is one of the best-studied agents in humans, animals, and *in vitro* systems. For these reasons, various authors and committees have recommended using radiation as a standard in assessing the risk of genotoxic chemicals (Bridges, 1973; Crow, 1973; Ehrenberg, 1974; Committee 17, 1975). In these methods, the dose-response relationship of a chemical agent is compared to radiation for the purposes of risk extrapolation, as discussed below (Section IV.B.1.3). Several different approaches have been taken to develop a unit of radiation dose equivalence as a general unit of chemical dose with which to express potency and ultimately, risk (Ehrenberg, 1979; ICPEMC Committee 4, 1983). The rem-equivalent chemical (REC) unit was introduced in 1975 (Committee 17) as a dose or product of concentration multiplied by the time that produces an amount of genetic damage equal to that produced by 1 rem of chronic irradiation. The rad-equivalent unit developed by Ehrenberg (1974, 1979) is based on dose and DNA reaction rates for alkylating agents.

All radiation-equivalence approaches assume that the relative response between radiation and chemicals is the same between animals and humans. This hypothesis is difficult to test for heritable genetic damage due to uncertainty concerning the genetic consequences of human radiation exposure. For disease outcomes associated with somatic-cell genetic damage (e.g., cancer), this critical assumption has been explored with some success (Ehrenberg, 1979). The concept of using radiation equivalence for heritable genetic risk estimation has been criticized on the grounds that the ionizing radiation and chemicals induce different patterns of response and different mutational spectra in laboratory animals.

1.2 Low-Dose Extrapolation Methods

For estimation of risk due to environmental exposures to genotoxic agents, the human exposures and associated target doses are generally much lower than the doses administered in laboratory animal studies. The most commonly assumed model for low-dose extrapolation in estimating potency by the methods discussed above is the linear nonthreshold dose-response. Low-dose extrapolations of germ-cell mutation rates in laboratory animals has relied on the linear or linear quadratic model (Ehling, 1991) until recently, when Rhomberg et al. (1990) applied both the multistage and Weibull models used previously in low-dose extrapolation for cancer risk assessment. In the case of ethylene oxide, Rhomberg et al. (1990) demonstrated that these nonlinear models fit the translocation dose response in mice better than the linear extrapolation model. Favor et al. (1990) have shown a threshold model to be the best fit of ENU-induced specific-locus mutations in spermatogonia of mice. It may be preferable to employ low-dose extrapolation models that are theoretically based on the biological mechanism of the measured response rather than being purely empirical and that fit the experimental data in the low-dose region of the dose response. In the case of cancer risk assessment, biologically based dose-response models have been proposed that invoke an initial interaction of genotoxic chemicals with DNA, followed by mutation, cell death, cell growth, and proliferation (Moolgavkar and Knudson, 1981). We expect such models to become increasingly important in risk assessment; however, their use will often require additional data on both biological dosimetry and the mechanism at the molecular level.

1.3 Human Risk Extrapolation Methods

Two extrapolation problems usually are encountered in estimating the human potency of a genotoxic agent: the extrapolation from high- to low-dose exposures encountered in the environment, discussed above, and the species extrapolation from estimations of potency in laboratory animal bioassays to

RISK CHARACTERIZATION STRATEGIES 151

human potency. A number of different approaches to the extrapolation from laboratory bioassay data to humans are described here, beginning with the more definitive probabilistic methods and ending with several methods that provide a relative ranking of genetic risk.

This problem is most serious when the objective of the risk assessment is to estimate the probability of human genetic disease occurring when one or both parents are exposed to genotoxic agents. Using even the best possible data, mutagenicity in the germ cells of rodents, it becomes problematic to extrapolate from a mutation incidence in mouse germ cells to disease incidence in humans because the fraction of mutations that result in human disease are not known.

1.3.1 Probabilistic Methods

Estimation of the probability of heritable genetic disease from exposure to a genotoxic agent currently relies on experimental *in vivo* mammalian germ-cell mutagenicity data. If such data are available for a mutagen in question, a human genetic risk estimation may be made using either the indirect or direct risk-estimation methods or a combination of both (Ehling, 1991). Human risk is usually expressed as the expected number of cases of genetic disease due to exposure to a genotoxic agent. Extrapolation methods to this end have been developed based upon the radiation genetic experience.

Indirect risk estimation uses the doubling dose. This method assumes a linear nonthreshold dose-response relationship and that the genetic damage from exposure to genotoxic agents is similar in nature to spontaneous mutation (Sobels, 1989). The method relies on an experimental estimate of the *in vivo* mutagenic activity of a genotoxic agent in the mouse as well as human genetic data (spontaneous mutation rates, number of loci). This dependence upon human genetic data for the risk extrapolation procedures may be avoided by expressing the expected risk due to exposure to a genotoxic agent as a relative risk (Ehling, 1991; ICPEMC, 1983). In this case, risk is calculated as the exposure dose divided by the doubling dose and expresses the increased mutation rate as a multiple of the spontaneous mutation rate. Estimates of radiation-induced genetic risk by the United Nations Scientific Committee on the Effects of Atomic Radiation (UNSCEAR, 1982) and BEIR Report (1990) use the indirect method and also assume that the doubling dose calculated for selected recessive mutations may be used to quantify dominant mutations, chromosomal diseases, and mutations with irregularly inherited diseases (Ehling, 1991).

The direct risk-estimation method, which uses dominant mutations to estimate genetic damage in the first generation, has been advocated by Ehling (1991). Ehling argues that this method avoids the problem of basing an estimate of induced mutations for one genetic endpoint upon experimental data from a different genetic endpoint. This method has relied on the use of dominant skeletal or dominant cataract mutations and also requires a number of assumptions, including (1) the dose response is linear, (2) the sensitivity of the dominant gene being used (e.g., cataract gene) is representative of all dominant genes, and (3) the ratio of the dominant mutation being measured (e.g., cataract mutations) to the total number of well-established dominant mutations in the mouse is the same as for humans.

These probabilistic risk-estimation procedures have the advantage that they are based upon the most appropriate experimental data for the human situation, and risk is expressed as an easily conceived quantity, i.e., probability or incidence of genetic disease. However, additional specific information is required, or critical assumptions become inherent to the extrapolation procedure. For example, pharmacokinetic data must be employed for species scaling or the assumption that sensitivity to mutation induction due to exposure to a compound is equal in man and mouse.

1.3.2 Comparative Methods

The comparative methods all rely on comparing the relative potency of the genotoxic agent in question either to a standard agent (e.g., radiation or a well-studied chemical) or other genotoxic agents. The underlying assumption in all comparative methods is that the relative potency in humans is either equivalent to or may be predicted from the relative potency in a laboratory bioassay.

1.3.2.1 Radiation equivalence method. The use of ionizing radiation as a standard in the evaluation of risk from exposure to environmental chemicals was the first comparative-risk method proposed in the 1970's (Bridges, 1973, 1974; Crow, 1973; Ehrenberg, 1974, 1979; Committee 17, 1975). The potency unit used in this method is either the rec-equivalent or the rad-equivalent discussed above. Törnqvist and Ehrenberg (1985) have used rad-equivalent risk model most extensively to estimate the cancer risk in Sweden from a series of urban air pollutants, particularly gaseous alkenes (e.g., ethene) (Törnqvist and Ehrenberg, in press). This method is based on the target doses of chemicals, determined by hemoglobin and/or DNA adducts and of gamma-radiation inducing the same response (Törnqvist et al., 1991).

1.3.2.2 Comparative potency method. The comparative potency method was developed to estimate human cancer risk of a series of related combustion emissions where there was no human cancer information for the specific complex mixture being assessed (e.g., diesel emissions) but there was human cancer data for similar mixtures (e.g., coke ovens and tobacco smoke) (Lewtas et al., 1981; Albert et. al., 1983; Lewtas et al., 1983). The underlying assumption in this method is the constant relative potency hypothesis. This is the hypothesis in which there is a constant relative potency across different bioassay systems (e.g., human and rodent) so that:

RISK CHARACTERIZATION STRATEGIES 153

$$\frac{\text{human potency carcinogen}_1}{\text{human potency carcinogen}_2} = (k) \frac{\text{bioassay potency carcinogen}_1}{\text{bioassay potency carcinogen}_2}$$

The potency in this method directly employs the slope that may be expressed in response per unit exposure or unit dose.

The constant relative potency assumption is implicit in any comparison that utilizes the relative toxicity of two substances in animals to estimate their relative toxicity in humans. This constant relative potency assumption is an experimentally testable hypothesis if the relative potency of two mixtures or components in one bioassay (e.g., humans) can be determined and compared to the relative potency in a second bioassay. The test of this hypothesis is whether there is a constant relationship (k) between the relative potencies in the two bioassays being compared. This method has been used in estimating the carcinogenic risk of complex mixtures of air pollutants containing polycyclic organic matter, such as combustion, pyrolysis, and incineration emissions (Lewtas, 1985, 1988, 1992, 1993).

A related approach is being applied to the risk assessment of individual polycyclic aromatic hydrocarbons (PAHs) and mixtures of PAHs by using benzo[a]pyrene as the reference chemical (Thorslund, 1988). Each PAH is expressed in equivalent units of B[a]P in a manner similar to the rad-equivalent approach. The human lung cancer risk from B[a]P has been extrapolated from rodents to humans using traditional scaling factors applied to the cancer data from a chronic rodent inhalation carcinogenesis study. The human lung cancer risk from the other PAHs are then determined in the same manner, as illustrated here for the comparative potency method.

1.3.2.3 Relative ranking. For the purposes of prioritizing or ranking genotoxic agents, the consistent use of one potency scale could serve as a relative ranking scale in a comparative risk-extrapolation model. The utility of such a ranking scale for estimating human risk from exposure to genotoxic agents would also implicitly invoke the constant relative potency hypothesis as discussed above. The use of the genotoxicity score (Sa) discussed above (see Section 1.1.4) for this ranking scale has as an advantage the ability to combine data from different genetic toxicity bioassays into one score.

2. *Quantitative Exposure Assessment*

The basic objectives of a quantitative assessment of human exposures to a genotoxic agent are to determine, in a cost-effective manner, the concentrations of that substance in contact media such as air, water, soils, foods, and beverages as well as the contact rates with those media (e.g., daily volume of water consumed and air inhaled). In this context, successful exposure-assessment strategies will be those that focus on the most important pathways of human exposure to the genotoxic substance(s) of concern.

One assessment approach includes an initial screening-level analysis for identifying the key exposure pathways, followed by more detailed exposure analyses that rely on measurement and/or modeling-based methods to characterize human exposures (Figure 7). The basic approach begins with a source-term analysis that involves the characterization of how a substance is released into the environment (e.g., atmospheric emission, waste water discharge, etc.), the physicochemical properties of the released material, and pertinent temporal aspects of a release (e.g., transient vs. chronic releases to the environment). This source-term analysis can be performed for one or more substances emitted from a specific industrial operation or for more generic release modes, such as releases to soils in a given geographic region. Releases can be based on actual measurements of emissions via sampling and analysis protocols, or estimated, based on emission factors (i.e., mass of genotoxicant emitted per unit of output or input of a given process) published in the literature or derived from analyses of the processes.

Once the preliminary source-term analysis is completed, a screening analysis is then prepared that identifies which environmental media are likely to have the highest concentrations of the contaminant and the pathways that could result in the highest exposures. An effective way of completing this type of screening-level analysis is to develop a multimedia environmental model that explicitly accounts for the movement of chemicals between the media at a given locale or perhaps a reference site that is indicative of a region. Mackay (1991) provides a thorough review of multimedia models that account for the partitioning of compounds between media boundaries (e.g., air-water and water-soil) based on the concept of chemical fugacity. These models can range in sophistication, have modest data input requirements, and are easily implemented on personal computers. In order to implement this type of model, parameters must be quantified dealing with selected properties of the chemical(s) released into an environmental compartment as well as properties of the receiving environment. For a given release scenario (e.g., a continuous release onto surface soils), a multimedia model will give the associated concentration in the various compartments (e.g., soil, air, water, or groundwater).

The media-specific concentrations predicted by the model can then be used to estimate the magnitude of potential human contacts with the genotoxicant via different exposure pathways. One convenient way of assessing exposures is to use pathway exposure factors (PEFs), which translate a contaminant's concentration in a given medium to a daily contact rate (expressed in terms of mass per kilogram of body weight per day) (see McKone and Daniels, 1991). Exposure factors can be developed for organic chemicals using physicochemical parameters such as their Henry's law constant, octanol-water partition coefficient, and diffusivities in water, air, etc., along with physiological parameters and lifestyle factors. The application of these exposure factors provides a matrix of exposures as a function of media and route of exposure (i.e., inhalation, ingestion, or dermal contact). This type of screening-level analysis will identify which contact media should be the subject of more detailed

RISK CHARACTERIZATION STRATEGIES

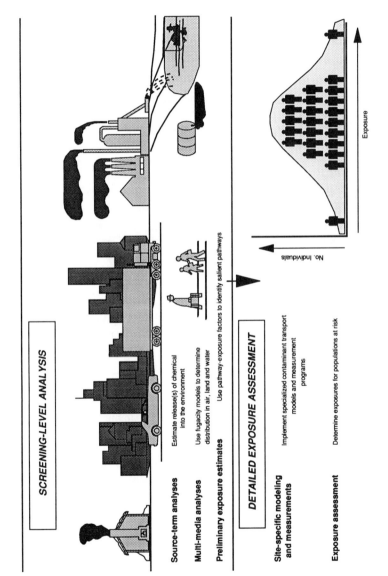

Figure 7. General strategy for conducting an exposure assessment of a genotoxic substance released into the environment.

measurement and modeling efforts. The final phase of the exposure-assessment methodology is to conduct site-specific analyses of exposure.

Once the key contact media have been identified in the screening analysis, various approaches can be adopted to determine the spatial and temporal variations in the concentration of a genotoxic substance in air, water, soil, and biota. There are many models available that simulate the transport and transformation of contaminants in groundwater, surface water, soil, and the atmosphere (see Jones et al., 1991). Depending on the sophistication of the models, the input requirements can be extensive. For example, when a Gaussian dispersion model is used to calculate the annual average concentration of an airborne substance around an industrial facility, it is necessary to have seasonal meteorological data on wind speeds, mixing heights, and atmospheric stabilities. Similarly, if annual concentrations are to be measured (instead of modeled) at fixed locations downwind from the facility, monitoring equipment must be operated for sufficient periods of time and in the case of particulate matter, subsequently analyzed in a laboratory for specific substances of concern. It may also be necessary to determine concentrations of the genotoxicant in indoor air, as people in many climates spend the majority of time indoors. Groundwater models require input data on the hydrologic properties of an aquifer, which in most cases must be determined by installing sampling wells for use in conducting applicable aquifer tests. For developing countries, there are clearly many tradeoffs that must be considered carefully when deciding on which approach (modeling or measurements) to adopt at a given site. Population exposures to a genotoxicant can be estimated as the product of the number of people exposed to a given contact medium and the concentration of the genotoxicant in the medium (exposure units of milligrams per kilogram of body weight per day).

Although point estimates of contact rates to a contaminant are frequently the product of exposure assessments, there are usually various kinds of uncertainties that affect the predicted results. Therefore, it is desirable to conduct even a rudimentary uncertainty analysis of the assumptions, parameters, and scenarios used to assess exposures. Formal uncertainty analyses using Monte Carlo techniques can also be conducted on personal computers, but input distributions must be specified for parameters; these may not be easily obtained.

3. Risk Quantitation

The final step in risk assessment is combining the extrapolated potency with the estimated individual and population exposure estimates to determine either the probability of an effect outcome or the comparative risk. As illustrated in Figure 8, this step can be done on either a comparative scale or a probability scale. It is possible to estimate the risk on an individual basis or for a population. The exposure assessment is usually based on either exposure concentration, daily intake, or dose. Favor et al. (Chapter 3) discuss the issues associated with the uptake, distribution, metabolism and dose in more detail.

RISK CHARACTERIZATION STRATEGIES

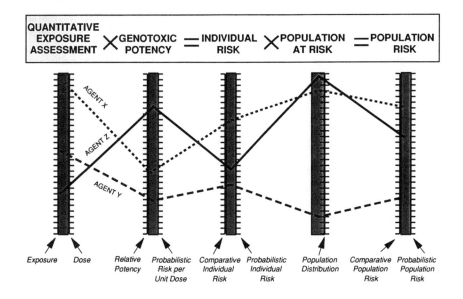

Figure 8. Quantitative risk characterization.

3.1 Probabilistic Risk Assessment

Ideally, one would have sufficient information about an agent to understand its dose-response relationships in several organisms, including ones closely related to the target organism such as the human. One might also have detailed environmental information about inventories, transport and fate of the agent, and its immediate access to the target species. Given these data elements, one can multiply the slope of the dose response by the exposure estimate to calculate the probability of a change in mutation rate at the somatic or germinal level. Such estimates, along with the necessary statistical uncertainties and caveats, constitute a probabilistic risk assessment. In principle, the risk assessments for any agents that have been evaluated by this process should be directly comparable to each other. As our understanding of the health significance of a change in mutation rate increases, the calculations for risks for such agents will also be comparable to other types of risk, such as fire, automobile accidents, or infectious disease. One could carry the argument one step further by reducing all such risks to a common metric, thus allowing their comparison to other societal options, including the economic benefits that may be associated with the use of the agents. We anticipate many practical problems with this approach, the most immediate being the remarkable sparsity of data with which to begin such analyses. Nevertheless, it is important to understand this approach, if for no other reason than to understand the limitations of the less-rigorous approaches. Hopefully, the commitment to continue collecting the requisite quantitative data will help achieve the practicality of a quantitative risk assessment for genotoxins.

3.2 Comparative Risk Assessment

For most environmental agents, and especially complex mixtures, it will often not be feasible to relate the genotoxic endpoints directly to events per unit dose for use in a numerical estimate of human or other biotic mutational risk. In many cases, the only data available will be gene mutation data in bacteria (e.g., *Salmonella typhimurium*), and in other cases, multiple short-term genotoxicity endpoints may have been studied. The Ames *S. typhimurium* histidine reversion assay has been the only genetic bioassay used widely in evaluating the genotoxic activity of complex environmental mixtures (DeMarini, 1991; Lewtas, 1991). Data are either available or readily collected on the relative genotoxic activity of emissions from many sources (e.g., industrial, automotive, and personal habits) as well as many environmental media (e.g., air, industrial effluents, drinking water, and food). It is even possible to estimate the total human exposure to genotoxic activity from air, water, and food by combining the data on the genotoxicity of air, water and food with the total intake from each of these routes of exposure. This approach has been used to prioritize the various air pollution sources of potential human genotoxic risk in various geographical areas (Lewtas and Gallagher, 1990).

There are, of course, important limitations to the use of short-term genetic bioassay data in comparing the potential genotoxic risk of agents in humans. Examples of such limitations would be endpoints in organisms remote from humans (such as prokaryotes), endpoints that are difficult to generalize quantitatively (such as yield of erythrocytic micronuclei), and endpoints that are not direct measures of gene mutation (such as strand breaks, sister-chromatid exchanges, or DNA adducts). A useful alternative is to stop short of absolute potency, and to use the various endpoints for genotoxicity in a composite weighting scheme, as described here for the genotoxicity score (Sa), which provides an ordering of agents by genotoxic potency, thus making it possible to do a comparative risk assessment. Similar constructions can apply to exposure data where, for example, one might have a class of agents that can be ranked by solubility or volatility, but where the detailed knowledge of their quantitative environmental behavior is lacking.

A set of related comparative-risk assessments will provide a list of the involved agents ordered by their cause for concern. This could be a very useful tool for committing resources to the necessary data accumulation to permit a probabilistic risk assessment. It can also be sufficiently convincing to allow risk managers to take regulatory action in a prioritized strategy, beginning at the top of the list with the most potent mutagens.

V. FACTORS THAT AFFECT RISK

A range of factors can influence the risk incurred by an individual or a population of organisms upon exposure to a mutagen.

RISK CHARACTERIZATION STRATEGIES

A. Endogenous Factors

Different species, and even individual organisms within a species, may exhibit different susceptibilities to the mutagenic effects of a given agent. The basis for a differential susceptibility may reside in the presence or absence or levels of various types of enzymes present in the organism.

Various genes that control the repair of DNA damage have been identified in many organisms — from bacteria to humans. In addition, different types of DNA repair activities are controlled by these genes, such as excision repair, demethylation, and glycosylation. Different individuals as well as different tissues within an individual may have different DNA repair activities, which will affect the ability of an organism to repair DNA damage or to process such damage into a mutation.

Many agents are not mutagenic per se but require conversion to a chemical form that can interact with DNA and cause mutations. This conversion, called metabolic activation, is accomplished by a variety of enzymes, e.g., P-450 enzymes. These enzymes vary in type and amount among species, individual organisms within a species, and within different tissues of an organism. In addition, the complex interplay among these enzymes as well as other metabolic capacities of an organism, including the tissue distribution of the activated mutagen, all influence the susceptibility of an organism to a given mutagen.

Many biochemical reactions within an organism produce highly reactive molecules, such as free radicals, that can damage DNA. In addition, mistakes in normal DNA metabolism can also lead to mutation, such as the endogenous deamination of 5-methyl cytosine (Youssoufian et al., 1986). Thus, a variety of endogenous factors can influence the susceptibility of an individual to risk from mutation.

B. Background Mutation Rate

All organisms or genes have an inherent background mutation rate, i.e., the occurrence of a mutation/cell/generation. Recent analyses suggest that this background mutation rate is more or less constant among organisms, although it varies per unit of DNA among organisms (Drake, 1991). It is possible that mutations in certain genes (DNA polymerases, DNA topoisomerases, and DNA repair genes) might increase or decrease the background mutation rate (Loeb, 1991).

Mutations can also result from dramatic changes in deoxyribonucleotide pools induced in microorganisms as well as in mammalian cells in culture (Kunz, 1982). That such mechanisms may be effective *in vivo* in mammals is exemplified by the acute and/or chronic administration of orotate (Rao et al., 1987) or galactosamine (Lesch et al., 1973), which results in increased neoplastic development via nucleotide pool imbalances. Changes in the nutritional states of mammals result in qualitative and quantitative changes in DNA adducts (Li and Randerath, 1990), which are potential promutagenic sites in DNA that could contribute to the background mutation rate. This process has been postulated to play a major role in the increased incidence of cancer seen in situations of increased cell replication induced by specific chemicals (Cohen and Ellwein, 1991).

C. Cell Replication

Recently, there has been a renewed emphasis on the role of cell division in toxicology, especially carcinogenesis (Cohen and Ellwein, 1991; Preston-Martin et al., 1990; Ames and Gold, 1990). As indicated above, the background rate of mutation is a function of the cell division rate. The "fixation" of mutations and the initiation stage of carcinogenesis require cell division, as does the expansion of initiated cell clones during the stage of promotion in carcinogenesis (Pitot et al., 1991). An even more important role of cell division in mutagenesis and carcinogenesis is the formation of structural changes in chromosomes that occur as "mistakes" of the mitotic phase of cell division (Ames and Gold, 1990).

D. Age

It is not yet clear how the age of an organism might influence its susceptibility to mutational damage induced by a mutagen. It is known that mutations may accumulate during the lifetime of an organism (Cole et al., 1988; Vijayalaxmi and Evans, 1984), making an older organism more vulnerable to the effects of a greater body burden of mutations than a younger organism that has a lower accumulation of mutations. There is also theoretical and some experimental evidence that aging itself is a consequence, to some extent, of accumulated mutations. It is also known that both cell replication and the immune system become impaired with age.

E. Diet

Diet may also influence the risk of incurring mutational damage from exposure to mutagens (Sugimura, 1982; Wakabayashi et al., 1991). Limited evidence suggests that at least in some species, a low-calorie diet or a diet low in fat or protein may alter the levels of certain enzymes involved in metabolic activation and may reduce the risk for mutation induction by exposure to mutagens. In addition, deficiencies in the levels of some vitamins, such as folate and Vitamin C, may increase the susceptibility to chromosomal mutations. Many foods contain both mutagens and antimutagens, and some epidemiological evidence suggests that consumption of some mutagenic foods in

RISK CHARACTERIZATION STRATEGIES 161

large quantities may account for certain types of cancers in some populations. Thus, diet may be an important factor affecting risk for mutation produced by mutagens.

F. Economic and Social Factors

Related to diet are economic and social factors that might also influence a person's risk for incurring mutations from exposure to mutagens. Poor diet, inadequate health care, infectious diseases, and excess exposure to known environmental mutagens (cigarette smoke, sunlight, etc.) could all interact to increase a person's susceptibility to mutagens. The presence of so many additional factors (described above) that may influence the risk of incurring mutations produced by exposure to genotoxic agents even further complicates risk characterization and quantitation. Nonetheless, such factors should be considered to the extent possible when characterizing the risk of a genotoxicant.

G. Duration of Exposure

The duration of exposure to a mutagenic substance, chronic vs acute, may affect the resulting risk incurred depending on the form of the dose-response curve in relation to exposure levels as well as the cell stage specificity of the mutagenic activity of the compound. For a strictly linear dose response, the genetic risk incurred would be additive and directly related to the cumulative exposure level regardless of exposure duration. For a nonlinear relationship of mutagenic effect to dose, the exposure duration would be expected to affect the mutagenic response and, therefore, the risk incurred. For example, for a threshold dose-response, the application of a total dose delivered chronically would result in a lower mutagenic effect than if the same total dose was delivered acutely. For a saturable metabolic activation, this difference would be reversed, i.e., a chronic exposure could be associated with a higher mutagenic effect than an acute exposure if the total dose is in the asymptotic region of the dose response.

The specificity of germ cell-stage sensitivity to mutation induction is also important when considering the genetic risk incurred by exposure to a chemical mutagen. For an acute exposure, chemicals with a mutagenic effect in postspermatogonial stages would result in a transitory genetic risk confined to conceptions resulting from sensitive gametes. For chemicals with a mutagenic effect in stem-cell spermatogonia, a permanent genetic risk will be incurred following acute exposure. For a chronic exposure to chemicals with mutagenic effects only in postspermatogonial stages, the genetic risk incurred will be equal to the genetic risk in the sensitive postspermatogonial germ-cell stage rather than the magnitude of the genetic risk incurred being associated with the cumulative exposure. Furthermore, upon cessation of exposure, the genetic risk will return to zero. For chemicals with mutagenic effects in stem-cell spermatogonia, the genetic risk incurred will increase constantly and will be related directly to the cumulative exposure. Finally, upon cessation of exposure, the genetic risk will remain at the final level attained, if the occurrence of induced mutations in stem cells is not associated with a decrease in the survival of the carrier cells.

VI. SIGNIFICANCE OF COMPARATIVE GENOTOXIC RISK ASSESSMENT

Components of this general process are used in many countries to determine whether new chemicals will be allowed into commerce and to set acceptable exposure limits for chemicals. Quantitative risk assessment, following this general framework has been used most widely to estimate human cancer risk. This chapter describes a more generic application of the risk-assessment framework, beginning with hazard identification and progressing to risk characterization by estimating the relative or comparative risk associated with the exposure of human and nonhuman biota to genotoxic agents and ionizing radiation. The progression of steps from hazard identification through exposure and dose-response assessment to risk characterization is illustrated in Figure 9. There are many potential applications for this methodology, including

1. Ranking the potential genetic hazard of new chemicals for the purpose of limiting the introduction of hazardous chemicals into commerce, the environment, food, and the workplace.
2. Ranking emissions from urban, rural, industrial, or energy-related activities to determine which emissions and human exposures have the highest potential genetic risk.
3. Ranking the genetic risk in geographical areas and regions for the purpose of intervention and control of the release of high levels of genotoxic agents.

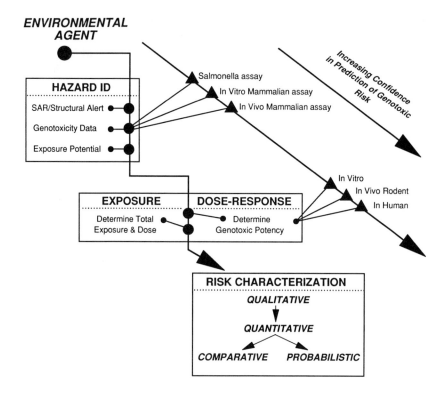

Figure 9. Scheme illustrating the progression of steps from hazard identification through exposure and dose-response assessment to risk characterization.

REFERENCES

Albert, R.E., R. Train, and E. Anderson (1977), Rationale developed by EPA for the assessment of carcinogenic risks, *JNCI,* 58, 1537–1541.

Albert, R., J. Lewtas, S. Nesnow, T. Thorslund, and E. Anderson (1983), Comparative potency method for cancer risk assessment: application to diesel particulate emissions, *Risk Anal.,* 3, 101–117.

Albertini, R.J., J.A. Nicklas, J.P. O'Neill, and S.H. Robinson (1990), In vivo somatic mutations in humans: measurement and analysis, *Annu. Rev. Genet.,* 24, 305–326.

Albertini, R.J. and S.H. Robison (1990), Human population monitoring, in *Genetic Toxicology,* A.P. Li and R.H. Heflich, Eds., CRC Press, Boca Raton, FL, 375–420.

Ames, B.N. and L.S. Gold (1990), Chemical carcinogenesis: too many rodent carcinogens, *Proc. Natl. Acad. Sci. U.S.A.,* 87, 7772–7776.

Ashby, J. and R.S. Morrod (1991), Detection of human carcinogens, *Nature,* 352, 185–186.

Aust, A.E. (1991), Mutations and cancer, in *Genetic Toxicology*, A.P. Li and R.H. Heflich, Eds., CRC Press, Boca Raton, FL, 93–117.
Barbacid, M. (1985), Human oncogenes, in *Important Advances in Oncology 1986*, V.T. Devita, S. Hellman, and S. Rosenberg, Eds., Lippincott, Philadelphia, 3–22.
BEIR III (1980), National Academy of Sciences/National Research Council, Advisory Committee on the Biological Effects of Ionizing Radiations (BIER). *The Effects on Populations of Exposure to Low Levels of Ionizing Radiations*, National Academy of Sciences Press, Washington, D.C.
BEIR Report (1990), *Health Effects of Exposure to Low Levels of Ionizing Radiation: BIER V*, National Academy of Sciences Press, Washington, D.C., 4, 73.
Bernstein, L., J. Kaldor, J. McCann, and M.C. Pike (1982), An empirical approach to the statistical analysis of mutagenesis data from the Salmonella test, *Mutat. Res.*, 97, 267–281.
Bishop, J.M. (1987), The molecular genetics of cancer, *Science*, 235, 305–310.
Brewen, J.G., J.R. Preston, and N. Gengozian (1975), Analysis of X-ray-induced chromosomal translocations in human and marmoset spermatogonial stem cells, *Nature*, 253, 468–470.
Bridges, B.A. (1973), Some general principles of mutagenicity screening and a possible framework for test procedures, *Environ. Health Perspect.*, 6, 221–227.
Bridges, B.A. (1974), The three-tier approach to mutagenicity screening and the concept of radiation-equivalent dose, *Mutat. Res.*, 26, 335–340.
Bridges, B.A., D.E. Bowyer, E.S. Hansen, A. Penn, and K. Wakabayashi (1990), The possible involvement of somatic mutations in the development of atherosclerotic plaques, *Mutat. Res.*, 239, 143–148.
Brooks, P. (1966), Quantitative aspects of the reaction of some carcinogens with nucleic acids and the possible significance of such reactions in the process of carcinogenesis, *Cancer Res.*, 26, 1994–2003.
Cohen, S.M. and L.B. Ellwein (1991), Genetic errors, cell proliferation, and carcinogenesis, *Cancer Res.*, 51, 6493–6505.
Cole, J., M.H.L. Green, S.E. James, L. Henderson, and H. Cole (1988), Human population monitoring: a further assessment of factors influencing measurements of thioguanine-resistant mutant frequency in circulating T lymphocytes, *Mutat. Res.*, 204, 1456–1460.
Committee 17 (1975), Environmental mutagenic hazards, *Science*, 187, 503–514.
Crow, J.F. (1973), Impact of various types of genetic damage and risk assessment, *Environ. Health Perspect.*, 6, 1–5.
Crump, K.S. (1981), An improved procedure for low-dose carcinogenic risk assessment from animal data, *J. Environ. Pathol. Toxicol.*, 5, 675–684.
Crump, K.S., D. Hoel, C. Langley, and R. Peto (1979), Fundamental carcinogenic processes and their implications for low dose risk assessment, *Cancer Res.*, 36, 2973–2979.
Dean, M. (1988), Molecular and genetic analysis of cystic fibrosis, *Genomics*, 3, 93–99.
Dellarco, V.L., Ed. (1990), United States Environmental Protection Agency genetic risk assessment on ethylene oxide, *Environ. Mol. Mutagen.*, 16, 81–134.
DeMarini, D.M, (1991), Environmental mutagens/complex mixtures, in *Genetic Toxicology*, A.P. Li and R.H. Heflich, Eds., CRC Press, Boca Raton, FL, 285–302.

Doll and Peto (1981), The causes of cancer: quantitative estimates of the available risks of cancer in the United States today, *J. Natl. Cancer Inst.,* 66, 1191–1308.
Drake, J. (1991), Spontaneous mutation, *Annu. Rev. Genet.,* 25, 125–146.
Drinkwater, N.R. and J.H. Klotz (1981), Statistical methods for the analysis of tumor multiplicity data, *Cancer Res.,* 41, 113–119.
Ehling, U.H. (1991), Genetic Risk Assessment, *Annu. Rev. Genet.,* 25, 255–280.
Ehrenberg, L. (1974), Genotoxicity of environmental chemicals, *Acta Biol. Iugoslav., Ser. B. Genet.,* 6, 367–398.
Ehrenberg, L. (1979), Risk assessment of ethylene oxide and other compounds, in: V.K. McElheny and S. Abrahamson, Eds., Banbury Report. I. Assessing Chemical Mutagens: The Risk to Humans, Cold Spring Harbor Laboratory, New York, 157–190.
Ehrenberg, L. (1985), Covalent binding of genotoxic agents to proteins and nucleic acids, in: Berlin, A., Draper, M., Hemminki, K., and Vainio, H., Eds., Monitoring Human Exposure to Carcinogenic and Mutagenic Agents, IARC Scientific Publications, Lyon, France, Publication No. 59, 107–114.
Evans, H.J. (1988), Mutation cytogenetics: past, present and future, *Mutat. Res.,* 204, 355–363.
Generoso, W.M., K.T. Cain, C.V. Cornett, N.L.A. Cacheiro, and L.A. Hughes (1990), Concentration-response curves for ethylene-oxide-induced heritable translocations and dominant lethal mutations, *Environ. Mol. Mutagen,* 16, 126–131.
Graedel, T.E., D.T. Hawkins, and L.D. Claxton (1986), *Atmospheric Chemical Compounds: Sources, Occurrence, and Bioassay,* Academic Press, New York.
Hartman, P.E. (1983), Mutagens: some possible health impacts beyond carcinogenesis, *Environ. Mutagen,* 5, 139–152.
HHS (1986), The Health Consequences of Involuntary Smoking, a report of the Surgeon General, U.S. Gov. Printing Office, Washington, D.C.
IARC (1987), IARC Monograph on the evaluation of carcinogenic risks to humans. Carcinogenicity: an update of selected IARC Monographs, Vols. 1–42, Supplement 7, Int. Agency for Research on Cancer, Lyon, France.
IARC (1991), Mechanisms of carcinogenesis in risk identification, IARC Tech. Rep. No. 91/002, Int. Agency for Research on Cancer, Lyon, France.
ICPEMC Committee 4 (1983), Estimation of genetic risks and increased incidence of genetic disease due to environmental mutagens, *Mutation Res.,* 115, 255–291.
IRLG (Interagency Regulatory Liaison Group), Work Group on Risk Assessment (1979), Scientific basis for identification of potential carcinogens and estimation of risks, *Natl. Cancer Inst.,* 63, 241–268..
Jones K.C., T. Keating, P. Diage, A.C. Chang, (1991), Transport and food chain modeling and its role in assessing human exposure to organic chemicals, *J. Environ. Quality,* 20, 317–329.
Kunz, B.A. (1982), Genetic effects of deoxyribonucleotide pool imbalances, *Environ. Mutag.,* 4, 695–725.
Lenski, R.E. (1987), The infectious spread of engineered genes, in *Application of Biotechnology: Environmental and Policy Issues,* AAAS Selected Symposium 106, J.R. Fowle, III, Ed., AAAS, Washington, D.C.

Lesch, R., Ch. Bauer, and W. Reutter (1973), The development of cholangiofibrosis and hepatomas in galactosamine induced cirrhotic rat livers, *Virchows Arch. Abt. B Zellpath,* 12, 285–289.

Lewtas, J. (1985), Development of a comparative potency method for cancer risk assessment of complex mixtures using short-term in vivo and in vitro bioassay, *Toxicol. Ind. Health,* 4, 193–203.

Lewtas, J. (1988), Genotoxicity of complex mixtures: strategies for the identification and comparative assessment of airborne mutagens and carcinogens from combustion sources, *Fundam. Appl. Toxicol.,* 10, 571–589.

Lewtas, J. (1991), Environmental monitoring using genetic bioassays, in *Genetic Toxicology,* A.P. Li and R.H. Heflich, Eds., CRC Press, Boca Raton, FL, 359–374.

Lewtas, J. (1992), Airborne carcinogens, *Pharmacol. Toxicol.,* 72SI, s55–s63.

Lewtas, J. (1993), Complex mixtures of air pollutants: characterizing the cancer risk of polycyclic organic matter, *Environ. Health Perspect.,* 100, 211–218.

Lewtas, J. and J. Gallagher (1990), Complex mixtures of urban air pollutants: identification and comparative assessment of mutagenic and tumorigenic chemicals and emission sources, in *Complex Mixtures and Cancer Risk,* H. Vanio, M. Sorsa, and J.J. McMichael, Eds., IARC Scientific Publ., No. 104, IARC, Lyon, France, 252–260.

Lewtas J., R. Bradow, R. Jungers, B. Harris, R. Zweidinger, K. Cushing, B. Gill, and R. Albert (1981), Mutagenic and carcinogenic potency of extracts of diesel and related environmental emissions: study design, sample generation, collection and preparation, *Environ. Int.,* 5, 383–387.

Lewtas, J., S. Nesnow, and R. Albert (1983), A comparative potency method for cancer risk assessment: clarification of the rationale, theoretical basis, and application to diesel emissions, *Risk Analysis,* 3, 133–137.

Li, D. and K. Randerath (1990), Association between diet and age-related DNA modifications (I-compounds) in rat liver and kidney, *Cancer Res.,* 50, 3991–3996.

Loeb, L.A. (1991), Mutator phenotype may be required for multistage carcinogenesis, *Cancer Res.,* 51, 3075–3079.

Lohman, P.H.M., M.L. Mendelsohn, D.H. Moore, II, M.D. Waters, D.J. Brusick, J. Ashby, and W.J.A. Lohman (1992), A method for comparing, combining and interpreting short-term genotoxicity data: the basic system, *Mutation Res.,* 266, 7–25.

Mackay, D. (1991), *Multimedia Environmental Models, The Fugacity Approach,* Lewis Publishers, Chelsea, MI.

Martin, R.H., K. Hildebrand, J. Yamamoto, A. Rademaker, M. Barnes, G. Douglas, K. Arthur, T. Ringrose, and I.S. Brown (1986), An increased frequency of human sperm chromosomal abnormalities after radiotherapy, *Mutat. Res.,* 174, 201–225.

McKone, T.E. and D.J.I. Daniels (1991), Estimating human exposure through multiple pathways from air, water, and soil, *Regul. Toxicol. Pharmacol.,* 13, 36–61.

McKusick, V.A. (1990), *Mendelian Inheritance in Man: Catalogs of Autosomal Dominant, Autosomal Recessive and X-linked Phenotypes,* 9th ed., John Hopkins University Press, Baltimore.

Mendelsohn, M.L., D.H. Moore, II, and P.H.M. Lohman (1992), A method for comparing, combining and interpreting short-term genotoxicity data: results and interpretation, *Mutation Res.,* 266, 45–60.

Meselson, M. and K. Russell (1977), Comparisons of carcinogenic and mutagenic potency, in *Origins of Human Cancer,* Book C, H.H. Hiatt, J.D. Watson, and J.A. Winsten, Eds., Cold Spring Harbor Laboratory, New York, 1473–1481.

Miller, E.E. and J.A. Miller (1966), Mechanisms of chemical carcinogenesis: nature of proximate carcinogens and interactions with macromolecules, *Pharmacol. Rev.,* 18, 805–835.

Miller, J.A. and E.C. Miller (1977), Ultimate chemical carcinogens as reactive mutagenic electrophiles, in *Origins of Human Cancer,* Book B, H.H. Hiatt, J.D. Watson, and J.A. Winsten, Eds., Cold Spring Harbor Laboratory, New York, 605–627.

Mohrenweiser, H.W. (1991), Germinal mutation and human genetic disease, in *Genetic Toxicology,* A.P. Li and R.H. Heflich, Eds., CRC Press, Boca Raton, FL, 67–92.

Moolgavkar, S.H. and A.G. Knudson (1981), Mutation and cancer: a model for human carcinogenesis, *JNCI,* 66, 1037–1052.

Moore II, D.H., M.L. Mendelsohn, and P.H.M. Lohman (1992), A method for comparing, combining and interpreting short-term genotoxicity data: the optimal use of dose information, *Mutation Res.,* 266, 27–42.

National Academy of Sciences (NAS) (1982), Committee on Chemical Environmental Mutagens (CCEM), *Identifying and Estimating Genetic Impact of Chemical Mutagens,* National Academy of Sciences Press, Washington, D.C.

National Academy of Sciences (NAS) (1983), Committee on the Institutional Means for Assessment of the Risks to Public Health, *Risk Assessment in the Federal Government: Managing the Process,* National Academy Press, Washington D.C.

National Research Council (NRC), (1983), *Risk Assessment in the Federal Government: Managing the Process,* National Academy Press, Washington D.C., 1–189.

National Research Council (NRC), (1986), *Environmental Tobacco Smoke, Measuring Exposures and Assessing Health Effects,* National Academy Press, Washington, D.C., 337.

Neel, J.V. and S.E. Lewis (1990), The comparative radiation genetics of humans and mice, *Ann. Rev. Genet.* 24, 327–362.

Pitot, H.C., Y.P. Dragan, M.J. Neveu, T.A. Rizvi, J.R. Hully, and H.A. Campbell (1991), Chemicals, cell proliferation, risk estimation, and multi-stage carcinogenesis, in *Chemically Induced Cell Proliferation: Implications for Risk Assessment,* Wiley-Liss, New York, 517–532.

Preston-Martin, S., M.C. Pike, R.K. Ross, P.A. Jones, and B.E. Henderson (1990), Increased cell division as a cause of human cancer, *Cancer Res.,* 50, 7415–7421.

Rao, P.M., E. Lanconi, S. Vasudevan, A. Denda, S. Rajagopal, S. Rajalakshmi, and D.S.R. Sarma (1987), Dietary and metabolic manipulations of the carcinogenic process: role of nucleotide pool imbalances in carcinogenesis, *Toxicol. Pathol.,* 15, 190–193.

Rattan, S.I.S., Ed. (1991), Cellular aging, *Mutation Res.,* 256 (2–6), 261.

Rhomberg, L., V.L. Dellarco, C. Siegel-Scott, K.L. Dearfield, and D. Jacobson-Kram (1990), Quantitative estimation of the genetic risk associated with the induction of heritable translocations at low-dose exposure: ethylene oxide as an example, *Environ. Mol. Mutagen.,* 16, 104–125.

Setlow, R. (1979), DNA repair, in Banbury Report. I. Assessing Chemical Mutagens: the Risk to Humans, V.K. McElheny and S. Abrahamson, Eds., Cold Spring Harbor Laboratory, New York, 81–95.

Sobels, F.H. (1989), Models and assumptions underlying genetic risk assessment, *Mutat. Res.,* 212, 77–89.

Stead, A.G., V. Hasselblad, J.P. Creason, and L. Claxton (1981), Modeling the Ames test, *Mutat. Res.,* 85, 13–27.

Sugimura, T. (1982), Mutagens, carcinogens, and tumor promoters in our daily food, *Cancer,* 49, 1970–1984.

Thorslund, T. (1988), Comparative potency approach for estimating the cancer associated with exposure to mixtures of polycyclic aromatic hydrocarbons. Report to U.S. EPA by ICF-Clement Associates, Fairfax, VA.

Törnqvist, M. and L. Ehrenberg (1985), Risk estimation of urban air pollution: information sources and methods, *Environ. Int.,* 11, 401–406.

Törnqvist, M., D. Segerback, and L. Ehrenberg (1991), The "rad-equivalence approach" for assessment and evaluation of cancer risks, exemplified by studies of ethylene oxide and ethene, in *Human Carcinogen Exposure, Biomonitoring and Risk Assessment,* R.C. Garner, P.B. Farmer, G.T. Steel, and A.S. Wright, Eds., IRL Press, Oxford, 141–155.

Törnqvist, M. and L. Ehrenberg (in press), Cancer risk estimation of urban air pollution, *Environ. Health Perspect.*

UNSCEAR (1977), Sources and Effects of Ionizing Radiation, United Nations, New York.

UNSCEAR (1982), Ionizing Radiation: Sources and Biological Effects, United Nations, New York.

U.S. EPA (1986a), U.S. Environmental Protection Agency, Guidelines for mutagenicity risk assessment, *Fed. Regist.,* 51(185):34006–34012.

U.S. EPA (1986b), U.S. Environmental Protection Agency, Guidelines for carcinogen risk assessment, *Fed. Regist.,* 51(185):33992–34003.

U.S. EPA (1986c), U.S. Environmental Protection Agency, Guidelines for exposure assessment, *Fed. Regist.,* 51(185):34042–34054.

U.S. EPA (1992), U.S. Environmental Protection Agency, Health Assessment of Environmental Tobacco Smoke, U.S. EPA, Washington, D.C.

Vijayalaxmi and H.J. Evans (1984), Measurement of spontaneous and X-irradiation induced 6-thioguanine resistant human blood lymphocytes using a T-cell cloning technique, *Mutat. Res,* 125, 878–894.

Wakabayashi, K., T. Sugimura, and M. Nagao (1991), Mutagens in food, in *Genetic Toxicology,* A.P. Li and R.H. Heflich, Eds., CRC Press, Boca Raton, FL, 304–338.

Waters, M.D., H.F. Stack, A.L. Brady, P.H.M. Lohman, L. Haroun, and H. Vainio (1988), Use of computerized data listings and activity profiles of genetic and related effects in the review of 195 compounds, *Mutat. Res.,* 205, 295–312.

Waters, M.D., H.F. Stack, N.E. Garrett, and M.A. Jackson (1991), The genetic activity profile database, *Environ. Health Perspec.,* 96, 41–45.

Wügler, F.E. and P.G.N. Kramers (1992), Environmental effects of genotoxins (Ecogenotoxicology), *Mutagenesis,* 2, 321–327.

Wyrobeck, A.J. (1983), Identifying agents that damage human spermatogenesis: abnormalities in sperm concentration and morphology, in *Monitoring Human Exposure to Carcinogenic and Mutagenic Agents,* A. Berlin, M. Draper, K. Hemminki, and H. Vanio, Eds., IARC Sci. Publ., No. 59, Oxford Univ. Press, Oxford, 387–402.

Wyrobeck, A.J., M. Currie, J.L. Stilwell, R. Balhorn, and L.H. Stanker (1990), Detecting specific-locus mutations in human sperm, in *Biology of Mammalian Germ Cell Mutagenesis,* Banbury Report 34, J.W. Allen, B.A. Bridges, M.F. Lyon, M.J. Moses, and L.B. Russel, Eds., Cold Spring Harbor Laboratory Press, New York, 93–111.

Youssoufian, H., H.H. Kazazian, D.G. Phillips, S. Aronis, G. Tsiftis, V.A. Brown, and S.E. Antonarakis (1986), Recurrent mutations in haemophilia A give evidence for CpG mutation hotspots, *Nature (London),* 324, 380–382.

CHAPTER 5

Monitoring Environmental Genotoxicants

J. T. MacGregor, L. D. Claxton, J. Lewtas, R. Jensen,
W. R. Lower, and G. G. Pesch

TABLE OF CONTENTS

I. Introduction 173

II. Methods for Monitoring Genotoxic Activity in
 the Environment 175
 A. Laboratory Bioassays 175
  1. Microbial Assays 176
   1.1 Gene Mutation 176
   1.2 Repair Assays 177
   1.3 Mitotic Recombination 177
  2. Plants 178
  3. Aquatic Organisms 179
   3.1 Vertebrates 179
   3.2 Invertebrates 182
  4. Mammalian Assays 184
   4.1 Mammalian Cell Culture 184
   4.2 *In Vivo* Mammalian Assays 185
 B. Methods for Sampling and Analysis for
  Laboratory Bioassay 186
 C. *In Situ* Environmental Bioassay of Air, Water, and Soil 186
  1. Microorganisms 187
  2. Terrestrial Plants 187
   2.1 *Tradescantia* 187
   2.2 *Zea maize* (Corn) 189
   2.3 Other Terrestrial Species 189
  3. Aquatic Species 189
   3.1 Mudminnow *(Umbra)* 190
   3.2 Mussels *(Mytilus)* 190

  4. Mammals 191
    4.1 Rodents 191
 D. Human Monitoring 191
  1. Somatic Cells 191
    1.1 Cytogenetic Damage 191
    1.2 Direct Measures of Mutation 194
    1.3 DNA Damage 196
    1.4 Presence of Mutagens or Mutagen Adducts in
      Blood, Urine, or Tissues 197
  2. Germ Cells 198
    2.1 Sentinel Phenotypes 198
    2.2 Protein Variants 198
    2.3 Sperm Analysis 198
    2.4 Methods under Development 199
  3. Experimental Design and Interpretation 199

III. Applications of Monitoring 200
 A. Laboratory Studies with Environmental Samples 200
  1. Emissions and Effluents 201
  2. Ambient Environment: Air and Water 202
  3. Microenvironments 203
  4. Monitoring Personal Exposure 203
  5. Transport and Transformation of Mutagens
    in the Environment 204
 B. *In Situ* Monitoring 205
  1. Plants 205
  2. Aquatic Species 206
  3. Mammals 208
 C. Human Monitoring 209

IV. Recommendations 211
 A. Definition of Objectives 211
 B. Sampling Design and Statistical Analysis 212
  1. Sampling Design 212
    1.1 Field Reference 212
    1.2 Laboratory Control 213
    1.3 Transport Control 213
  2. Data Management 213
 C. Criteria for Selection of Organisms 213
  1. Habitat of Concern 213
  2. Surrogates for Species of Concern 214
  3. Availability and Validation State of System 214
  4. Laboratory/Field "Transplantability" 214
  5. Ability to Measure Endpoint of Interest 214

D. Criteria for Endpoint Selection 215
   1. General Considerations 215
   2. Resources 215
   3. Species Limitations 216
   4. Mutational Specificity 216
   5. Metabolism 216
   6. Purpose of the Monitoring Effort 217
   7. Tiers and Matrices 217

V. Future Needs 218

References 219

I. INTRODUCTION

The environment contains a wide variety of man-made and naturally occurring genotoxic agents, including mutagens and carcinogens. The objective of this chapter is to summarize and evaluate existing methods of monitoring for the presence of these genotoxicants in the environment, and for genetic damage in humans and nonhuman biota. We define environmental monitoring as the practice of identifying and tracking either genetic damage in environmental organisms or the presence in the environment of agents with the potential to induce genetic damage. This can be accomplished either by direct measurement of known genotoxic substances or by using bioassays for genotoxic activity. The human health effects and/or ecological hazard of pollutants and natural constituents can be assessed by bioassay of samples brought to the laboratory from the environment of interest, or by *in situ* bioassay of biological effects on indigenous species and introduced indicator organisms present in the environment. The *in situ* bioassay is a cost-efficient method of hazard identification and is useful for short-term or long-term monitoring. If no hazard is identified biologically, the level of concern may be downgraded without costly follow-up studies. If the major genotoxicants present in a complex mixture are known, then chemical measurement of these agents may provide a rapid and cost-effective method of monitoring environmental samples. However, bioassays of environmental samples often show the presence of mutagenic activity that cannot be accounted for by the substances identified by chemical analysis (Pitts et al., 1979; Chriswell et al., 1979; Lewtas et al., 1982, 1988; Waters et al., 1990). The complexity of the direct analytical approach is illustrated by the survey of Graedel et al. (1986), which found that information about genotoxic activity was available for less than 11% of the approximately 2800 compounds that have been identified in ambient air. Additionally, chemical analysis alone does not provide information about the effect on biological activity of interactions among components of chemical mixtures. Bioassays for genotoxic activity provide a more general method of identifying genotoxicants in environmental samples, and bioassay-directed fractionation and chemical analysis is an effective approach for identifying mutagenic constituents in environmental samples (Claxton, 1983a; Schuetzle and Lewtas, 1986).

From the genetic point of view, there are two principal target-cell populations in environmental organisms: somatic and germ cells. Somatic alterations are associated with the health or viability of individuals in a population, but are not passed to subsequent generations, while germ-cell alterations are passed on to descendants of the affected individuals. Bioassays for genetic damage can be based on many endpoints, including gene mutations, metaphase chromosomal aberrations, DNA damage (such as strand breakage), DNA repair activity, the presence of covalent DNA adducts or modified DNA bases, and micronuclei. These genetic endpoints can be evaluated in organisms exposed *in situ* in the environment and can also be evaluated in laboratory bioassays that employ microbial organisms, higher plants, aquatic organisms, mammalian cells *in vitro,* and mammals as test organisms. The organisms used for *in situ* studies may be surrogate species introduced into the environment by the investigator or "sentinel" species naturally present in the environment of interest.

Data obtained by monitoring can be used in several important ways as part of the genotoxic risk-assessment process. Specific applications of monitoring include the following:

Identification of genotoxic hazards in a population or environment of interest — Monitoring for genotoxic activity provides information about the presence of genotoxicants in the environment. Mutagenic activity can be monitored in environmental organisms without knowledge of the specific chemical agents present, and the characteristics of the activity observed may provide clues to the identity of the responsible agents. Biological endpoints are especially useful for monitoring environmental contamination because such contamination usually consists of complex mixtures or unknown degradation products. Such information can be used to provide a warning of the presence of potentially hazardous genotoxic activity, to track mutagenic contamination to its sources, or to monitor emissions from known sources. Laboratory assays for mutagenic or genotoxic activity in extracts of environmental samples (such as air, water, or soil) and the measurement of genotoxic damage in sentinel and surrogate species have been used widely to obtain such information.

Direct measurement of damage in a population of interest — Direct measurement of genetic damage is of particular value when a specific potential hazard (capacity to induce a particular type of damage) has been identified in laboratory or *in situ* studies and a direct measure of the risk of such damage in a population of interest is needed.

Dosimetry and determination of dose-response relationships —When the relationship between exposure and a specific biological response is known, monitoring a population for that response allows estimation of exposure.

Evaluation of the relationship between exposure, genetic effect, and biological consequence — One of the most important objectives of risk assessment is confirmation of the link between exposure and biological outcome (disease or adverse population effects). Population studies provide information essential to establish this link.

Risk confirmation through association and/or intervention — The quantitative impact of suspected environmental risk factors on relevant genetic endpoints can be determined by population studies in which these risk factors are either associated with observed genetic damage or specifically modified in intervention trials while monitoring genetic damage in the population.

This chapter focuses on approaches for monitoring the presence of genotoxic agents in the environment and the occurrence of genetic damage in environmental organisms, including man. The following sections summarize the currently available assays applicable to monitoring for genetic alterations in laboratory and field settings, and the approaches to environmental monitoring which have been used, or which could be used, for the above purposes. The advantages and disadvantages of specific test systems for these various objectives are discussed, approaches to monitoring that are currently considered practical and scientifically justified are recommended, and needs for the development of specific new methodologies are identified.

II. METHODS FOR MONITORING GENOTOXIC ACTIVITY IN THE ENVIRONMENT

A. Laboratory Bioassays

The development of short-term genetic bioassays in the mid-1970s led rapidly to the use of these assays in environmental monitoring (Lewtas, 1991). The essential elements that facilitated the application of these bioassays to environmental monitoring were (1) their relative simplicity, rapidity, and low cost and (2) the small sample size requirements. The genetic bioassay used most widely has been the *Salmonella typhimurium his⁻* reversion assay described by Ames et al. (1975). This bioassay was adopted rapidly throughout the world and applied to environmental studies of complex mixtures. Within 2 years after the description of this bacterial mutagenesis bioassay, scientists in the U.S., Japan, and Germany reported monitoring studies of environmental pollutants in which this bioassay was used to detect and quantitate the mutagenic activity associated with environmental samples (Waters et al., 1979). In 1978, a Symposium on the Application of Short-Term Bioassays to the Analysis of Complex Environmental Mixtures reported on the use of short-term genetic bioassays to monitor air, water, industrial effluents, and commercial products, including synthetic petroleum products, cigarette smoke, and food (Waters et al., 1979). Since 1979, five additional books that address the use of genetic bioassays in the analysis of complex environmental mixtures have been published in this series (Waters et al., 1981, 1983, 1985, 1990a, b; and Sandhu et al., 1988).

Many genetic assay systems that could be applied to environmental monitoring are now available. In practice, only a limited number of these systems are used for this purpose. Environmental monitoring studies usually require that a relatively large number of measurements be made over time or across locations. This requirement for a large number of analyses is the principal reason that microbial assays have most often been used in monitoring programs. In one of the earliest reviews of the use of microbial assay systems in the detection of environmental mutagens in complex mixtures, Rosenkranz (1979) stated that "It is our opinion that of all the systems available to date, the *Salmonella* mutagenicity assay procedure is the most versatile and is adaptable to a number of experimental situations that reflect environmental situations." Ten years later, in a recent review of environmental monitoring using genetic bioassays, Lewtas (1991) drew the same conclusions. Therefore, this section summarizes bioassay systems currently available but emphasizes the use of the *Salmonella* mutagenicity assay in environmental monitoring. Plants and aquatic organisms, though useful as laboratory assays for monitoring activity in environmental samples, are most useful in *in situ* studies directly monitoring mutagens in the environment. These systems are therefore discussed more extensively in the section on *in situ* assays.

1. Microbial Assays

1.1 Gene Mutation

A number of microbial assays using bacteria, bacteriophage, fungi, or yeast are available. The most widely used is the *S. typhimurium his⁻* reversion assay developed by Ames and co-workers (Ames et al., 1975; Maron and Ames, 1983). *S. typhimurium, Escherichia coli, Bacillus subtilis* (Kada and Watanabe, 1983), the yeast *Saccharomyces cereviscae* (Constantin et al., 1980), and the plant *Tradescantia* (Schairer, 1983) have commonly been used (Graedel et al., 1986).

Guidelines for conducting the *S. typhimurium his⁻* reversion assay have been published (Maron and Ames, 1983; Claxton et al., 1987). This assay has recently been used in two international collaborative studies to evaluate the inter- and intralaboratory variation in the mutagenicity of complex environmental mixtures: one using NIST Standard Reference Mixtures (Claxton et al., 1992a, b), and a similar collaborative study from Japan monitoring the mutagenicity of airborne particulate matter (Matsushita et al., 1992).

The relatively small sample mass requirements (milligram quantities) of microbial genetic bioassays is one of the important attributes that facilitated

MONITORING ENVIROMENTAL GENOTOXICANTS 177

their application to environmental monitoring. Recent advances in exposure monitoring have resulted in the need for assay methods that could be applied to even smaller (microgram) quantities of environmental samples. Microsuspension bacterial mutation assays have been developed to meet these needs through simple modifications of the *S. typhimurium* preincubation assay in which the assay is conducted in a 10- to 100-fold smaller volume (Kado et al., 1986). The spiral *Salmonella* assay (Houk et al., 1989, 1991) also provides a rapid way to bioassay small amounts of sample. This assay has been applied to both pure compounds (Houk et al., 1989) and complex mixtures (Houk et al., 1991).

1.2 Repair Assays

Very few of the assays developed for detecting nonmutational DNA damage have been applied to environmental monitoring. The induction of prophage λ in *E. coli* WP2 (Rossman et al., 1985) and the induction of SOS repair of DNA in either the SOS/umu test (Oda et al., 1985; Ong et al., 1988) or the SOS chromatest (Quillardet et al., 1982) have been applied to monitoring environmental samples. These assays and their mechanisms have been reviewed by Elespuru (1984, 1987). The prophage λ induction assay in *E. coli* has been applied to the monitoring of ambient air (Rossman et al., 1985) and more recently to the detection of genetic activity of hazardous wastes and chlorinated compounds, some of which are not mutagenic in the *Salmonella* assay but are carcinogenic in rodents (Houk and DeMarini, 1987, 1988). These investigators have proposed this assay as an adjunct to the *Salmonella* assay for industrial effluents, waste waters, and solid wastes, where the *Salmonella* assay may fail to detect genotoxic chlorinated compounds. This assay appears to detect a broader range of genetic damage than the *Salmonella* mutation assay. In particular, agents that induce single-strand breaks, such as free radicals that induce oxidative DNA damage, may be monitored by SOS/phage-induction assays (Elespuru, 1984; Battista et al., 1990).

The SOS assays for induction of DNA repair (SOS repair) in bacteria have been applied to a number of complex environmental samples (Whong et al., 1986). These assays are even more rapid than the *Salmonella* mutation assay. The SOS chromatest, for example, can be completed in 1 d, and the assay has been automated. However, this assay may in some cases be less sensitive than the *Salmonella* mutation assay (Nylund et al., 1992).

1.3 Mitotic Recombination

Among other endpoints that have been used in microbial systems is the measurement of mitotic recombination. Mitotic recombination cannot be measured in bacteria because it requires the presence of more than one chromosome. *Saccharomyces cerevisiae* has been used widely to screen chemicals and complex environmental mixtures for mitotic recombination (Zimmerman, 1975, 1985).

2. Plants

Specific-locus assay — Two assays have been used most widely for specific-locus tests: (1) mutation at the waxy locus of *Zea maize* (corn) (Plewa and Gentile, 1976; Lower et al., 1983a, b) and (2) mutation at the pink/blue stamen hair locus of *Tradescantia* clone 4430 or clone 03 species hybrid heterozygotes (Underbrink et al., 1973; Schairer, 1983; Lower et al., 1983a, b). Currently, the *Tradescantia* system is the most convenient due to the ease of scoring stamen hair cells with a dissecting microscope. The corn system is very labor intensive at present, but may become more practical with automation of the scoring of pollen by flow cytometry or image analysis.

Multilocus assay — *Arabidopsis* has had limited use as a multilocus assay scoring for chlorophyll mutations or death of seeds in the silique (seed pod), but more research is required to expand its potential use (Redei et al., 1980).

Cytogenetic assay — Several species have been used to monitor cytogenetic changes. Micronucleus (MN) formation in pollen mother cells of *Tradescantia* (Ma, 1979; Ma et al., 1981, 1983; Lower et al., 1984) and, more recently, sister chromatid exchange in root tips of *Tradescantia* (Yan and Ma, 1990) have been used. Micronuclei are chromatin bodies found in the cytoplasm of a cell as a consequence of chromosome breakage or spindle apparatus malfunction during mitosis (Schmid, 1976). The *Tradescantia* micronucleus assay system can be used to monitor air, water, or soil in laboratory bioassays. Root tip chromosomal aberrations have been studied in onion *(Allium sepa)* (Grant and Zura, 1992) and broad bean *(Vicia faba)* (Ma, 1982). The root tip systems are generally used by exposing the roots to aqueous solutions of mutagens or other test chemicals.

DNA adduct analysis — Analysis of DNA adducts can provide a direct measure of DNA damage to both germinal and somatic tissues. The increased formation of 8-hydroxy-2′-deoxyguanosine in the chloroplast DNA of 21-d-old beans *(Phaseolus vulgaris)* and peas *(Pisum sativum)* exposed to ozone and detected by high-performance liquid chromatography (HPLC) with electrochemical detection has been reported by Floyd et al. (1989). Parallel genetic and molecular dosimetry experiments were conducted by Schy and Plewa (1989), in which the frequency of forward mutation at the yellow g 2 (yg 2) locus of *Z. mays* and the level of covalently bonded ethyl methanesulfonate (EMS)-induced methyl DNA adducts were determined. The adducts were extracted from plant embryonic tissue containing leaf primordia, identified by HPLC, and a linear relationship was reported between molecular dose, mutation induction, and EMS concentration. Rether et al. (1990) induced nuclear DNA adducts in plant cell suspensions of *Echinacea purpurea* (purple coneflower) treated with benzo [a] pyrene (B[a]P) and detected by ^{32}P postlabeling. Alex and Dupuis (1989) reported DNA binding to Cd^{2+} through electrostatic interactions *in vitro* under physiological conditions, but in real biological systems the DNA is intact and the effect of Cd^{2+} may not be directly related to its action on extracted DNA. Other additional work with ^{32}P postlabeling also provides some of the early information that DNA adducts are formed in plants exposed to known mutagens/carcinogens (Ireland et al., 1990; Lower et al., 1991). *In vivo* and *in vitro* tests of the direct-acting mutagen *N*-methyl-*N*-nitrosourea (MNU) on growing barley *(Hordeum vulgare)* leaves, and DNA isolated from barley, identified by chromatography and ^{32}P postlabeling indicated the presence of normal DNA constituents as well as additional spots considered to be adducts. There are unpublished data by Lower on leaf tissue of *H. vulgare, Atriplex semibacata* (Australian saltbrush), *Nicotiana tabacum* (tobacco), *Lycopersicon esculentum* (tomato), and bulbs or tubers of *Tulipa gesneriana* (tulip), *Pachyrhizus erosus* (Jicama), and *Helianthus tuberosus* (Jerusalem artichoke) with MNU and B[a]P. At the concentrations used, the bulbs and tubers, but not leaves, showed DNA adducts to B[a]P (B[a]P requires activation to be mutagenic). Leaves exposed to MNU (a direct-acting mutagen) showed DNA adducts in barley, tobacco, *Tradescantia,* and Australian saltbrush but not tomato. Metabolic variations between leaf tissue that is rich in chlorophyll and relatively poor in mitochondria, and root tissue lacking chlorophyll and rich in mitochondria probably largely explain this difference.

3. Aquatic Organisms

Aquatic organisms may be used in at least three ways to detect the presence of mutagenic compounds in aquatic habitats: (1) as laboratory bioassays, (2) as surrogates placed *in situ,* and (3) as sentinel species indigenous to a study area. Most studies with aquatic species have involved laboratory exposures. Applications of the same species to a combination of field and laboratory experiments is particularly useful. Field tests alone may serve as monitoring tools in areas stressed by mutagens, but do not permit easy identification of cause-effect relationships. These relationships are best identified in well-designed laboratory experiments where, e.g., dose-dependent responses may be demonstrated.

3.1 Vertebrates

Cytogenetic assays: anaphase observations — Historically, the first detailed experimental studies of genotoxic responses in aquatic organisms, conducted in the 1950s and 1960s, examined radiation damage to early developmental stages of freshwater fish. Most of this initial research was done in the Soviet Union (for a review, see Kligerman 1980, 1982). The rapidly dividing cells of fish embryos make them ideal for anaphase observations of chromosomal bridges and fragments, and multipolar mitoses. The techniques used for making these observations on fish embryos are well developed and relatively easy to apply (Longwell and Hughes, 1980; Liquori and Landolt, 1985). This assay has been applied in the laboratory exposure of several species of fish to complex mixtures (Landolt and Kocan, 1987). However, anaphase abnormalities do not necessarily reflect heritable damage and should be interpreted with caution.

Cytogenetic assays: metaphase observations — Metaphase observations are recommended for clastogenicity studies (Evans, 1976), although many species have karyotypes that are not amenable to classic metaphase analysis (e.g., many species of fish have large numbers of small chromosomes). In 1975, Kligerman et al. proposed the mudminnow *Umbra limi* as a model for cytogenetic studies in fish. The karyotype of this species consists of 22 large meta- and submetacentric chromosomes, ideal for metaphase analysis. Suitable chromosome preparations are found in a variety of tissues in adult specimens. Metaphase chromosome analysis has been used to examine the effects of radiation on the mudminnow (Kligerman et al., 1975), including studies of the dose-response curve for X-irradiation (Mong and Berra, 1979). Chemically induced chromosomal observations have also been studied (Kligerman and Bloom, 1975; Sugatt, 1978). Kligerman (1982) recommends fishes of the genus *Umbra* for clastogenicity studies because of their favorable karyotypes and because they are hardy, easy to handle, relatively small in size, and sensitive to the induction of chromosomal aberrations.

Sister-chromatid exchange analysis — Metaphase preparations allow accurate observation of chromosome structure and permit sister-chromatid exchange (SCE) analysis. Measurement of SCE is a useful cytogenetic technique because of its sensitivity and ease of scoring. It involves a four-strand exchange within the DNA of a chromosome. This phenomenon was first visualized by Taylor (1958) using autoradiography and tritium-labeled DNA. However, Taylor's method was laborious and difficult to score accurately. Development of differential staining methods to study SCE with light microscopy transformed SCE from a limited research tool into one that could be applied extensively to the study of environmental mutagenesis. Several studies have shown that SCE is a more sensitive method for detecting many genotoxicants than traditional chromosome and chromatid observations (Latt, 1974; Perry and Evans, 1975; Bloom, 1978). The SCE technique was reviewed by the U.S. Environmental Protection Agency (EPA) Gene-Tox program and recommended for environmental application (Latt at al., 1981). It has been applied to several species of marine and freshwater fish, including the mudminnow (Kligerman and Bloom, 1976), a flatfish from Puget Sound (Stromberg et al., 1981), and the toadfish (Maddock and Kelly, 1980). Bishop and Valentine (1982) studied SCE responses in the mudminnow and produced dose-response curves for several mutagens. Like metaphase analysis, SCE analysis requires a suitable karyotype and actively growing tissues. In addition, labeling of the DNA for a defined number of cell cycles is required. Interpretation of effects on this endpoint is limited by the uncertain relationship between SCE and genetic effect.

Micronucleus test — The use of micronuclei as a marker of genetic damage has had some limited application to fish. Hooftman and de Raat (1982) elicited a dose response in the mudminnow with the mutagen ethyl methanesulfonate. Hose et al. (1987) scored micronuclei in two fish species, white croaker and kelp bass, from contaminated areas off southern California. This test is attractive because it is not dependent on the quality of the karyotype and is easily scored. However, because most micronuclei arise during the mitosis immediately preceding their observation, knowledge of cell kinetics and the relationship of exposure to cell kinetics can be a factor in interpreting results. This is probably not a significant issue when long-term exposure has occurred in the field, but must be taken into account when conducting laboratory studies.

Specific-locus assays — Specific-locus assays that score alterations in phenotypic characteristics in progeny of exposed parents have been developed in several species. In fish, these characteristics include the number of vertebraae, spinal curvature, color patterns, aggressiveness, eye formation, sex ratio, scale formation, and electrophoretic variants of polymorphic proteins (Kligerman, 1982; Anderson and Harrison, 1986). To date, aquatic studies of this type have been done with aquarium species such as guppies and goldfish. These species are easily bred, have relatively short generation times, and have known genetic markers. Most of these studies considered the effects of radiation (Schroeder, 1969, 1979; Holzberg and Schroeder, 1972; Purdom and Woodhead, 1973; Anders et al., 1973a, b; Dubinin et al., 1975). Only a few studies have considered the effects of chemicals on mutations in fish (e.g., Tsoi, 1971; Tsoi et al., 1976). Specific-locus assays demonstrate heritable gene damage in organisms, but are costly because of the large sample sizes needed to obtain statistically significant data.

Dominant-lethal assays — Dominant-lethal studies are based on the scoring of embryonic death due to chromosomal damage induced by the exposure of one parent. For technical reasons, this usually involves exposure of males to test agents, with subsequent mating to unexposed females. The number of normal, abnormal, and dead progeny are scored and compared with controls. (See Kligerman, 1982, for a review of studies of dominant-lethal mutations in fish.) Chemically induced, dose-related, dominant-lethal responses have been demonstrated in *Tilapia mossambica* (Hemsworth and Wardhaugh, 1978) and in guppies, *Poecilia reticulata* (Mathews et al., 1978). Although dominant-lethal tests detect agents and exposures that cause clastogenic damage, they tend to be insensitive to low-dose effects.

DNA strand breaks — DNA strand breaks are produced by direct chemical or physical damage to the DNA or via secondary effects during the DNA repair process. One sensitive method for measuring DNA strand breaks is the alkaline unwinding assay. This technique subjects DNA to conditions of pH and temperature that foster unwinding at sites of single-strand breaks. The amount of double-stranded DNA remaining is then measured fluorometrically using a DNA-binding dye, Hoechst Dye 33258 (Kanter and Schwartz, 1982), that exhibits a marked differential fluorescence with single- and double-stranded

DNA. Shugart (1988a, b) modified this technique, originally applied *in vitro*, for application to environmental samples. Shugart (1988a, b) detected an increase in strand breaks in two freshwater fish species, the bluegill sunfish and the fathead minnow, exposed in the laboratory to benzo[a]pyrene in their water. This technique has been applied to turtles (Meyers et al., 1988) and to several marine species (Nacci and Jackim, 1989). The DNA alkaline unwinding assay is sensitive and useful as a screening tool, but must be used with caution because nongenotoxic factors may confound the response (Nacci and Jackim, 1989).

Unscheduled DNA synthesis — This technique measures DNA synthesis resulting from DNA damage. The damage elicits cellular excision and repair processes. Tritiated thymidine is used to label DNA during repair synthesis. Labeled DNA is measured by autoradiography or liquid scintillation counting. This method was reviewed by the U.S. EPA (Mitchell et al., 1983) and recommended for use in genotoxicity screening. It has had limited application in aquatic species.

DNA adduct analysis — Analysis of chemical adducts in DNA is highly relevant to risk assessment of mutagens because it provides a direct measure of DNA damage in the tissue of interest. Several methods for DNA adduct analysis have been applied to fish. Shugart et al. (1987) analyzed benzo[a]pyrene adducts on DNA of bluegill sunfish. They used acid hydrolysis of DNA isolated from the livers of exposed fish and analyzed released tetrols by high-performance liquid chromatography (HPLC) or fluorescence detection. The DNA adducts formed were similar to those reported in other organisms. Randerath et al. (1985) developed the ^{32}P postlabeling technique for analysis of DNA adducts. This technique incorporates radioactive phosphate into DNA after the DNA has been exposed, *in vitro* or *in vivo*, to (nonradioactive) covalently binding chemicals. The labeled adducts are then separated by thin-layer chromatography (TLC) and analyzed by autoradiography or quantitated by HPLC. The adducts from TLC can be quantified by scintillation counting. Varanasi et al. (1989) used this technique to study benzo[a]pyrene-diolepoxide-DNA adducts in liver of English sole. The fish were exposed by injection and the consequent DNA adducts were detected in liver.

3.2 Invertebrates

Cytogenetic assays: anaphase observations — Anaphase observations can be made on rapidly growing invertebrate embryos. This assay has been used in sea urchins to evaluate the cytogenetic activity of arsenic and mercury (Pagano et al., 1982a, b). Hose (1985) offers a detailed description of this assay as applied to sea urchins. This assay should be viewed as a screening tool. Bivalve mollusks are particularly suitable for use in this assay because they also have karyotypes amenable to classic metaphase and SCE analysis.

Cytogenetic assays: metaphase observation — Two species of aquatic invertebrates have been tested for chromosomal effects associated with genotoxic agents: the aquatic larval stages of a chironomid midge, *Chironomus tentans,* and a marine worm, *Neanthes arenaceodentata*. Chironomid midge larvae have large polytene chromosomes in which inversions and small deletions are visible without any staining. Blaylock (1966a, b) studied the effects of chronic radiation exposure in White Oak Lake at the Oak Ridge National Laboratory. These larvae also were used in laboratory exposures to tritium in water. Responses were detected at dose rates of 38 rad/d, with a minimum accumulated dose of 760 rad. Pesch et al. (1981) proposed the marine worm *N. arenaceodentata* as a cytogenetic model for marine genetic toxicology. Marine worms have excellent karyotypes for metaphase observations. Pesch et al. (1981) demonstrated dose responses for chromosomal aberrations in worms exposed to chemical mutagens and promutagens. Harrison et al. (1984, 1987) and Anderson et al. (1987) used *N. arenaceodentata* in a series of experiments that examined the effects of radiation on chromosomes and the consequences of such damage to life-span and reproductive success. Pesch (1990) summarized cytogenetic research on marine worms in a review article and concluded that this test system is potentially useful but requires further development before it is broadly applied.

Sister-chromatid exchange analysis — SCE analysis has been applied to two groups of aquatic invertebrates, marine worms and marine bivalves (Pesch, 1990). All of the results published to date on SCE analysis in marine worms have come from two laboratories, the U.S. National Laboratory at Livermore, CA (Harrison and co-workers) and the U.S. EPA Laboratory at Narragansett, RI (Pesch and co-workers). Marine worms represent useful test organisms for evaluating the bioavailability and toxicity of chemicals in sediments. Contaminated sediments represent a major management problem in marine environments. Pesch et al. (1981) demonstrated dose responses for SCE and mutagenic compounds and for SCE and highly contaminated sediments (Pesch et al., 1987). Harrison et al. (1984, 1987) demonstrated SCE responses to radiation exposures. SCE response in the chromosomes of the bivalve mussel *Mytilus edulis* was demonstrated in adults (Dixon and Clarke, 1982) and in larvae (Harrison and Jones, 1982) exposed to known mutagens. This bivalve is capable of metabolizing promutagens, and there is evidence that this capability is inducible (Dixon et al., 1985). The SCE response has been observed in other species of mussel, e.g., *M. galloprovincialis* (Brunetti et al., 1990). Chromosomes must be labeled with 5-bromodeoxyuridine (BrdUdr) for two cell cycles before they can be differentially stained to visualize SCEs. Slower-growing adult animals require more time for incorporation of the BrdUdr label than larval stages. For adult mussels, labeling is administered through the seawater for periods of several days (Dixon and Clarke, 1982) vs. labeling for several hours for larvae of the mussel (Harrison and Jones, 1982). The mussel larval assay is the simplest and most rapid aquatic invertebrate SCE assay.

Micronucleus test — Micronuclei have been used as a test endpoint in two species of marine invertebrates, mussels and sea urchins. Brunetti et al. (1990)

and Migliore et al. (1990) exposed adult mussels, *M. galloprovincialis,* to chemicals and complex wastes in laboratory experiments and observed micronuclei in gill tissue. Hose (1985) proposed the use of several endpoints in sea urchin larval assays, including micronuclei. Larval assay systems are suited for the micronucleus test because these assays are rapid, easily manipulated, and cell division is rapid and observable.

Specific-locus assays — Almost no research has been done on specific-locus responses to mutagens with aquatic invertebrates. Squire (1973) irradiated brine shrimp, *Artemia salina,* and measured sterility in F_1 sons of irradiated males. Aquatic invertebrates are well suited for the specific-locus assay. The culture, spawning, and early development of bivalves and sea urchins are well studied. The high fecundity and short development times are well suited to the experimental needs for specific-locus assays. Unfortunately, there is a paucity of information about the genetics of these species, and appropriate genetic markers are needed to make specific-locus assays practical.

Dominant-lethal assays — Aquatic invertebrates such as shellfish and sea urchins have several favorable characteristics for use in dominant-lethal assays. The culture, spawning, and early development of bivalves and sea urchins are well studied. Hose (1985) recommends bioassay procedures for dominant-lethal assays in sea urchins. Metalli and Ballardin (1962, 1972) and Ballardin and Metalli (1965) studied the dominant-lethal effects of irradiation on brine shrimp, *A. salina.* They used hatching success as an index of dominant lethality. Although just a few studies have been done to date, the use of aquatic invertebrates as subjects for dominant-lethal assays has strong potential as a screening tool for mutagens.

DNA strand breaks — The DNA alkaline unwinding assay has had very limited application in aquatic invertebrates for the detection of DNA damage. Nacci and Jackim (1989) reported the successful applications of this assay to sperm from sea urchins and mussels. They observed DNA strand breaks in sperm from blue mussel, *M. edulis,* exposed to methyl methanesulfonate and from sea urchin, *Arbacia punctualata,* exposed to N-methyl-N'-nitro-N-nitrosoguanidine. They suggest that this assay may be useful for screening potential genotoxicants and for investigating relationships between DNA processes such as damage, synthesis, and repair.

4. Mammalian Assays

4.1 Mammalian Cell Culture

Mammalian cell assays have been used in hazard identification of some environmental mixtures. Since a large range of endpoints can be measured in mammalian cells, including human cells (e.g., gene mutations, DNA damage, chromosomal effects, etc.), these assays have been used to confirm that bacterial mutagens are mammalian cell mutagens and, in some specific chemical cases, to measure genetic effects in mammalian cells that are not detectable in bacteria. In the case of complex environmental samples of air, water, waste, or specific environmental source emissions, bacterial gene mutation assays in *Salmonella* have been more sensitive in detecting the presence of mutagens in these samples than mammalian cell assays (Lewtas, 1983, 1991).

Mammalian cell mutation assays in Chinese hamster ovary (CHO) cells and mouse lymphoma cells have been the most widely used to assess the induction of gene mutation by complex mixtures (Lewtas, 1983; Li et al., 1983). Human lymphocytes in culture have been used to detect cytogenetic effects, including sister chromatid exchanges, chromosomal aberrations, and induction of micronuclei (Barale et al., 1990). Because of sensitivity problems in the CHO mammalian cell mutation assay, Li et al. (1983) have developed a specific protocol with increased efficiency and sensitivity for application to environmental samples. Examples of publications of mammalian cell assays include studies of environmental samples from different sources, including automotive engines (Lewtas, 1983; Li et al., 1983), and from the seasons of summer and winter (Barale et al., 1990).

Although mammalian cell assays for cellular toxicity, genotoxicity, morphological transformation, and other genetic endpoints (e.g., cell-cell communication, DNA adducts, etc.) have been applied to complex environmental samples for the purpose of hazard identification (see Chapter 1 of this volume), these assays have been used less than microbial assays for environmental monitoring because (1) the mass of environmental sample required is generally 10 to 100 times that required for laboratory measurements in microbial systems, (2) the time required to conduct mammalian cell assays is generally several weeks rather than the several days for microbial assays, (3) mammalian cell assays are more costly in labor hours, cell culture materials, and equipment, and (4) with some important exceptions, the results from these assays generally agree with the microbial assays.

4.2 In Vivo Mammalian Assays

Mammalian assays conducted *in vivo* have also been used in hazard identification of environmental mixtures. Whole-animal mammalian assays have used primarily mice, although other small rodents such as hamsters can be used . The endpoints most commonly used in these studies are cytogenetic changes such as chromosomal aberrations, sister chromatid exchanges, and micronuclei induction. As with the mammalian cell assays, these assays have been used to confirm that complex environmental mixtures exhibiting bacterial mutagenic activity are genotoxic *in vivo* in mammals. Examples of applications of *in vivo* assays include evaluation of the genotoxicity of diesel particulate matter (Pereira, 1982) and water chlorination products (Meier et al., 1990).

Germ-cell mutagenesis studies have also been done with rodents, but this is not common due to the expense of the test.

B. Methods for Sampling and Analysis for Laboratory Bioassay

The methods for sampling air, water, soil, and other environmental media (e.g., waste sites) for mutagenicity studies have been developed through adaptations of methods developed for analysis of trace organics (e.g., polycyclic aromatic hydrocarbons) in the environment. In the case of air sampling, high-volume sampling has been used traditionally; however, massive volume samplers and, more recently, small personal samplers have been applied to air sampling (Claxton, 1981, 1983b; Jungers et al., 1981; Kado et al., 1987). Water and soil sampling have often involved the collection of grab samples and samples using special collection systems (Claxton et al., 1991; Donnelly et al., 1983; Loper and Tabor, 1983).

Extraction or some other method for separating the organic mutagens from the inorganic, carbonaceous, or other nonmutagenic matrices is usually required prior to bioassay. This step may be required in order to concentrate the mutagens and thereby improve the bioassay detection limit. In the case of insoluble particulates, such as air particles or soil, this step is required to facilitate exposure of the cell to the mutagens. Although the first methods used were those developed for extracting trace organics, biologists soon found that mutagens were also present in the more polar organic matter that was not well extracted by solvents, such as cyclohexane, which were used in many early studies. One of the most widely used solvents for extraction is dichloromethane, because it extracts both nonpolar and polar organics and does not extract the inorganic matter. Although soxhlet extraction was originally the most widely used extraction method, many investigators have recently replaced this procedure with the simpler and more rapid sonication method (Savard et al., 1992; Claxton et al., 1992b).

Fractionation is usually required for the separation and identification of specific mutagenic components (Claxton, 1982; Austin et al., 1985; Schuetzle and Lewtas, 1986; Alfheim et al., 1983). Separation of highly toxic components from the mutagenic components in samples is also often necessary to allow the measurement of mutagenicity (Epler et al., 1979; Claxton, 1983a). The fractionation methods used include all of the methods classically used to isolate and identify trace organics (e.g., solvent partitioning and chromatographic separations) as well as newer high-resolution HPLC and solid-phase extraction methods.

C. *In Situ* Environmental Bioassay of Air, Water, and Soil

In the natural environment, no biological system is acted upon by a single chemical. The adverse biological effects induced by environmental chemicals result, for the most part, from complex interactions among various chemical and physical components of a given environment. These toxic effects are often the result of additive, synergistic, or antagonistic interactions among the many chemicals in complex mixtures as well as their interaction with physical factors. Such an interactive and dynamic milieu may be impossible to reproduce in the laboratory. Therefore, from the perspective of hazard identification and environmental assessment, it is useful to measure the biological effects as they exist in nature.

Relatively few species are currently used for the *in situ* environmental bioassay of mutagenicity. The organisms most used are domestic flowering plants, indigenous aquatic vertebrates and invertebrates, and small, terrestrial mammals. There are two approaches to the *in situ* environmental bioassay, based on the source of the organisms and how they are used: (1) the use of surrogate individuals or species and (2) the use of indigenous individuals and species. Surrogate organisms may be transported from the greenhouse, laboratory, or farm field to the site of interest, from the site of interest to the laboratory, or from one field site to another site. Indigenous species may include both species native to an area and species introduced either purposefully or accidentally at some time in the past and which have become indigenous to an area (e.g., many weeds, the starling bird, and the house mouse in the U.S. originally were introduced from Europe, Asia, or Africa).

1. Microorganisms

Microorganisms generally are not used for *in situ* monitoring methods because environmental samples have to be prepared (extracted, solvent exchanged, concentrated, etc.) before the samples can be used with the assay. One primary exception, however, is the use of microbial systems to monitor the mutagenicity of gaseous mutagens. Because many environmentally encountered gases are short-lived and involved in dynamic and rapid chemical interactions, it is not possible to collect some gaseous species for bioassay testing in the laboratory. It is, therefore, advantageous to be able to test gases *in situ*. Ong et al. (1985) developed a flow-through liquid impinger system for monitoring volatile mutagens in the workplace. Others (Claxton et al., 1990) used a microbial plate system in flow-through chambers to examine the mutagenicity of atmospheric transformation products for volatile wood smoke and automotive emissions in an *in situ*-simulating study. Although these studies are more difficult and less reproducible than typical microbial mutagenesis studies, they have provided information that could not have been obtained by other means (Ong et al., 1985; Claxton et al., 1990).

2. Terrestrial Plants

2.1 *Tradescantia*

The species used most widely to date as a surrogate is *Tradescantia* clone 4430. Two principal assay systems have been used: the stamen hair bioassay for detecting single-gene somatic mutation and the assay for chromosomal aberrations as micronuclei in pollen mother cells.

Methods are available that permit the *in situ* monitoring of mutations induced by air, water, soil/sediment, and their combinations. Air contamination can be monitored effectively with plants such as *Tradescantia*. Gases and particles of 0.5 µm or less pass the stomata readily during respiration and enter the plant. Larger particles are excluded. Water pollution can be monitored effectively by aquatic vertebrates such as fish, invertebrates such as mussels and marine worms, and also with aquatic or semi-aquatic plants such as the Royal Fern, and terrestrial plants such as *Tradescantia*. Soil and sediments are most obviously assayed by rooted plants, such as corn and *Tradescantia,* and indigenous plants. Cut flower stalks placed in soil and sediments saturated with water also provide an effective bioassay for monitoring genotoxic activity in soil and sediments.

Since the mid-1970s, plants have been used to determine *in situ* genotoxic activity in air pollution, sludges, the ambient environment of petrochemical complexes, a lead smelter, bodies of water, and obscurant smokes. Five recent reviews discuss *in situ* genotoxicity monitoring with examples of surrogate, indigenous, and domestic species (Grant and Zura, 1982; Couch and Harshbarger, 1985; Ma and Harris, 1985; Sandhu and Lower, 1989; Lower et al., 1990; Lower and Kendall, 1990).

From 1976 to 1980, ambient air quality at 18 U.S. sites was monitored by the use of a mobile monitoring vehicle in a series of *in situ* studies instituted by Brookhaven National Laboratory using the *Tradescantia* stamen hair bioassay to monitor for genetic hazards. Certain locations that had a high concentration of industries showed significantly higher levels of mutagenicity than the clean sites (Grand Canyon, AZ, and Pittsboro, NC) used as concurrent negative controls. This study was a costly undertaking, but demonstrated the successful application of both the *in situ* plant mutation bioassay and the use of a mobile laboratory to carry out the assay. A more common approach is to physically transport plants to each site by auto or airplane. This is efficient if only a few sites are to be examined for mutagenesis, and it has been employed successfully by several investigators with transport distances from tens to thousands of kilometers from the laboratory. An example of this approach is a study conducted from 1976 to 1978, involving monitoring the combination of air and soil at five sites ranging from 0.3 to 11.7 km from a primary lead smelter (Lower et al., 1983b). This study used three species of plants, *Tradescantia,* corn, and soybean, and one species of indigenous mammal, the white-footed mouse *(Peromyscus leucopus)*. Six endpoints were measured: stamen hair mutagenesis and photosynthetic electron transport of *Tradescantia,* pollen mutation and linear growth (as height) of corn, nitrogen fixation of soybean, and excretion of γ-amino levulinic acid by the white-footed mouse. The soil concentrations of Pb, Cd, Cu, and Zn were also measured. Significant correlations were found between distances from the smelter, soil metal concentrations, and the endpoints measured in the six bioassays. This approach to *in situ* bioassay has been used to determine the mutagenicity of other media, such as diesel smoke emissions, soil, water, and the environment in the vicinity of nuclear power plants (Ma et al., 1983; Lower et al., 1983a, b; Lower et al., 1984).

The *Tradescantia* micronucleus assay allows monitoring of chromosomal aberrations and anaphase chromosome loss by using micronuclei at the tetrad stage in meiotic pollen mother cells as a biomarker of those events. This sensitive assay has been used for testing more than 150 chemicals and in several *in situ* monitoring studies, including studies of sewage sludges, diesel exhaust, and obscurant smokes (Lower et al., 1984; Ma et al., 1983; Schaeffer et al., 1987).

2.2 Zea maize (Corn)

The corn pollen assay uses the expression of the gene for amylose synthesis in pollen cells of corn as a marker for induced mutation. Pollen with a dominant allele (Wx) can synthesize amylose and stains black with iodine, whereas pollen grains with the recessive allele (wx) cannot synthesize amylose and stains tan with iodine. Thus, the induction of both forward mutation (Wx to wx) and reverse mutation (wx to Wx) can be measured in the meiotic cells. This assay has been used to detect the mutagenicity of herbicides applied in the fields, to evaluate the mutagenicity of city sewage sludges, and to monitor the genotoxicity of the ambient environment near an oil refinery and a petrochemical complex (Hopke et al., 1982; Lower et al., 1983a, b).

2.3 Other Terrestrial Species

A natural population of *Osmunda regalis* (Royal Fern) growing in a river heavily contaminated with paper-recycling waste was reported to have a high incidence of chromosomal aberrations (paracentric inversions, reciprocal translocations, and ring chromosomes). Solvent extracts of samples of the paper-recycling waste collected from the wastewater treatment facility and tested in the laboratory also showed mutagenicity in the *Salmonella his$^-$* reversion test and in the soybean mitotic crossing-over assay.

Other plant assays that could be further developed for the *in situ* mutagenesis bioassay of soil, water, and air include the *Hordeum vulgare* (barley) chlorophyll-deficient mutation bioassay and chromosome aberration bioassay (Constantin and Nilan, 1982a, b), *Arabidopsis* embryo mutation bioassay (Redei, 1982), *Vicia faba* (broad bean) root tip cytogenetic bioassay (Ma, 1982), and *Allium cepa* (onion) root tip cytogenetic bioassay (Grant and Zura, 1982).

3. Aquatic Species

Only a few aquatic species have been used as surrogates to detect environmental mutagens *in situ*. In the marine environment, two species of mussels,

M. edulis (Nacci and Jackim, 1989) and *M. galloprovincialis* (Migliore et al., 1990) have been used. In freshwater, the eastern mudminnow *U. pygmaea* has been used (Alink et al., 1980; Prein et al., 1978; Sugatt, 1978). Cytogenetic methods for determining genetic damage to fish species and their application to evaluating environmental pollutants have recently been summarized by Al-Sabti (1991).

3.1 Mudminnow (Umbra)

Prein et al. (1978) and Sugatt (1978) scored chromosomal damage in eastern mudminnows exposed to contaminated river water. In fish exposed to Rhine River water, 28% of the metaphases had chromosomal damage (Prein et al., 1978). However, when the same species was held in the heavily polluted Maas and Lek Rivers for up to 4 weeks, very little chromosomal damage was observed. Alink et al. (1980) exposed the eastern mudminnow to Rhine River water and observed significant increases in SCE frequencies, supporting the observations of Prein et al. (1978).

The mudminnow assay has great promise as a model system for laboratory and *in situ* surrogate studies. Kligerman (1980) proposes this system as a tool for detecting mutagens in freshwater environments. Kligerman (1982) recommends the improvement and validation of existing systems along with the compilation of a database from which meaningful comparisons among organisms and test systems can be made.

3.2 Mussels (Mytilus)

Marine mussels of the genus *Mytilus* are well studied and found circumglobally in all but tropical environments. They have excellent karyotypes for traditional metaphase observations and SCE analysis. A variety of mutagenicity assays are made relatively easily on mussels. They are easily handled and cultured, and may be induced to spawn. Thus, heritable genetic damage could be studied. They are widely used for marine monitoring. Adult and larval mussels may be used for laboratory studies. Adults may be used as *in situ* surrogates or sampled from indigenous populations. These assays with mussels are established and useful tools for monitoring marine environments.

Migliore et al. (1990) sampled adult mussels, *M. galloprovincialis*, from a clean site, then held a subsample at the clean site and transferred the rest to two contaminated sites. The spontaneous frequency of micronuclei in gill tissue was observed in animals from each location over an 18-week period. There was a significant increase in micronuclei at both of the contaminated sites. Nacci and Jackim (1989) conducted a similar experiment with adult mussels, *M. edulis*. The animals were exposed *in situ* at an estuarine superfund site. The gill tissues were examined for DNA strand breaks using the alkaline unwinding assay. A significant increase in DNA strand breaks was seen in the animals held at the contaminated site for as little as 3 d.

4. Mammals

Wild, feral, pet, and laboratory mammals have been employed only recently for *in situ* environmental monitoring. Representative studies are summarized below.

4.1 Rodents

The use of mammals for *in situ* environmental mutagenesis has been primarily with indigenous species of terrestrial rodents, e.g., the white-footed mouse, the cotton rat, and the house mouse (Nayak and Petras, 1985; Thompson et al., 1986; McBee, 1987; Tice et al., 1988; Thompson et al., 1988; Lower et al., 1990; Lower and Kendall, 1990). The utility of small mammals is dependent largely on their ecological and geographic distribution, their occurrence at sites of interest, and their ease of capture. Small rodents are easily captured alive. *Peromyscus leucopus*, along with its often sympatric and often indistinguishable close relative, the deer mouse *(P. maniculatus)*, together range over three fourths of the U.S. in many habitats from oak forests, grasslands, and mountains to deserts, making these prime candidates for use in monitoring. In addition, they can readily be raised and propagated in the laboratory. The distribution of the cotton rat and kangaroo rat is restricted to the southeastern and western U.S., respectively. The house mouse is found commonly, perhaps ubiquitously, in areas inhabited by man. It is, therefore, also a prime candidate for genetic bioassay in and near man's habitats.

D. Human Monitoring

Methods useful for monitoring genetic damage in human somatic and germ cells have in the past been limited. The past decade, however, has witnessed the development of a number of new methods that are expected to open the way to the widespread use of direct human monitoring as a method of identifying and tracking the health effects of environmental agents. Markers of genetic damage in human populations that are currently available are summarized in Table 1.

1. Somatic Cells

1.1 Cytogenetic Damage

The first studies of genetic damage in human populations were carried out in the 1960s, following the development in the 1950s of the currently employed methods for human cell culture and chromosome preparation. The developments that made this possible include methods of culturing lymphocytes and the use of lectins to stimulate lymphocyte division *in vitro* (Nowell, 1960; Moorhead et al., 1960), the use of colchicine to accumulate metaphase chromosomes (Tjio and Levan, 1956; Ford and Hamerton, 1956), and hypotonic

Table 1. Markers of Genetic Damage or Exposure in Human Populations

Direct Measures of Mutation and Cytogenetic Damage
  Cytogenetic damage: Chromosomal aberrations, micronuclei, SCE, aneuploidy
  Mutation assays: *hprt*, glycophorin A, HLA, T cell receptor, Hb
DNA Damage
  Adducts: ^{32}P-postlabeling, immunochemical methods, direct methods
  Modified DNA bases: oh^8-dG, thymine glycol, etc.
  Strand breaks: Single-cell electrophoresis/comet assay
Markers of Exposure and Surrogate Markers
  Body fluid mutagenicity
  Hemoglobin adducts
Markers of Germ Cell Damage
  Sentinel phenotypes, inborn errors of metabolism, protein variants, gene monitoring, sperm aneuploidy

---

treatment to promote chromosome dispersion (Hsu, 1952). These methods have been applied extensively to study the genetic effects of radiation in human populations and for biodosimetry of radiation exposure (Evans and Lloyd, 1978; Bender et al., 1988; Evans et al., 1979; Buckton et al., 1978; Awa, 1983; Awa et al., 1987; Evans et al., 1979; Roesch, 1987). Cytogenetic endpoints are by far the most extensively applied in human studies of genetic damage, and there is an extensive literature on cytogenetic studies in human populations exposed to physical and chemical agents in occupational and environmental settings. This literature will not be reviewed comprehensively in this document. Ashby and Richardson have summarized 113 human surveillance studies conducted between 1965 and 1984 (Ashby and Richardson, 1985), an ICPEMC task group has recently addressed the methodology and criteria for population-monitoring studies using cytogenetic techniques (Carrano and Natarajan, 1988), Bender et al. (1988) and Wolff (1991) have reviewed the use of cytogenetic techniques for biological dosimetry, and Kucerova (1982) has summarized cytogenetic studies of chemical exposures in occupational populations up to 1982. The use of cytogenetic and other genetic endpoints for population monitoring and biological dosimetry studies has been the subject of a number of international conferences (see, e.g., Gledhill and Mauro, 1991; Bridges et al., 1982).

The classical technique for analysis of cytogenetic changes in human somatic cells is to enumerate chromosomal aberrations in peripheral lymphocytes that have been stimulated to grow *in vitro* (Bender et al., 1988). Most of the changes detected using this classical technique (chromosome or chromatid deletions, dicentrics, and rings) are unstable *in vivo* and are not related directly to mutational events that occur in peripheral lymphocytes or in other tissues. Balanced chromosomal translocations that result in stable genetic aberrations can result in alterations of genes in the immediate vicinity of the translocation site, but traditionally these have been detectable only by highly labor-intensive chromosome banding techniques (Awa et al., 1987). Because this analysis is very labor intensive, it has not been applied extensively to genotoxicity analysis. The recent development of the methodology of fluorescence *in situ* hybridization of DNA probes to chromosomes provides a much more efficient analytical procedure for the detection of stable chromosomal translocations (Lucas et al., 1989, 1992). This should be an exceptionally useful tool for future studies.

A second cytogenetic event that can be detected using *in situ* hybridization techniques is the occurrence of aneuploidy in individual cells. By performing hybridization with DNA probes that are specific for particular genes, one can monitor for the occurrence of an increase (or decrease) in the frequency of that gene from the diploid state. Partly because *in situ* hybridization can be performed on either metaphase chromosomes or interphase nuclei, this new technique allows detection of gene duplication or deletion, endpoints that have in the past been difficult to monitor (Kallioniemi et al., 1992a, b; Wyrobek et al., 1990). Expected improvements in automated microscopic image analysis should allow this to become an important method for human monitoring (Piper et al., in press).

The occurrence of micronuclei is often used as an indirect indicator of cytogenetic damage in cells exposed to genotoxic agents. Micronuclei occur when chromosomal fragments or whole chromosomes lag at anaphase and remain in the cytoplasm of daughter cells after the formation of the nuclear membrane of the daughter cell. Their occurrence therefore indicates either chromosomal damage or other events that lead to anaphase lag of chromosomes. Because micronuclei can be scored much more rapidly than chromosomal aberrations, and because they can be enumerated in interphase cells that are difficult to culture *in vitro* for chromosomal analysis, they are frequently used as a surrogate genotoxicity endpoint in monitoring studies. Micronuclei have been most commonly scored in lymphocytes from peripheral blood using cytochalasin B to restrict *in vitro* replication of stimulated cells to a single replicative cycle (Fenech and Morley, 1986). This endpoint has been used to demonstrate cytogenetic damage in workers exposed to cytostatic pharmaceuticals (Yager et al., 1988), in patients receiving radiographic contrast agents (Cochran and Norman, 1982), and in workers exposed to styrene in the boatbuilding industry (Brenner et al., 1991). This endpoint, however, is considered insensitive for radiation dosimetry because there is a much higher and more variable spontaneous background of micronuclei in cultured lymphocytes than of radiation-specific chromosomal damage such as dicentric formation (Prosser et al., 1988, 1989).

Micronuclei can also be scored in a variety of epithelial cells, including buccal cells and cells from the nasal cavity, bronchi, urinary tract and bladder, and cervix (Stich and Rosin, 1983a, b; Stich et al., 1982a, b, 1985; Stich, 1987).

These endpoints have been used to demonstrate genotoxic damage from tobacco and betel quid chewing (Stich et al., 1982a, b; Livingston et al., 1990), to show a synergistic genotoxic effect between smoking and alcohol consumption (Stich and Rosin, 1983b), to demonstrate reduced genotoxic damage in smokeless tobacco users given β-carotene (Stich et al., 1985; Stich, 1987), and to study occupational exposures to chromic acid and ethylene oxide (Sarto et al., 1990).

The incidence of micronuclei in red blood cells of asplenic individuals is an especially useful marker in intervention trials designed to quantitate the contribution of risk factors responsible for elevated cytogenetic damage in individuals (Schlegel et al., 1986; Smith et al., 1990). This endpoint has been used to demonstrate the importance of folate status in spontaneous cytogenetic damage, and to identify dietary and lifestyle factors associated with elevated micronucleus frequencies (Everson et al., 1988; Smith et al., 1990).

As with the other cytogenetic measures, the major drawback of these micronucleus assay techniques is their indirect relationship with mutagenesis.

Another cytogenetic technique that has been used for human monitoring is the measurement of sister-chromatid exchange (SCE) frequencies (Carrano and Natarajan, 1988). Although SCE frequency and mutation frequency are both induced by many chemicals, the SCE endpoint itself is not a mutation (Yager et al., 1990). Although this endpoint has been used in human population studies — e.g., chemotherapy patients (Hongslo et al., 1991), pesticide workers (Carbonell et al., 1990), and pentachlorophenol exposure (Seiler, 1991) — it is considered less relevant than chromosomal aberrations because it has not been linked directly to human disease (DeMarini et al., 1994).

1.2 Direct Measures of Mutation

New technologies developed during the last 10 years allow detection and analysis of mutations that occur in human cells *in vivo*. Detection of somatic cell mutations in humans has been developed for five different genes (for reviews, see Albertini et al., 1990; Jensen et al., 1990): hypoxanthine phosphoribosyl transferase (HPRT), glycophorin A (GPA), histocompatibility locus (HLA), hemoglobin (Hb), and T cell receptor (TCR). The basis of all of these techniques is the enumeration of the frequency of cells that display an altered phenotype that results from mutational changes *in vivo*. Differences in tissue type, cell type, and gene locus make these assays complementary to one another, so multi-endpoint somatic mutation analysis on each individual is desirable.

The HPRT assay uses autoradiographic or clonogenic techniques to detect peripheral blood lymphocytes that fail to express this purine-metabolizing enzyme (as a result of mutation). While it is the most commonly used of the somatic mutation techniques, its validity is subject to several caveats. The gene is located on the X chromosome, so there is only one functional copy in each cell. Therefore, if large deletions occur at this locus, any nearby genes that are critical for cell viability would be lost, resulting in cell death and no phenotypic expression of the mutational event. Also, it has been shown that cells deficient in *hprt* appear to be selected against *in vivo,* so that their frequency in peripheral blood decreases with time after mutagenic exposure (Caskey and Stout, 1989). Another problem is that subpopulations of T cells may proliferate and clones of mutant T cells can be amplified *in vivo* after the mutagenic event has already occurred (Nicklas et al., 1988). Thus, a high observed mutant frequency may not be indicative of a high rate of mutation in a given individual. In spite of these limitations, the HPRT assay has proved to be a useful assay and has been used to demonstrate a relationship between the exposure of individuals to both chemical and physical mutagens and the frequency of mutant T cells. A major advantage of the clonogenic HPRT assay is the capability to recover mutant cells and to obtain detailed characterization of each clone of mutant T cells using restriction enzyme analysis of the T cell receptor gene (Nicklas et al., 1986).

The GPA assay uses immunolabeling of this cell-surface antigen on peripheral blood erythrocytes and flow cytometry to enumerate cells which have lost one of the two allelic forms of GPA. It detects cells that have lost the expression of one allele of the GPA gene locus (gene loss mutations) and also those that have lost one of the allelic forms and gained a copy of the other (by mitotic recombination or chromosome segregation errors). These two phenotypes are induced by different kinds of genetic effects on somatic cells, so their frequencies measure unrelated kinds of genotoxicity. A recent simplification of the GPA technique (Langlois et al., 1990) allows it to be performed on commercially available single-beam flow cytometers. It is the most easily applied of the five somatic-cell genotoxicity analytical techniques, but it also has several drawbacks. First, it can only be applied to individuals who are heterozygous at the glycophorin A locus, so only 50% of the human population can be analyzed. Also, this antigen is expressed only on erythroid cells, which in the peripheral blood do not carry DNA. Thus, the genetic changes that result in changes in cellular phenotype cannot be assessed directly. Nevertheless, it is proving to be a useful screening assay for defining cohorts that have been exposed to mutagenic agents and for monitoring the extent of genotoxic exposures.

The HLA assay is based on the analysis of cell-surface antigens in peripheral blood lymphocytes using immunolabeling techniques. In concept, it is similar to the GPA assay. It was first performed in a clonogenic assay using complement fixation (McCarron et al., 1989) and subsequently has been adapted to flow cytometry (Kushiro et al., 1992). The types of genetic changes detected in this assay are similar to those detected with the GPA assay, including both mutational and segregational alterations. Like the GPA assay, it suffers from the limitation that it can only be applied to individuals who are heterozygous at the HLA locus being analyzed. Since all HLA genes are very polymorphic, no allelic form can be analyzed that occurs in a majority of the human population. This assay is still untested in monitoring exposed populations, but appears to be a promising approach.

The fourth of the human somatic mutation assays uses immunolabeling to detect erythrocytes that express a very specific amino acid substitution that results from a mutation in the hemoglobin gene. To date, this assay has been performed by semiautomated microscopic analysis (Tates et al., 1989), although flow cytometric techniques also have been attempted (Bigbee et al., 1984). Since it measures the occurrence of a single base substitution, it analyzes for a very specific genotoxic event. This affords both advantages and disadvantages. It can be used to monitor for particular types of chemical effects when exposure is suspected for base-specific chemicals. However, many exposures result in multiple genotoxic effects, and the hemoglobin-based assay would not detect many of these events. Another disadvantage is that the gene target is very small, so the frequency of variant cells is correspondingly small. This makes the analysis very difficult and requires that a large number of cells in each individual (approximately 10^8 cells) must be scored in order to obtain a statistically significant number of mutation events. If the hematopoietic cell population that is affected is less than 10^8 cells, the assay will be ineffective. As with the HLA assay, this assay has not yet been applied to many cohorts of exposed individuals.

Recently, an assay has been developed which detects the loss of T cell receptors from peripheral T lymphocytes (Nakamura et al., 1991). A defect in either the α- or β-chains results in a failure to express the CD3 molecules on the cell surface. The assay immunologically measures the frequency of T cells which do not express this marker and enumerates these as mutated cells. The frequency of these cells is much higher than are measured for HLA, HPRT, or GPA in similar populations ($\sim 10^{-4}$). Thus, there may be a background of nonmutagenically derived variant cells. Nonetheless, this assay shows many sensitivities similar to the other somatic mutation assays, e.g., the variant frequency is increased in individuals exposed to ionizing radiation and increased with age. However, when atomic-bomb survivors were tested, they showed no increase in mutant cell frequency with dose (Kyoizumi et al., 1992). Thus, the assay appears to be transient in response. This lack of persistence makes it a transient biodosimeter. Results demonstrating the response of this biomarker to mutagenic chemical exposure have not yet been published.

1.3 DNA Damage

Several methods are available that allow analytical detection of chemical alterations induced in human cells by mutagens. These techniques include the detection of DNA adducts by ^{32}P-postlabeling, monoclonal antibody-based immunoassay, quadruple mass spectroscopy, and methods for detecting DNA strand breaks and alkali-labile lesions (Shields et al., 1991).

^{32}P-postlabeling is a sensitive and versatile method of monitoring DNA adducts in human tissues (Gupta et al., 1982; Reddy and Randerath, 1986;

Randerath et al., 1989; Schoket et al., 1991). It has been applied in a variety of occupational and environmental exposure settings to demonstrate DNA adducts in smokers (Randerath et al., 1989), foundry workers (Phillips et al., 1988; Perera et al., 1988), coke oven workers (Harris et al., 1985; Haugen et al., 1986), aluminum production plant workers (Schoket et al., 1991), environmental industrial exposures (Hemminki et al., 1990), and in other populations. The tissue studied most commonly is the peripheral blood lymphocyte population, but the method is suitable for any accessible tissue from which DNA can be isolated. It has been applied, for example, to estimate fetal exposure by measuring placental adducts (Everson et al., *Science,* 1986).

The use of immunological techniques for quantitation of DNA adducts was pioneered by the radiation genetics group at the University of Leiden, and these techniques have found significant application in biomonitoring studies (Baan et al., 1982, 1985a, b, 1986; Fichtinger-Schepman et al., 1982, 1984, 1985a, 1987a, b). Lohman et al. (1984, 1985) have discussed these and other methods of identifying DNA damage in exposed humans.

Methods have also been developed to measure oxidatively damaged and alkylated DNA bases and deoxyribonucleosides excreted in urine, and the excretion of these products has been used to estimate oxidative or alkylation damage to DNA. These markers include thymine and thymidine glycol (Adelman et al., 1988), 8-hydroxydeoxyguanosine, 7-methylguanosine, ^6O-methylguanine, and 5-methyldeoxycytidine (Park et al., 1992; Shigenaga et al., 1989). To date, these methods have been used mainly to study endogenous oxidative DNA damage (Ames, 1989).

Another indicator of DNA damage that has been used widely in mutagenesis research is the presence of DNA single- or double-strand breaks. Various methods of detecting strand breakage have proved very useful. One limitation of these methods is that monitoring radiation-induced damage is limited by the rapid rejoining of strand breaks following termination of exposure *in vivo* (Olive et al., 1991). A new method, based on alkaline microgel electrophoresis (Singh et al., 1988), that allows rapid and sensitive detection of DNA single-strand breaks, alkali-labile sites, DNA cross-linking, and excision repair sites (Singh et al., 1988; Tice et al., 1991) is the single-cell gel (SCG) or comet assay. This technique allows detection of intercellular differences in DNA damage and repair in virtually any eukaryotic cell population, requires only small numbers (a few thousand) of cells, and is rapid (approximately 3 h for processing and analysis) and sensitive.

1.4 Presence of Mutagens or Mutagen Adducts in Blood, Urine, or Tissues

Protein adducts formed by exposure to mutagenic substances have also been used as an indicator of exposure to genotoxicants. The use of this marker of exposure has been reviewed by Skipper and Tannenbaum (1990). Because analyses for protein adducts and direct chemical analysis of mutagens and their metabolites are generally specific to the type of chemical mutagen, the more general technique of using simple mutagenicity tests, such as the *Salmonella* assay (Ames, 1978; Yamasaki and Ames, 1977), to measure the presence of mutagenic chemicals in body fluids has been used more commonly. This provides a means by which exposure to genotoxicants can be detected even when the type of exposure is unknown or identification of the mutagenic agent cannot be performed easily. This monitoring technique has proved to be quite useful in many exposure situations (Ahlborg et al., 1985; Albertini and Robison, 1991), and new tester strains and methods have been developed to increase the sensitivity of these methods (Einistö et al., 1990).

2. Germ Cells

2.1 Sentinel Phenotypes

A number of studies have been performed to link the frequency of appearance of "sentinel phenotypes" with the exposure of a parent to environmental mutagens. A sentinel phenotype can be defined as a clinical disorder or syndrome that occurs as a consequence of a single, highly penetrant mutant gene, i.e., a dominant or X-linked trait that is uniformly expressed and accurately diagnosable near birth (Mulvihill and Czeizel, 1983). Large studies in which sentinel phenotype monitoring has been performed are (1) offspring of atomic-bomb survivors (Neel, 1981; Schull et al., 1981; Albertini et al., 1990), (2) children of cancer patients who were treated with radiation and/or chemical mutagens (Brewen et al., 1975), and (3) offspring of people who attempted suicide with large doses of chemicals (Czeizel et al., 1984). These studies all failed to demonstrate induction of germ cell mutations as a result of the exposure studied. The only possible indication of a human germ cell effect in man as a result of genotoxic exposures is that of abnormal meiotic chromosomes in testes of individuals exposed to radiation (Brewen et al., 1975).

2.2 Protein Variants

Another attempt to show heritable effects of genotoxic exposure of humans is the study of protein variants in offspring of atomic-bomb survivors (Neel et al., 1988). Protein variants can be detected by electrophoretic analysis of specific sets of proteins present in each person's cells or serum. This large, careful study showed no statistically significant increase in genetic effects in the progeny of atomic-bomb survivors.

2.3 Sperm Analysis

Several indirect endpoints have been used to detect genotoxicity in germ cells. For example, sperm morphology is known to be affected by exposure of the male testes to mutagenic substances (Moruzzi et al., 1988). Sperm morphology analyses have been performed on dry cleaning workers (Eskenazi et al., 1991), shipyard painters (Welch et al., 1988), and on individuals who smoke cigarettes (Oldereid et al., 1989). In recent years, a technique which uses *in situ* DNA hybridization on human sperm has been developed and is being used to detect aneusomies in sperm (Han et al., 1992; Wyrobek et al., 1990). This technique is so recent that few applications have been attempted, but it promises to be very useful.

Cytogenetic analysis of sperm has been performed by *in vitro* fertilization of hamster eggs with human sperm and subsequent microscopic analysis of paternal chromosomes formed in the first few mitoses (Brandriff et al., 1990). These studies have indicated an increase in germinal chromosomal aberrations when males were exposed to ionizing radiation (Martin et al., 1986). However, these measurements have proved to be extremely difficult because of the cross-species fertilization procedures, since males with significant testicular exposure often become sterile for extended periods after exposure, and because the procedure and scoring are laborious.

2.4 Methods under Development

Direct molecular analysis procedures are now being applied directly to DNA from parents and offspring so as to determine DNA-damaging effects directly (Mohrenweiser et al., 1989). In addition, other new molecular analytical techniques show the possibility of detailed analysis of DNA sequences in sperm as a monitoring technique (Honghua et al., 1990). These techniques are still too new to evaluate.

3. Experimental Design and Interpretation

Because many genotoxicants exhibit selectivity with respect to both target tissue and type of damage induced, a combination of several of the somatic mutagenic monitoring techniques is the best means for performing a complete analysis of mutational occurrence in humans. For example, the lack of a mutagenic effect in bone marrow erythropoietic cells as measured by the GPA assay does not assure the lack of somatic cell damage in liver. For practical purposes, however, technical limitations and the availability of resources make it necessary to restrict most studies to one or two endpoints. The choice of endpoint will depend to some extent on the degree of knowledge about the potential exposure. For example, dicentric formation in peripheral lymphocytes is a sensitive biomarker of exposure to ionizing radiation, but these unstable aberrations disappear with time after exposure. This is therefore the endpoint of choice immediately following acute radiation exposure, but balanced translocations are more useful for retrospective dosimetry a year or more after exposure because they persist for very long periods in the exposed cell population (Littlefield et al., 1991). On the other hand, long-lived markers are not very useful for intervention studies because their incidence does not decrease rapidly when genotoxic exposure is removed. Because human monitoring is usually conducted only when a genotoxic risk has been at least partly characterized, these and other known factors should be taken into account when designing human monitoring studies.

III. APPLICATIONS OF MONITORING

Short-term genetic bioassays are increasingly, and rightly, being used as an integral part of environmental assessment studies. No practical set of chemical analyses can fully characterize or monitor all of the potential genotoxic agents in complex mixtures. The use of short-term genetic bioassays complements chemical analysis to directly provide data on the genotoxic activity of complex mixtures. A series of case studies describing integrated assessments is found in Waters et al. (1983a, b, 1985). These studies include studies of indoor and outdoor air pollution, industrial effluents, and occupational environments.

An important aspect of environmental assessment studies using genetic bioassays should be the assessment of total human exposure, where human monitoring methods discussed elsewhere in this chapter are combined with environmental and personal exposure monitoring. Integrated multidisciplinary studies of this type make possible direct measurement of exposure, dosimetry, and genetic effects. An example of the use of biomonitoring of both environmental and human exposure is a study being conducted to assess exposure and lung cancer risk in a population in Xuan Wei Provence in China (Mumford et al., 1990). This study involves three different cohorts exposed to combustion emissions in their home from either smoky coal, smokeless coal, or wood. Biomonitoring of the air was conducted using bacterial genetic bioassays. Mutagenesis studies in mammalian cells and animal tumor studies are being used to assess the comparative cancer potency of these three emissions. Human biomonitoring of DNA adducts in tissues from exposed individuals are being assessed, as well as the cancer mortality in the populations. The combined use of genetic toxicology methods to assess environmental and human exposure, dosimetry, and genotoxic effects provides a powerful set of data for assessment of human risk from exposure to genotoxic agents.

A. Laboratory Studies with Environmental Samples

Although the utilities of the methods vary, each of the methods described in Section II can be used to determine whether or not genotoxic hazards exist within the environment. This can involve samples taken directly from the environment, or individual compounds or fractions from environmental samples. In order to use bioassays to identify environmental hazards, the following factors are very important: (1) sampling methods, (2) experimental design, both spatially and temporally, (3) relevance of the bioassay to the likely genotoxicants present (e.g., sensitivity, response to the type or class of substances), and (4) technical feasibility for the desired goals. These factors are discussed in Section IV below.

Benefits of using direct bioassay of environmental samples, or crude extracts of environmental samples, include simplicity and inclusion of the role of interactions and ecological transformation. There are, however, some potential disadvantages with using direct bioassay of environmental samples: (1) the preparation of samples for bioassay may produce artifactual results by either destroying or creating genotoxic components, (2) the chosen bioassay may not detect certain classes of genotoxicants, or (3) a complex mixture may contain components that mask the genotoxicity of other components. Because of these advantages and disadvantages, it is desirable to combine information from the bioassay of environmental chemicals with that from analytical techniques and *in situ* monitoring.

A discussion of the application of bioassays for differing monitoring strategies and media follows.

1. Emissions and Effluents

The application of bioassay methods to monitoring the mutagenicity of industrial and other source emissions and effluents was one of the earliest applications of genetic bioassays (Waters et al., 1979). Short-term bioassays have proven useful in evaluating the effects of engines, fuels, and control technologies on the mutagenicity of emissions from combustion sources (Lewtas, 1982). Emissions from combustion sources were collected, extracted, and tested in the *Salmonella his⁻* reversion assay in order to quantify and compare the mutagenic activity of emissions. Aqueous effluents from industrial sources have been collected from the plant outlet pipes in order to monitor the emissions of mutagens from industries (McGeorge et al., 1985). These earliest studies showed that bacterial mutagenicity assays were useful for detecting the emission of mutagens from many of these sources.

Mutagenic emission factors have been calculated to compare the mutagenicity of emissions from different sources. When making such comparisons, it is critical to compare mutagenic emission factors on a common basis such as fuel consumption, distance driven, energy consumption, or yield. For example, mutagenic emission factors of mobile sources can be expressed as mutagenic activity per kilometer, or those from stationary sources can be expressed on the basis of fuel or energy consumption (revertants per kilogram of fuel or joule of heat). This direct comparative analysis of mutagenic emission factors provides a simple means of comparing activity from various sources. Typical mutagenic emission factors from stationary and mobile sources range over three to five orders of magnitude (Lewtas, 1985, 1991).

The Ames *Salmonella*/microsome assay has been applied extensively to evaluations of the mutagenic activity of aqueous effluents from municipal and industrial wastewaters. Studies of industrial effluents have concentrated on either specific manufacturing process types, such as coal gasification processes and textile mill effluents, or have surveyed the mutagenicity of effluents discharged from various industrial processes. Generally, aqueous effluents must be concentrated prior to mutagenicity testing. Therefore, to compare the mutagenic emission rate from effluents, the mutagenic activity of the concentrated organics (e.g., in revertants per microgram of organics) is multiplied by the organic discharge rate (in terms of the amount of organics per volume of effluent or per time) to determine the mutagenic emission rate. The range of mutagenic emissions from various industrial effluents has ranged from less than 100 to over 100,000 revertants per liter (Lewtas, 1991). A thorough review of the genotoxicity of industrial wastes and effluents is available (Houk, 1992).

Plant systems have been used to study soils and air (Lower et al., 1983a, b) and waters. Diesel exhaust and obscurant smokes have been evaluated in the laboratory setting using cuttings of the flower stalks of *Tradescantia* placed in glass containers or chambers. Cut flower stalks can be exposed by placing them, cut surface down, in water or soil samples. Plants are also suitable for the study of gasses and aerosols, which can enter through the stomata. Exposure to air:diesel exhaust mixtures, for example, have been shown to produce a significant increase in the micronucleus frequencies in *Tradescantia* (Ma et al., 1981, 1983). The cut surface of the flower stalk breaks the root barrier and readily permits the translocation of substances up the stalk and to the developing flower buds. Other plants can be used similarly. Corn or *Arabidopsis* seeds can be exposed by soaking in water or soil containing a chemical or extracts.

2. Ambient Environment: Air and Water

Ambient air particles contain extractable organic matter that was first shown to be carcinogenic in rodents more than 45 years ago and later was shown to induce oncogenic transformation in mammalian cells in culture (see review by Lewtas, 1990). Studies in several countries have shown that the extractable organics from air particles were mutagenic in the *Salmonella his*⁻ reversion bioassay. These studies showed that mutagenic activity was usually concentrated in the smaller particles and was more concentrated at industrial and urbanized sites than at rural sites (Lewtas et al., 1982). Monitoring studies of mutagenic activity over time demonstrated higher concentrations of mutagens in the winter months and during certain times of the day related to traffic (Alfheim et al., 1983). Evidence that airborne mutagenic activity is derived mainly from combustion emission sources is based on studies showing that the concentration of mutagenic activity was highest downwind from combustion sources, as well as on modeling studies of mutagenic activity in air, in which specific sources of airborne mutagenicity were identified and quantitated by biomonitoring (Lewis et al., 1988; Pierson et al., 1983). Bioassay studies of the organics emitted from a variety of combustion sources (e.g., automobile and residential oil emissions) have confirmed the presence of mutagens in these sources.

Drinking water is another major ambient environment through which humans are exposed to mutagens and carcinogens. Chemical analysis has shown that drinking water contains hundreds to thousands of chemicals in low concentrations. Many of these chemicals result from disinfection of drinking water with chlorine. The first such chemicals identified were the trihalomethanes. Concern over these observations was increased by the finding that chloroform is carcinogenic in rats and mice. Results from animal studies were subsequently supported by a number of studies indicating a correlation between drinking water chlorination and gastrointestinal- and-urinary tract cancer mortality. In addition, drinking water concentrates and chemicals found in drinking water have been identified as mutagens in the *Salmonella* test. It is now thought that the trihalomethanes represent only a fraction of the products of chlorination. Many substances produced by chlorination of humic and fulvic acids and isolated from water remain to be identified. Several reviews are available (Meier, 1988; Alink, 1982; Bull, 1985; Loper, 1980; Nestman, 1983).

3. *Microenvironments*

Monitoring exposure in various microenvironments (i.e., automobiles, homes, and work places) has been used to determine total human exposure to environmental pollutants (Sexton and Ryan, 1988). The concentration of pollutants in indoor microenvironments is often higher than outdoors due to numerous indoor sources and the small equilibration volume of air. Since homes have a small air volume and low air-exchange rates, indoor air samplers are limited to sampling less than 10% of the total volume per hour to prevent disturbing the normal air flow and concentrations. Microsuspension and spiral system mutagenicity bioassay methods make it possible to work with the small mass quantities recovered with these sampling conditions (Lewtas et al., 1987; Houk et al., 1989, 1991).

In environmental monitoring studies, unvented combustion sources have been found to make a major contribution to indoor mutagenicity. In developed countries, the highest concentration of mutagenic activity observed indoors or outdoors has been in indoor environments where tobacco smokers are present. Unvented kerosene heaters have also resulted in elevated levels of mutagenic activity. Environmental monitoring studies using genetic bioassay analysis of indoor air particulate matter have identified environmental tobacco smoke, emitted as sidestream smoke, to be the major source of mutagens associated with airborne particles indoors (Lewtas et al., in press).

4. *Monitoring Personal Exposure*

Until very recently, genetic bioassays and monitoring personal exposure have not been coupled. The development of sampling and bioassay methods for application to personal monitoring is currently in progress (Lewtas et al., 1993). The development of such methods will result in a better understanding of external exposure to genotoxic agents. Exposure monitoring can also be approached from the biomonitoring of human populations, as discussed in

Section II.D. The combination of environmental and personal exposure monitoring with human population monitoring yields important information for understanding the relationship between exposure, dose, and effects.

5. Transport and Transformation of Mutagens in the Environment

Recognition that transport and transformation of chemicals in the environment can result in substantial chemical changes has led to the demonstration that these chemical changes can substantially change the genetic activity of pollutants. Claxton et al. (1990) have recently reviewed the effects of atmospheric transformation on the mutagenicity of volatile organic air pollutants. In the case of gaseous pollutants, there is evidence that nonmutagenic gases, e.g., toluene, are transformed in the presence of ozone and nitrogen oxides to mutagenic compounds (Shepson et al., 1985). In the case of particulate air pollutants, Claxton and Barnes (1981), Alfheim et al. (1983), Kamens et al. (1984), Gibson et al. (1985), and others have demonstrated significant changes in the character of mutagens present in the air after transport and transformation.

Plants and the immediate environment with which plants interact transform many chemicals. Some of these transformations can differ from those that occur in animals. Plants commonly take up chemicals through the foliage and roots; uptake may also occur through seeds and the cut surface of a stem or stalk. Alteration of the chemical may occur before and during penetration of the plant cuticle of leaves, fruit, stems, roots, and seeds. Prior to uptake by plants, the chemicals may undergo transformations due to photolytic and hydrolytic processes and by microbial activity, particularly in the soil. Once deposited on a plant surface, a chemical may be absorbed by and stored in cuticular waxes and may undergo change before penetration into cells.

Different parts of a plant may have very different metabolic potentials (Baldwin, 1977), and different species may have differences in metabolism (Mumm and Hamilton, 1979, 1983). Tissues rich in chloroplasts, such as leaves, are generally poor in mitochondria, while root tissues of various kinds, e.g., tubers, bulbs, etc., and flower heads, such as cauliflower, are poor or lacking in chloroplasts but rich in mitochondria. Whereas conjugated compounds are commonly excreted in animals, in plants they are commonly immobilized and stored.

There are significant differences between plants and animals in the amount of mitochondria and range of specificities of cytochrome P-450 to activate promutagens: higher plants contain small amounts of cytochrome P-450 with narrow substrate specificities (Higashi et al., 1983, 1985; O'Keefe and Leto, 1989; Wagner et al., 1990), whereas animals generally have much larger concentrations of cytochrome P-450 with broad substrate activity.

B. *In Situ* Monitoring

1. Plants

The species of plants and the bioassays discussed in Section II have different characteristics that influence their utility in various monitoring situations. *Tradescantia* is a particularly versatile species. It can be used for short-term exposure (minutes) to long-term chronic exposure (months), and can be used either as a whole rooted plant or cut flower stalks to monitor air, water, soil, or any combination of these media. The continuous production of new flower stalks by each plant and the rapid production of flowers (up to 20 flowers per inflorescence at the rate of one flower every 2 d per inflorescence) facilitates monitoring on even a daily basis. *Tradescantia* is cultured easily in the laboratory and can be maintained under field conditions for months. At least six bioassay endpoints can be measured, including the two measures of mutagenesis discussed above, sister chromatid exchange of root tips, photosynthesis electron transport, flower production as a measure of growth, and pollen abortion. One drawback of *Tradescantia* is the lower increase in the mutation frequencies of the two principal mutation systems, micronuclei production and, in particular, stamen hair mutagenesis, compared to corn. This can be compensated for by an increase in the number of observations. Very preliminary work indicates that *Tradescantia* pollen can be used for a mutagenesis assay in the same way that corn pollen is used. This may well be the case for pollen of many species of plants. If this proves to be true, the utility of *Tradescantia* and other plant species will be further enhanced. Although the purpose of this chapter is to comment on bioassays for mutagenesis, other nonmutagenic bioassays in the same organism (and often in the same individual) may be of use, too, particularly if they can be measured easily and cheaply at the same time.

Corn is less versatile compared to *Tradescantia*. A given individual corn plant provides a measure of mutation only once, at the time of anthesis (the release of pollen). The developing corn plant has only a small number of primordial pollen stem cells which eventually give rise to pollen mother cells and pollen. A mutation can occur at any stage of the mitotic division of these cells, as well as at meiosis, and can give rise to variable and often very large numbers of mutant pollen grains. The use of samples of pollen from 10 to 20 separate corn plants will partly overcome this problem and provide an estimated mean mutation frequency and a measurement of error. Somatic leaf mutations, such as yellow:green sectors, have been found to be of questionable use for *in situ* bioassay due to the fact that leaf damage by insects and weather obscures the correct scoring of sectors. The same difficulty is also encountered in similar somatic mutation assays in other plant species, such as soybean. The time to anthesis in corn takes from about 1 month (the most rapid) in an early synthetic strain, to 3 or more months in the standard corn varieties. Corn is not as easily grown in the laboratory as *Tradescantia,* it requires more care, and it is more vulnerable than *Tradescantia* because the long corn stalk can be easily damaged and broken. The low-growing *Tradescantia* is less likely to be damaged by man or other animals. The utility of the corn pollen bioassay, or the bioassay using the pollen of any plant, can only be realized if an automated method, such as flow cytometry, is used; manual scoring by microscope is too labor intensive to be practical.

The Royal Fern has had very limited use for *in situ* environmental bioassay under restricted ecological and geographic conditions, where it occurs endogenously along streams in the eastern U.S. It should be possible, however, to transport the fern from one site to another in the field, so this species could be used more widely.

The other species of plants discussed above — barley, *Arabidopsis,* broad bean, and onion — have not been used widely for *in situ* environmental bioassay, although they have potential for such use. All are easily grown in the laboratory and in the field.

Plants are obviously well suited for terrestrial monitoring *in situ* and have been used successfully for this purpose (Schairer, 1983; Lower et al., 1983a, b; Ma and Harris, 1985; Hopke et al., 1982). Air, water, and soil, individually or in any combination, can be assessed with a combination of *in situ* field assays or field and laboratory assays. To assess the components of a field site contaminated with toxic chemicals, the following scheme may be illustrative. Plants planted in the soil over weeks or months will record and integrate soil, water, and air components. Plants planted in pots of clean soil or a potting mixture at the same location will record the effects of airborne contamination plus whatever washes out with the rain. Plants planted in pots of soil transported from the test site to a field comparison site or the laboratory will record the effects in the soil alone. Plants transported to a field comparison site or laboratory and placed in contaminated water or planted in clean potting soil and watered with contaminated water will record the effects of the water. Plants as cuttings or whole rooted plants can be used to monitor sludges, seepages, etc. right in the field. If cuttings are used, wetting the ground with water to a soupy consistency and inserting the cuttings is sufficient.

2. Aquatic Species

Aquatic species have been used successfully to identify mutagenic hazards in aquatic environments. In a classic study, Longwell and Hughes (1980, 1981) examined mackerel eggs sampled from surface waters of the New York Bight. They observed anaphase chromosome irregularities, variable development rates, and differences in viability. These effects were found in impacted areas of the Bight close to the coast and ocean dumping grounds. In addition to the New York Bight, mutagens and mutagenic damage have been detected in many other polluted marine environments, including Long Island Sound (Pesch et al.,

1987), Chesapeake Bay (Pittinger et al., 1987), Puget Sound (Kocan et al., 1985), and sites in Canada (Payne, 1976) and Europe (Parry et al., 1976; Dixon, 1982; Brunetti et al., 1986).

The aquatic organisms used most extensively for studies of mutagenic damage in marine environments are fish. Studies of marine fish have used two types of observation: (1) anaphase aberrations in embryo/larval stages and (2) micronuclei in the circulating erythrocytes of adult fish. The largest database on cytogenetic abnormalities in embryo/larval stages of fish has been compiled by Longwell (1976) and Longwell and Hughes (1980, 1981). Observations on mackerel eggs and larvae were collected at an array of stations in the New York Bight sampled for several years. They found a close association between contaminant concentrations in the water and anaphase aberrations in the organisms. Hose and Smith (1986) and Hose et al. (1987) reported on the frequencies of micronuclei in circulating erythrocytes of two marine fish species collected in contaminated areas off southern California. They found the frequencies of micronuclei to be elevated in fish from the contaminated sites relative to fish from less contaminated sites. For example, the frequencies of micronuclei in white croaker were four times higher, and in kelp bass 11 times higher, in contaminated than in clean areas. These increases were related to measured environmental contaminant concentrations. Longwell and Hughes (personal communication) have made a similar observation on the frequencies of micronuclei in erythrocytes of winter flounder, *Pseudopleuronectes americanus,* collected at an extensive array of stations in the New York Bight and Long Island Sound. These results support the use of the micronucleus test as a rapid monitoring tool to detect the presence of environmental mutagens. Hose et al. (1987) recommend that this test be developed similarly to that proposed for use in laboratory animals (MacGregor et al., 1980).

A few studies have examined mutagenic damage in aquatic invertebrates. Blaylock (1966a, b) examined the effects of irradiation on large, polytene chromosomes of chironomid midge larvae, *Chironomus tentans.* This insect has aquatic larval stages which Blaylock sampled from White Oak Lake over a 5-year period. White Oak Lake is located at the Oak Ridge National Laboratory. It received low-level radiation wastes for many years. These insects were exposed to chronic radiation doses of approximately 0.63 rad/d. Their chromosomes were examined for evidence of damage, including inversions and small deletions. During the entire 5-year period, ten inversions and one deletion unique to the exposed population were observed. These aberrations were observed in only one of the annual collections. It was concluded that these types of aberrations may occur but would be eliminated by natural selection. This approach does not appear to be useful for routine monitoring.

Brunetti et al. (1990) and Dixon (1982) have examined indigenous populations of marine mussels *(Mytilus)* for mutagenic damage. Brunetti et al. (1990) observed micronuclei in gill tissue of the mussel *M. galloprovincialis.* They offered a series of recommendations for the use of this test as a monitoring tool.

Dixon (1982) observed aneuploidy in mussel embryos *(M. edulis)* spawned from stock sampled from a polluted dock. Brunetti et al. (1986) examined SCE frequencies in the eggs of *M. galloprovincialis* spawned from adults collected at two stations with markedly different levels of contamination. The SCE frequencies in eggs from the contaminated site were significantly higher, more than a doubling.

A great variety of mutagenicity tests have been applied to marine mussels. The early developmental stages are easily obtained and are useful for laboratory testing. Adults may be used as surrogates or sampled as indigenous populations. These adults may be examined directly or spawned to identify mutagenic damage transmitted to subsequent generations.

Cancerous diseases in aquatic animals have also been proposed to be indicators of contamination with mutagenic chemicals. However, there are very few cases where cause-effect relationships have been demonstrated (Mix, 1986). Viral and parasitic etiologies undoubtedly also contribute to the high incidence of cancer in some aquatic organisms, making it very difficult to demonstrate a causal relationship with chemical exposure. Therefore, although the epidemiology of cancer in aquatic animals may indicate environmental problems, environmental monitoring for mutagens should rely on methods that detect mutagenic activity directly.

3. Mammals

Feral, wild, pet, and laboratory mammals can be monitored to obtain an indication of the presence of environmental mutagens and carcinogens (Sandhu and Lower, 1989; Lower et al., 1990). Nayak and Petras (1985) studied SCEs in *Mus musculus* (house mouse) caught in corn cribs at various distances from industrial sites in southern Ontario, Canada. An inverse relationship was observed between SCE frequency and the distance of the sample collection site from the nearest major industrial center, suggesting that atmospheric pollutants may induce significantly higher levels of SCEs. From a cytogenetic study of bone marrow and germ cell damage in wild populations of *Peromyscus leucopus* from a hazardous waste site and uncontaminated areas, Tice et al. (1988) concluded that these mice can be used to monitor the toxicity of waste sites. A similar but larger study in Texas used *Sigmodon hispidus* (cotton rat) and *P. leucopus* to demonstrate significant differences in the incidence of chromosomal aberrations between a control site and two petrochemical waste disposal sites (Thompson et al., 1986; McBee, 1987; Thompson et al., 1988). From work at a military facility in California, Schaeffer et al. (1987) detected significantly higher frequencies of SCEs in the native rodents, *Dipodomys merriami* (kangaroo rat), exposed to obscurant smokes than in unexposed animals.

An interesting work has been reported on the use of *in situ* monitoring with rodents to confirm the observations from a preliminary human epidemiological study and to identify carcinogenic agents in an industrial environment in Shanghai, People's Republic of China (Wang et al., 1984). Preliminary studies on workers from factories in Shanghai showed an increased incidence of lung cancer among workers from rubber factories, especially among those who worked in areas where compounding, mixing, and milling took place. Chronic *in situ* monitoring with albino rats and inbred mice yielded a strong positive correlation between lung cancer in the test animals and air concentrations of N-phenyl-2-napthylamine and antioxidant D.

The single-cell gel (SCG) electrophoresis assay is a sensitive technique for evaluating the effect of environmental pollution on levels of DNA damage in cells collected from any organ or tissue (e.g., blood, brain, liver, ovaries, spleen, and testis) of virtually any animal. This technique has been used to assess the extent of DNA damage in blood, brain, liver, and spleen of rodents *(Ochrotomys nuttalli)* collected at a hazardous waste site (Nascimbeni et al., 1991), in fish exposed to polluted water (Tice et al., unpublished data), and in earthworms inhabiting contaminated soil (Vankerkom et al., 1993). Given the ability to utilize small numbers of cells from any organ/tissue sampled from virtually any eukaryotic species, this technique will be very valuable for environmental biomonitoring.

Domestic animals, such as dogs, can provide useful information on the potential health effects of environmental contaminants. Their use, however, appears to have been limited to investigations on the incidence of neoplasia. Pets share their owner's environment (particularly that of the young child) and are not subject to confounding factors such as alcohol consumption and active smoking. Pet dogs have been used to evaluate environmental influences on the occurrence of breast cancer (Schneider et al., 1969; Schneider, 1970) and osteosarcoma (Tjalma, 1966). Glickman et al. (1983) screened pet dogs for mesothelioma and found significant correlations between disease incidence in pet dogs and the owner's asbestos-related occupation and hobby.

In large-scale studies of dogs (Reif and Cohen, 1970; Hayes et al., 1981), the incidence of bladder cancer was significantly correlated with the overall level of industrial exposure. These studies found similar correlations between industrial activity and mortality from bladder cancer in humans, suggesting that canine bladder cancer could be used as a sentinel condition for the identification of carcinogenic hazard in the general environment.

C. Human Monitoring

Human monitoring has been used to determine the extent of exposure to specific genotoxicants and to perform risk assessment on the exposed populations. Cytogenetic analysis has been used more extensively than the somatic-cell mutation assays because the latter have become available only recently. For example, the analysis of chromosomal aberrations in peripheral blood lymphocytes has long been accepted as an international standard for determining the extent of exposure to ionizing radiation that an individual has received (Bender et al., 1988). Cytogenetic studies have demonstrated clastogenicity in human populations exposed to radiation (Tough et al., 1960; Evans and Lloyd, 1978; Evans et al., 1979; Lloyd et al., 1980), industrial chemicals (Kucerova, 1982), and drugs used in chemotherapy (Sorsa et al., 1985). The frequency of SCEs in peripheral blood lymphocytes has been used to monitor exposure of patients to chemotherapeutic agents and to study exposure to cigarette smoke and occupational exposures (Wasserman et al., 1990; Carrano and Moore, 1982; Sorsa et al., 1985). Micronucleus assays have been used to demonstrate that genetic damage may be related to lifestyle and nutritional factors, and can be reduced by interventions in which nutritional balance is improved (Everson et al., 1988) or chemopreventive agents are employed (Stich, 1987; Stich et al., 1985).

Of the somatic-cell mutation assays, only the HPRT and GPA assays have been used to a significant extent. The HPRT somatic mutation assay has been used to show an increased *in vivo* mutant T lymphocyte frequency in cancer patients treated with ionizing radiation (Sala-Trepat et al., 1990) or with chemotherapy (Dempsey et al., 1985; Ammenheuser et al., 1988). HPRT mutant frequencies also appear to be increased in blood samples from cigarette smokers (Cole et al., 1988, 1991) and from atomic-bomb survivors exposed to ionizing radiation (Kyoizumi et al., 1992). In subjects given chemotherapy for breast cancer, mutant frequency appeared to be inversely related to folic acid status (Branda et al., 1991).

The GPA assay has shown an increased frequency of variant erythrocytes after exposure of individuals to ionizing radiation (Straume et al., 1991; Jensen et al., 1990) or mutagenic chemicals (Bigbee et al., 1990), and also has been used to show high variant cell frequencies in sensitive individuals who are more susceptible to somatic cell mutation than the general population (Langlois et al., 1989; Bigbee et al., 1989). It also showed an increased trend in people who smoke cigarettes as compared to nonsmokers (Jensen, unpublished data).

The HLA or hemoglobin-based somatic mutation assays have not been applied to many cohorts of exposed individuals, but the HLA assay is well characterized and the occurrence of HLA mutations has been studied in normal populations (Morley et al., 1990; Turner and Morley, 1990; McCarron et al., 1989). The hemoglobin assay has shown an indication of higher frequencies of mutant erythrocytes in cigarette smokers (Tates et al., 1989).

Direct measurements of DNA damage, such as detection of DNA adducts or the occurrence of DNA repair, are now being applied rather extensively. (For a recent review of the status of these studies, see Gledhill and Mauro, 1991.) Adducts of polycyclic aromatic hydrocarbons, for example, have been found in DNA from the placental tissue of mothers who had been heavy cigarette smokers during their pregnancy (Everson et al., 1988b) and in occupationally exposed foundry workers (Perera et al., 1988; Phillips et al., 1988), and occupational exposure to aromatic amines has been associated with adducts in exfoliated urothelial cells (Kaderlik et al., 1993).

The SCG alkaline electrophoresis assay for DNA strand breaks and alkali-labile lesions has been employed to evaluate the effect of donor age on *in vitro* sensitivity to X-ray-induced DNA damage (Singh et al., 1990), to demonstrate increased levels of DNA damage in lymphocytes sampled from chemotherapy-treated breast cancer patients (Tice et al., 1992), and to evaluate levels of damage in epithelial cells collected from the urine of bladder cancer patients (McKelvey et al., 1992). It is currently being used to evaluate DNA damage in blood cells obtained by fingerprick and in cells lavaged from the lungs of smokers and nonsmokers, and in alveolar macrophages and lung epithelial cells obtained from volunteers exposed to ozone (Rice, unpublished data).

Similarly, the measurement of mutagenic chemicals directly in body fluids is now being used regularly. Adducts to proteins, such as hemoglobin or serum albumin, are most commonly applied. A number of examples are given by Gledhill and Mauro (1991).

It appears that a combination of the above endpoints would be the best way to monitor individuals for somatic mutations. For example, a combination of the GPA somatic-cell mutation assay, chromosomal analysis by fluorescence *in situ* hybridization, micronucleus frequency, and body fluid analysis for mutagenic compounds could be applied to a large population in a screening mode with relative ease. This combination would allow evaluation of damage in different tissues (e.g., lymphocytes and erythrocytes), different modes of action (e.g., mutagenicity and clastogenicity), and different targets (e.g., DNA and proteins). The results from this combination of endpoints would provide much more significant information than would a single endpoint, and would provide more power for risk analysis.

IV. RECOMMENDATIONS

As discussed above, monitoring can be used to achieve different objectives at various stages of risk assessment. Monitoring programs will vary with the specific objective and environment under consideration, and it is therefore not possible to recommend a simple scheme that would meet all possible needs and applications.

In this section, we attempt to identify the factors that influence the experimental design, organisms, and endpoints to be used when developing specific monitoring programs that meet the objectives described above, and to define the process required to implement a specific program.

A. Definition of Objectives

The objectives of monitoring were defined in Section I, and specific applications of monitoring at the various levels of risk assessment were discussed more fully in Sections II and III. Selection of organisms, endpoints, and experimental design requires a clear definition of the specific objective to be achieved. Once the specific objectives of a program are defined, the appropriate organisms and endpoints to be monitored may be selected and the experimental and sampling design devised. The factors that must be considered when making these selections and devising specific experimental protocols, and recommendations for the implementation of the processes, are discussed in the following sections.

B. Sampling Design and Statistical Analysis

1. Sampling Design

Green (1984) emphasizes that in the design of any study there should be a logical flow of purpose to question to hypotheses to model to sampling (or experimental) design to statistical analysis to tests of hypotheses to interpretation and presentation of results. The sampling, experimental, and statistical procedures to be used must be in place before the actual monitoring begins. "Biologically defined objectives should dominate and determine the statistics...". Monitoring may be conducted for a number of different purposes. For example, compliance monitoring is used to ensure that characteristics of an effluent are within standards specified by legal regulations. Status and trends monitoring is used to establish baselines from which trends in environmental quality may be assessed, and monitoring may be used to identify problems or "hot spots" such as superfund sites or point source discharges. Each of these applications has a unique purpose which leads logically to unique sampling designs. Suggested criteria for deciding whether or not a hazard exists are as follows: (1) a simple comparison of biological responses at a study site vs. the same responses at a clean reference site, (2) a demonstrated statistical relationship between exposure and response data, with the high exposure being found in the vicinity of the study site, and (3) a risk analysis using exposure/response measures from the study site supported by an available dose-response database. The purpose of this section is simply to call attention to the need for careful planning before initiating an extensive monitoring program. An excellent discussion on the statistics and experimental design of environmental monitoring is given by Gilbert (1987).

1.1 Field Reference

Field reference or field control sites are an integral part of any *in situ* environmental monitoring, and the data from such sites are used to determine whether the effects at the test sites are above ambient levels. A field reference site should be, as nearly as possible, identical to the test site except for the variable under investigation. Close geographic proximity is also advantageous. Field reference sites may be difficult to find because of the prevalence of contamination unrelated to that under investigation, and several field reference sites may be useful in case one proves to be contaminated with other xenobiotics which could confuse the interpretation of data from a test site.

1.2 Laboratory Control

A laboratory control may also be useful in addition to, but never in place of, a field reference to provide information on the ambient or background levels of effects by comparison to the field reference.

1.3 Transport Control

At times, a transport control may be of value if the test organism is to be transplanted from one location to another and if the transport requires some length of time or the organism is unavoidably exposed to some stress or condition that could confuse the measurements taken on the organisms that have been transplanted to the test site or field reference site.

2. Data Management

Data management is a significant problem not to be overlooked. Monitoring programs tend to generate large databases. If these data are to be analyzed in useful ways, it is likely the data will be stored on computers. This necessitates careful design for data storage, and data management systems to retrieve, manipulate, analyze, and graphically display data to ensure their full utilization. For small monitoring programs, personal computer technology will suffice, and readily available programs such as dBase, Lotus 1-2-3, Excel, or other similar products will serve. Larger programs should include consultation with experts on this topic during the planning phase.

C. Criteria for Selection of Organisms

1. Habitat of Concern

If the purpose of a monitoring effort is to evaluate a specific environmental site, the sampling design should be tailored specifically for that site. The order of preference for selecting test organisms would be (1) the species of primary concern in the environment under study (e.g., an endangered or resource species), (2) an indigenous sentinel species, (3) a surrogate species placed *in situ*, and (4) laboratory species to evaluate environmental samples.

The habitat of concern will thus define the organisms to be used for monitoring. Various components of a habitat may also require specific assessment, e.g., the air, water, or soil components at a given site may need to be assessed to determine their contribution to the overall effects of contamination. When possible, the same organism and the same biological response should be used to evaluate the various components. If different organisms and biological responses are used, a known correlation should exist between the biological responses.

2. Surrogates for Species of Concern

Surrogate species should experience exposures similar to the species of concern and have similar sensitivities. The surrogate species should also be selected to provide an economy of time and effort. A surrogate is used when the species of interest is unavailable or difficult or too costly to use. Surrogates are commonly used for the species of concern when that species is man.

3. Availability and Validation State of System

Developing and validating new test systems is very expensive. Further, it takes years for a new test system to gain acceptance in the scientific community. Until acceptance is achieved, the information provided by the new test system is of little value for risk-management purposes. Existing, well-accepted test systems should be used to monitor programs if at all possible. Some degree of validation is desirable under the specific conditions in which the system is to be used. A system available and validated for the laboratory may not function well in the field, and field validation is desirable before field implementation.

4. Laboratory/Field "Transplantability"

A combination of field and laboratory experiments is particularly useful Field tests may serve to identify areas exposed to mutagens. However, field studies alone do not permit easy identification of cause-effect relationships. These relationships are best identified in well-designed laboratory experiments where, for example, dose-dependent responses may be demonstrated. Therefore, the capability of making the same measurements on the same species in both laboratory and field studies greatly facilitates risk assessment. The transplantability or transport of species between field and laboratory and between sites within the field greatly enhances their usefulness.

5. Ability to Measure Endpoint of Interest

Two factors need to be considered when selecting a species to support measurements of a particular endpoint. First, the species needs to be suitable for that endpoint. For example, for SCE or metaphase analysis, it must possess a suitable karyotype. Second, it must be possible to handle and manipulate the species to provide suitable tissues for measurements. For example, rapid and predictable cell division is required for cytogenetic observations. Measurability can be a problem, particularly for *in situ* field monitoring. Many uncontrolled environmental conditions can render an organism or a biological response useless. For example, the yellow/green leaf spots in corn leaves that result from somatic mutation frequently cannot be scored in the field because of insect and heat damage to the leaf, although the scoring is reliable in the laboratory. Field testing of a system is always desirable before actual use to ensure that the biological response can be measured under the ambient conditions where the test is applied.

D. Criteria for Endpoint Selection

1. General Considerations

The purpose of monitoring is to provide information for a risk-analysis process. This process should define the level of concern appropriate for a particular problem or situation. Selection of endpoints to be measured should be based on the information needs of the risk-assessment process and, therefore, should be responsive to the issues raised in Chapters 1 and 4 of this volume. The organism chosen and the design of monitoring studies are highly influenced by the numbers and types of genotoxicity endpoints for which monitoring is required. The endpoints that can be monitored effectively also depend upon resources, time constraints, media(s) monitored, types of genotoxicants likely to be present, chemical and tissue specificity of the bioassay, and the type of genotoxicity damage for which monitoring is needed. For example, the endpoints chosen when the primary concern is a broad understanding of the types of genotoxicants present in an environment will be different than when the primary concern is a specific pollutant (Claxton and Barry, 1977).

2. Resources

The resources that affect the endpoint selection include monetary resources, available scientific expertise, availability of needed organisms and equipment, and sufficient time to complete the design, implementation, and analysis stages (Lave and Omenn, 1986). In relative terms, the least expensive type of monitoring would be the testing of a single environmental chemical in a simple *in vitro* microbial assay. Cost increases with the use of higher organisms and with increased complexity. For example, air-monitoring studies designed for comparative risk analysis are extremely expensive. Such studies have used samples from multiple sites taken over extended time periods and tested in multiple systems, including a whole-animal carcinogenicity bioassay. Extensive studies also may include efforts in analytical chemistry, meteorology, human surveys and/or epidemiology, and sophisticated modeling. Most studies will obviously fall between these extremes.

Studies also cannot be conducted satisfactorily without the appropriate scientific and technical expertise. Expertise for some bioassays, e.g., the *Salmonella* assay, can be found in many countries; however, expertise for other bioassays is less available. Information on available expertise and training can be found through contact with a variety of agencies (e.g., Environmental Mutagen Societies, International Commission for Protection Against Environmental Mutagens and Carcinogens, U.N. International Program for Chemical Safety, and U.N. Environmental Program).

3. Species Limitations

Many studies are limited to a specific environmental medium (e.g., air, water, or soil). In some cases, especially *in situ* studies, this will limit the species that can be used and thus limit the endpoints that can be considered. For laboratory studies, the amount of sample available often limits the endpoint considered (Claxton, 1983a, b; Claxton et al., 1988).

4. Mutational Specificity

Different chemical classes elicit different genotoxic responses. For example, the endpoints affected by combustion emissions, chlorinated wood preservatives, and textile-manufacturing effluents are quite different. When a monitoring effort is initiated, some information concerning the pollutants of concern is usually available. Because the *Salmonella* assay detects mutagens that are polycyclic aromatic hydrocarbons (PAHs) and nitrated and oxygenated derivatives of PAHs (Claxton et al., 1988), the assay is very useful for monitoring organic combustion material. However, because the *Salmonella* assay is not efficient at detecting chlorinated hydrocarbon mutagens, it would be less appropriate to use it for examining soil samples potentially contaminated with a wood preservative such as pentachlorophenol (Waters et al., 1990a, b). The same types of considerations apply to other bioassays.

5. Metabolism

Genotoxicants of different chemical classes will be metabolized by different metabolic pathways, and to different extents, by different species and tissues. Therefore, the species and tissue responses are influenced by the type of metabolism that is likely to occur. PAH compounds such as benzo[a]pyrene, common in combustion emissions, require microsomal metabolism before they can be detected by the *Salmonella* assay. This type of metabolism is usually supplied by a male rat liver homogenate (Ames et al., 1975; Maron and Ames, 1983). Some azo dyes that could be used by the textile industry require azo reduction to generate mutagenic derivatives (Esancy et al., 1990a, b; Freeman et al., 1990). In mammals, the microbial flora found in the intestinal tract contain azo reductase activity (Chadwick et al., 1992). Some azo reduction products are active in *S. typhimurium* TA1538 but not strain TA98. Different metabolism methods and bioassay endpoints would therefore be chosen for combustion products than for textile dye effluents. Preliminary chemical characterization of the monitored media will often help the genetic toxicologist select the appropriate endpoint.

6. Purpose of the Monitoring Effort

The choice of endpoint will be influenced by whether the effort is for hazard identification, exposure assessment, or risk assessment (Lewtas, 1988). For hazard identification, it is preferable to be able to examine multiple endpoints under a number of varying conditions using simple, rapid tests. For exposure assessment, the numbers and types of endpoints will depend upon whether a qualitative or quantitative assessment is needed, upon the size characteristics of the exposure population, and upon the route of exposure. For risk assessment, it is preferable to examine endpoints chosen in the most relevant and quantitative manner possible.

Monitoring can be used not only to document the presence and/or level of pollution by genotoxicants, but also in a preventative mode or to assess the level of remediation that has been achieved in response actions. In a preventative mode, monitoring can be used to determine whether a process will introduce genotoxicants to the environment. For example, whether an industrial process will add a genotoxicant to the environment may require only a simple screening test if that test is known to detect the genotoxicity characteristic of that particular industrial emission. On the other hand, the assessment of whether a remediation effort eliminated known pollutants without creating genotoxicants or contaminating additional media may require a range of bioassays with differing endpoints. The overall type of public health or ecological concern must also be considered. For example, monitoring a single environmental carcinogen and its metabolites would require different endpoints than would monitoring the heritable effects of a noncarcinogenic genotoxicant in an ecological setting.

7. Tiers and Matrices

If the type of genotoxicant or genotoxic activity present in an environment is not known, the use of "tier-testing" or "testing matrices" may be considered (Claxton and Barry, 1977; Waters et al., 1983a, b). In a tier-testing scheme, samples are first tested with less expensive, more rapid tests. Depending upon the outcome, another level of testing may or may not be undertaken in order to better characterize the genotoxicity of the substance. With a test matrix, a battery of tests established at several different levels is outlined. The selection of the battery that is used is based upon previous knowledge and the purpose of the study. (For a more thorough discussion of the types and uses of such systems, see the discussion of Waters et al., 1983a, b.)

V. FUTURE NEEDS

Environmental monitoring has evolved over the past decade toward providing data more useful for exposure and risk assessment. The objective of many monitoring studies in the 1960s and 1970s was to monitor concentrations of pollutants at ambient locations, such as rooftops and in large bodies of water, where the pollutants would be well mixed and represent a homogeneous sample. In the 1980s, a number of studies focused on assessing the emission of mutagens from point sources. In the future, the emphasis will likely shift to individual exposure monitoring, especially in the case of human exposure, and to understanding which sources and factors lead to increased exposure and increased risk.

Future needs include additional development of bioassays for *in situ* monitoring. To date, the bioassays and species used most commonly are those historically used for genetic research in the laboratory. It is timely to design assays specifically for the purpose of *in situ* monitoring, using actual species of concern rather than surrogates. Other needs include development of new analytical techniques that afford capabilities not currently available. The new molecular analytical capabilities that are rapidly becoming available can be applied to determining new endpoints in species different from those previously employed for monitoring. Practical germ cell specific-locus assays are particularly needed because of the importance of heritable genetic alterations to environmental populations. Methods that permit assessment of genetic damage in a wider variety of human tissues are badly needed, especially for those epithelial or endothelial cell populations that cannot presently be monitored for mutations and which are responsible for the most common human cancers. New endpoints for these analyses should include recombinatorial changes and aneuploidy. Molecular analysis of oncogenes or tumor suppressor genes that have direct relevance to risk estimation for carcinogenesis is needed, and can be expected to be developed in the near future. Many of the bioassays that are used presently have not been linked closely with specific genetic risks in the populations of biota currently used for monitoring studies. Research into the correlation of mutagenic effects with disease outcome and with deleterious heritable effects must be extended considerably.

REFERENCES

Adelman, R., R.L. Saul, and B.N. Ames (1988), Oxidative damage to DNA: relation to species metabolic rate and life span, *Proc. Natl. Acad. Sci. U.S.A.,* 85, 2706–2708.

Ahlborg, G. Jr., B. Bergstrom, C. Hogstedt, P. Einistö, and M. Sorsa (1985), Urinary screening for potentially genotoxic exposures in a chemical industry, *Br. J. Ind. Med.,* 42, 691–700.

Albertini, J., J.A. Nicklas, J.P. O'Neill, and S.H. Robison (1990), *In vivo* somatic mutations in humans: measurement and analysis, *Annu. Rev. Genet.,* 24, 305–326.

Albertini, R.J. and S.H. Robison (1991), Human population monitoring, in *Genetic Toxicology,* A.P. Li and R.H. Heflich, Eds., CRC Press, Boca Raton, FL, pp. 375–420.

Alex, A. and P. Dupuis (1989), FT-IR and Raman investigation of cadmium binding to DNA, *Inorg. Chim. Acta.,* 157, 271–281.

Al-Sabti, K. (1991), *Handbook of Genotoxic Effects and Fish Chromosomes,* J. Stefan Institute, Ljubljana, Yugoslavia.

Alfheim, I., G. Lofroth, and M. Moller (1983), Bioassay of extracts of ambient particulate matter, *Environ. Health Perspect.,* 47, 227–238.

Alink, G.M. (1982), Genotoxicants in water, in *Mutagens in our Environment,* M. Sorsa and H. Vainio, Eds., Alan R. Liss, New York, 261–276.

Alink, G.M., E.M.H. Frederix Wolters, M.A. van der Gaag, J.F.J. van de Kerkhoff, and C.L.M. Poels (1980), Induction of sister chromatid exchanges in fish exposed to Rhine water, *Mutat. Res.,* 78, 369–374.

Ames, B.N. (1978), Identifying environmental chemicals causing mutations and cancer, *Science,* 204, 587–593.

Ames, B.N. (1989), Mutagenesis and carcinogenesis: endogenous and exogenous factors, *Environ. Mutagen.,* 13, 1–12.

Ames, B.N., J. McCann, and E. Yamasaki (1975), Methods for detecting carcinogens and mutagens with the *Salmonella*/mammalian microsome mutagenicity test, *Mutat. Res.,* 31, 347–364.

Ammenheuser, M.M., J.B. Ward, Jr., E.B. Whorton, Jr., J.M. Killian, and M.S. Legator (1988), Elevated frequencies of 6-thioguanine-resistant lymphocytes in multiple sclerosis patients treated with cyclophosphamide: a prospective study, *Mutat. Res.,* 204, 509–520.

Anders, A., F. Anders, and K. Klinke (1973a), Regulation of gene expression in the Gordon Kosswig Melanoma system. I. The distribution of the controlling genes in the genome of the Xiphophorin fish, *Platypoecilus maculatus* and *Platypoecilus variatus,* in *Proc. Symp. Genetics and Mutagenesis of Fish,* J.H. Schroeder, Ed., Springer-Verlag, New York, 33–52.

Anders, A., J. Anders, and K. Klinke (1973b), Regulation of gene expression in the Gordon Kosswig Melanoma system. II. The arrangement of chromatophore determining loci and regulating elements in the sex chromosomes of Xiphophorin fish, *Platypoecilus maculatus* and *Platypoecilus variatus,* in *Proc Symp. Genetics and Mutagenesis of Fish,* J.H. Schroeder, Ed., Springer-Verlag, New York, 53–63.

Anderson, S.L. and F.L. Harrison (1986), Effects of Radiation on Aquatic Organisms and Radiobiological Methodologies for Effects Assessment, EPA 520/1-85-016, U.S. Environmental Protection Agency, Washington, D.C.

Anderson, S.L., F.L. Harrison, G. Chan, and D.H. Moore (1987), Chromosomal Aberrations, Reproductive Success, Life Span, and Mortality in Irradiated *Neanthes arenaceodentata* (Polychaeta), EPA 520/1-87-007, U.S. Environmental Protection Agency, Washington, D.C.

Ashby, J. and C.R. Richardson (1985), Tabulation and assessment of 113 human surveillance cytogenetic studies conducted between 1965 and 1984, *Mutat. Res.,* 154, 111–133.

Austin, A.C., L.D. Claxton, and J. Lewtas (1985), Mutagenicity of the fractionated organic emissions from diesel, cigarette smoke condensate, coke oven and roofing tar in the Ames assay, *Environ. Mutagen.,* 7, 471–487.

Awa, A. (1983), Chromosome damage in atomic bomb survivors and their offspring — Hiroshima and Nagasaki, in *Radiation-Induced Chromosome Damage in Man,* A.A. Sandberg, Ed., Alan R. Liss, New York, 433–453.

Awa, A.A., T. Honda, S. Nerlisishi, T. Sufuni, H. Shimba, K. Ohtaki, N.M. Nakano et al. (1987), Cytogenetic studies of the offspring of atomic bomb survivors, in *Cytogenetics: Basic and Applied Aspects,* G. Obe and A. Basler, Eds., Springer-Verlag, Berlin, 166–183.

Baan, R.A., M.A. Schoen, O.B. Zaalberg, and P.H.M. Lohman (1982), The detection of DNA damages by immunological techniques, (Proc. 12 EEMS meeting, Helsinki, Finland), in *Mutagens in Our Environment,* M. Sorsa and H. Vainio, Eds., Alan R. Liss, New York, 111–124.

Baan, R.A., O.B. Zaalberg, A.M.J. Fichtinger-Schepman, M.A. Muysken-Schoen, M. J. Lansbergen, and P.H.M. Lohman (1985a), Use of monoclonal and polyclonal antibodies against DNA adducts for the detection of DNA lesions in isolated DNA and in single cells, in DNA adducts: dosimeters to monitor human exposure to environmental mutagens and carcinogens, *Environ. Health Perspect.,* 62, 81–88.

Baan, R.A., P.H.M. Lohman, O.B. Zaalberg, M.A. Schoen, A.M.J. Fichtinger-Schepman, H.H. Schutte, and G.P. van der Schans (1985b), Future tools in biomonitoring, in *Carcinogens and Mutagens in the Environment, Vol. IV: The Workplace: Monitoring and Prevention of Occupational Hazards,* H.F. Stich, Ed., CRC Press, Boca Raton, FL, 101–116.

Baan, R.A., P.H.M. Lohman, A.M.J. Fichtinger-Schepman, M.A. Muysken-Schoen, and J.S. Ploem (1986), Immunochemical approach to detection and quantitation of DNA adducts resulting from exposure to genotoxic agents, in *Monitoring of Occupational Genotoxicants,* M. Sorsa and H. Norppa, Eds., Alan R. Liss, New York, 135–146.

Baldwin, B.C. (1977), *Xenobiotic Metabolism in Plants in Drug Metabolism: From Microbes to Man,* D.V. Parke and R.L. Smith, Eds., Taylor and Francis, London, 191–217.

Ballardin, E. and P. Metalli (1965), Sulla relazione fra poliploidia radiosensibilita in ovociti de *Artemia salina,* Leach, *Boll. Zool.,* 32, 613.

Barale, R., L. Migliori, B. Cellini, L. Francioni, F. Giorgelli, I. Barrai, and N. Loprieno (1990), Genetic toxicology of airborne particulate matter using cytogenetic assays and microbial mutagenicity assays, in *Genetic Toxicology of Complex Mixtures,* M.D. Waters, F.B. Daniel, J. Lewtas, M.M. Moore, and S. Nesnow, Eds., Plenum Press, New York, 57–71.

Battista, J.R., C.E. Donnelly, T. Ohta, and G.C. Walter (1990), The SOS response and induced mutagenesis, in *Mutation and the Environment, Part A: Basic Mechanisms,* M.L. Mendelsohn and R.J. Albertini, Eds., Wiley-Liss, New York, 169–178.

Bayley R., A. Carothers, X. Chen, S. Farrow, J. Ji. L. Gordon, J. Piper, D. Rutovitz, M. Stark, and N. Wald (1991), Radiation dosimetry by automatic image analysis of dicentric chromosomes, *Mutat. Res.*, 253(3), 223–235.

Bender, M.A., A.A. Awa, A.L. Brooks, H.J. Evans, P.G. Groer, L.G. Littlefield, C. Pereira, R.J. Preston, and B.W. Wachholz (1988), Current status of cytogenetic procedures to detect and quantify previous exposures to radiation, *Mutat. Res.*, 196, 103–159.

Bigbee, W.L., E.W. Branscomb, and R.H. Jensen (1984), Detection of mutated erythrocytes in man, in *Individual Susceptibility to Genotoxic Agents in the Human Population,* F.J. de Serres and R.W. Pero, Eds., Plenum Press, New York, 249–266.

Bigbee, W.L., R.G. Langlois, M. Swift, and R.H. Jensen (1989), Evidence for an elevated frequency of *in vivo* somatic cell mutations in ataxia telangiectasia, *Am. J. Hum. Genet.*, 44, 402–408.

Bigbee, W.L., A.W. Wyrobek, R.G. Langlois, R.H. Jensen, and R.B. Everson (1990), The effect of chemotherapy on the *in vivo* frequency of glycophorin A "Null" variant erythrocytes, *Mutat. Res.*, 240, 165–175.

Bishop W.E. and L.C. Valentine (1982), Use of the central mudminnow *(Umbra limi)* in the development and evaluation of a sister chromatid exchange test for detecting mutagens *in vivo,* in *Aquatic Toxicology and Hazard Assessment: Fifth Conference,* ASTM STP 766, J.G. Pearson, R.B. Foster, and W.E. Bishop, Eds., American Society for Testing and Materials, Philadelphia, 99–108.

Blaylock, B.G. (1966a), Chromosomal polymorphism in irradiated natural populations of *Chironomus, Genetics,* 53, 131–136.

Blaylock, B.G. (1966b), Cytogenetic study of a natural population of *Chironomus* inhabiting an area contaminated by radioactive waste, in *Proc. Symp. Disposal of Radioactive Waste into Seas, Oceans, and Surface Waters.* International Atomic Energy Agency, Vienna, 835–846.

Bloom, S.E. (1978), Chick embryos for detecting environmental mutagens, in *Chemical Mutagens: Principles and Methods for their Detection,* Vol. 5, A. Hollander and F.J. de Serres, Eds., Plenum Press, New York, 203–232.

Branda, R.F., J. P. O'Neill, L.M. Sullivan, and R.J. Albertini (1991), Factors influencing mutation at the *hprt* locus in T-lymphocytes: women treated for breast cancer, *Cancer Res.,* 51, 6603–6607.

Brandriff, B.F., L.A. Gordon, and A.V. Carrano (1990), Cytogenetics of human sperm: structural aberrations and DNA replication, *Prog. Clin. Biol. Res.,* 340B, 425–434.

Brenner, D., A.M. Jeffrey, L. Latriano, L. Wayneh, D. Warburton, M. Toor, R.W. Pero, L.R. Andrews, S. Walles, and F.P. Perera (1991), Biomarkers in styrene-exposed boat builders, *Mutat. Res.,* 261, 225–236.

Brewen, J.G., R.J. Preston, and N. Gengozian (1975), Analysis of X-ray-induced chromosomal translocations in human and marmoset spermatogonial stem cells, *Nature (London),* 253, 468–470.

Bridges, B.A., B.E. Butterworth, and I.B. Weinstein, Eds. (1982), *Indicators of Biological Exposure,* Cold Spring Harbor Laboratory, New York, 580.

Brusick, D.J., G.T. Arce, J.C. Bailar, R.C. Gupta, R. Herbert, P.H.M. Lohman, C.W. Moore, R.F. Murray, G.A. Sega, R.B. Setlow, and J.A. Swenberg (1988), *Drinking Water and Health: DNA Adducts,* Subcommittee on DNA Adducts, Safe Drinking Water Committee, National Research Council U.S.A., National Academy Press, Washington, D.C. 1–73.

Brunetti, R., I. Gola, and F. Malone (1986), Sister chromatid exchange in developing eggs of *Mytilus galloprovincialis* Lmk. (Bivalvia), *Mutat. Res.,* 174, 207.

Brunetti, R., F. Majone, M. Zordan, and A.G. Levis (1990), Cytogenetic alterations in *Mytilus galloprovincialis* as indicators of genotoxic pollutants in the marine environment: methodological aspects, in *Advances in Applied Biotechnology Series. Vol. 5, Carcinogenic, Mutagenic, and Teratogenic Marine Pollutants: Impact on Human Health and the Environment,* World Health Organization, United Nations Environment Program, Gulf Publishing, Houston, 101–110.

Buckton, K.E., G.E. Hamilton, L. Paton, and A.O. Langlands (1978), Chromosome aberrations in irradiated ankylosing spondolitis patients, in *Mutagen-Induced Chromosome Damage in Man,* Edinburgh University Press, Scotland, 142–150.

Bull, R.J. (1985), Carcinogenic and mutagenic properties of chemicals in drinking water, *Sci. Total Environ.,* 47, 385–413.

Carbonell, E., M. Puig, N. Xamena, A. Creus, and R. Marcos (1990), Sister chromatid exchange in lymphocytes of agricultural workers exposed to pesticides, *Mutagenesis,* 5, 403–405.

Carrano, A.V. and D.H. Moore (1982), The rationale and methodology for quantifying sister chromatid exchange in humans, in *Mutagenicity: New Horizons in Genetic Toxicology,* J.A. Heddle, Ed., Academic Press, New York, 267–304.

Carrano, A.V. and A.T. Natarajan (1988), Considerations for population monitoring using cytogenetic techniques, *Mutat. Res.,* 204, 379–406.

Caskey, C.T. and J.T. Stout (1989), Molecular genetics of HPRT deficiency. *Semin. Nephrol.,* 9(2), 162–167.

Chadwick, R.W., S.E. George, and L.D. Claxton (1992), Role of the gastrointestinal mucosa and microflora in the bioactivation of dietary and environmental mutagens or carcinogens, *Drug Metab. Rev.,* 24, 425–492.

Chriswell, C.D., B.A. Glatz, J.S. Fritz, and H.J. Svec (1979), Mutagenic Analysis of Drinking Water, in *Application of Short-Term Bioassays: the Fractionation and Analysis of Complex Environmental Mixtures,* M.D. Waters, S. Nesnow, J. Lewtas-Huisingh, S.S. Sandhu, and L. Claxton, Eds., Plenum Press, New York, 478–494.

Claxton, L. and P.Z. Barry (1977), Chemical mutagenesis: An emerging issue for public health, *Am. J. Publ. Health,* 67(11), 1037–1042.

Claxton, L.D. (1981), Mutagenic and carcinogenic potency of diesel and related environmental emissions: *Salmonella* bioassay, *Environ. Int.,* 5, 389–391.

Claxton, L.D. (1982), Review of fractionation and bioassay characterization techniques for the evaluation of organics associated with ambient air particles, in *Genotoxic Effects of Airborne Agents,* R.R. Tice, D. L. Costa, and K.M. Schaich, Eds., Plenum Press, New York, 19–34.

Claxton, L.D. (1983a), The integration of bioassay and physio-chemical information for complex mixtures, in *Short-Term Bioassays in the Analysis of Complex Environmental Mixtures III,* M.D. Waters, S.S. Sandhu, J. Lewtas, L. Claxton, N. Chernoff, and S. Nesnow, Eds., Plenum Press, New York, 153–162.

Claxton, L.D. (1983b), Characterization of automotive emissions by bacterial mutagenesis bioassay: a review, *Environ. Mutagen.,* 5, 609–631.

Claxton, L.D. and H.H. Barnes (1981), The mutagenicity of diesel exhaust particle extracts collected under smog-chamber conditions using the *Salmonella typhimurium* test system, *Mutat. Res.,* 88, 255–272.

Claxton, L.D., J. Allen, A. Auletta, K. Mortelmans, E. Nestmann, and E. Zeiger (1987), Guide for the *Salmonella typhimurium*/mammalian microsome tests for bacterial mutagenicity, *Mutat. Res.,* 189, 83–91.

Claxton, L.D., A.G. Stead, and D. Walsh (1988), An analysis by chemical class of Salmonella mutagenicity as predictors of animal carcinogenicity, *Mutat. Res.,* 205, 197–225.

Claxton, L.D., T.E. Kleindienst, E. Perry, and L.T. Cupitt (1990), Assessment of the mutagenicity of volatile organic air pollutants before and after atmospheric transformation, in *Genetic Toxicology of Complex Mixtures,* M.D. Waters, F.B. Daniel, J. Lewtas, M.M. Moore, and S. Nesnow, Eds., Plenum Press, New York, 103–111.

Claxton, L.D., V.S. Houk, R. Williams, and F. Kremer (1991), Effect of bioremediation on the mutagenicity of oil spilled in Prince William Sound, Alaska, *Chemosphere,* 23(5), 643–650.

Claxton, L.D., J. Creason, S. Bagley, D.W. Bryant, Y.A. Courtois, G. Douglas, C.B. Clare, S. Goto, P. Quillardet, D. Jagannath, K. Kataaoka, G. Mohn, P. Nielson, T. Ong, T. Pederson, H. Shimizu, L. Nylund, H. Tokiwa, G.J. Vink, Y. Wang, and D. Warshawsky (1992a), Results of the IPCS: a collaborative study on complex mixtures, *Mutat. Res.,* 276, 23–32.

Claxton, L.D., G. Douglas, D. Krewski, J. Lewtas, H. Matsushita, and H.S. Rosenkranz (1992b), Overview, conclusions, and recommendations of the IPCS collaborative study on complex mixtures, *Mutat. Res.,* 276, 33–60.

Cochran, S.T. and A. Norman (1982), Cytogenetic effects of contrast material, *Invest. Radiol.,* 17, 178–182.

Cole, J., M.H.L. Green, S.E. James, L. Henderson, and H. Cole (1988), A further assessment of factors influencing measurements of thioguanine-resistant mutant frequency in circulating t-lymphocytes, *Mutat. Res.,* 204, 493–507.

Cole, J., A.P.W. Waugh, D.M. Beare, M.S. Trepat, G. Stephens, and M.H.L. Green (1991), HPRT mutant frequencies in circulating lymphocytes: population studies using normal donors, exposed groups and cancer-prone syndromes, in *New Horizons in Biological Dosimetry,* B.L. Gledhill and F. Mauro, Eds., Wiley-Liss, New York, 319–328.

Constantin, M.J., K. Lowe, T.K. Rao, F.W. Larimer, and J.L. Epler (1980), The detection of potential genetic hazards in complex mutagenesis assays, in *Short-Term Bioassays in the Analysis of Complex Environmental Mixtures II,* M.D. Waters, S.S. Sandhu, J.L. Huisingh, L.D. Claxton and S. Nesnow, Eds., Environmental Science Research, Vol. 22, Plenum Press, New York, 253–266.

Constantin, M.J. and R.A. Nilan (1982a), The chlorophyll-deficient mutant assay in barley (*Hordeum vulgare*), a report of the U.S. EPA's GENE-TOX program, *Mutat. Res.,* 99, 37–49.

Constantin, M.J. and R.A. Nilan (1982b), Chromosome aberration assays in barley (Hordeum vulgare), a report of the U.S. EPA's GENE-TOX program, *Mutat. Res.,* 99, 13–36.

Couch, J.A. and J.C. Harshbarger (1985), Effects of carcinogenic agents on aquatic animals: an environmental and experimental overview, *Environ. Carcin. Rev.,* 3, 63–105.

Czeizel A., I. Szentesi, and G. Molnar (1984), Lack of effect of self-poisoning on subsequent reproductive outcome, *Mutat. Res.,* 127, 175–182.

De Ferari, M., M. Artuso, S. Bonassi, S. Bonatti, Z. Cavalieri, D. Pescatore, E. Marchini, V. Pisano, and Abbondandolo A. (1991), Cytogenetic biomonitoring of an Italian population exposed to pesticides: chromosome aberration and sister-chromatid exchange analysis in peripheral blood lymphocytes, *Mutat. Res.,* 260, 105–113.

DeMarini, D.M., A.M. Richard, M.D. Shelby, and M.D. Waters (1994), Comparative risk assessment associated with exposure of human and nonhuman biota to genotoxic agents and ionizing radiation I. Hazard identification, in *Methods for Genetic Risk Assessment,* D.J. Brosick, Ed., CRC Press, Lewis Pub., FL, Chap. 1.

Dempsey, J.L., R.S. Seshadri, and A.A. Morley (1985), Increased mutation frequency following treatment with cancer chemotherapy, *Cancer Res.,* 45, 2873–2877.

Dixon, D.R. (1982), Aneuploidy in mussel embryos (*Mytilus edulis* L.) originating from a polluted dock, *Mar. Biol. Lett.,* 3, 155.

Dixon, D.R. and K.R. Clarke (1982), Sister chromatid exchange: A sensitive method for detecting damage caused by exposure to environmental mutagens in the chromosomes of adult *Mytilus edulis, Mar. Biol. Lett.,* 3, 163–172.

Dixon, D.R., I.M. Jones, and F.L. Harrison (1985), Cytogenetic evidence of inducible processes linked with metabolism of a xenobiotic chemical in adult and larval *Mytilus edulis, Sci. Total Environ.,* 46, 1–8.

Donnelly, K.C., K.W. Brown, and B.R. Scott (1983), The use of short-term bioassays to monitor the environmental impact of land treatment of hazardous wastes, in *Short-Term Bioassays in the Analysis of the Complex Environmental Mixtures III,* M.D. Waters, S.S. Sandhu, J. Lewtas, L.D. Claxton, N. Chernoff, and S. Nesnow, Eds., Environmental Science Research, Vol. 27, Plenum Press, New York, 59–78.

Dubinin, N.P., Y.P. Altukhov, E.A. Salmenkova, A.N. Milishnikov, and T.A. Novikova (1975), Analysis of monomorphic markers of genes in populations as a method of evaluating the mutagenicity of the environment, (transl.), *Dokl. Biol.Sci.,* 225, 527–530.

Egami, N. and Y. Hyodo Tagachi (1973), Dominant lethal mutation rates in the fish, *Oryzias latipes,* irradiated at various stage of gametogenesis, in *Genetics and Mutagenesis of Fish,* J.H. Schroeder, Ed., Springer-Verlag, New York, 75–81.

Einistö, P., T. Nohmi, M. Watanabe, and M. Ishidate Jr. (1990), Sensitivity of *Salmonella typhimurium* YG1024 to urine mutagenicity caused by cigarette smoking, *Mutat. Res.,* 245, 87–92.

Elespuru, R.K. (1984), Induction of bacteriophage lambda by DNA interacting chemicals in *Chemical Mutagens: Principles and Methods for their Detection, Vol. 9,* F.J. deSerres, Ed., Plenum Press, New York, 213–231.

Elespuru, R.K. (1987), Inducible responses to DNA damage in bacteria and mammalian cells, *Environ. Mol. Mutagen.,* 10, 97–116.

Epler, J.L., B.R. Clar, C.-H. Ho, M.R. Guerin, and T.K. Rao (1979), Short-term bioassay of complex organic mixtures: II. Mutagenicity testing, in *Application of Short-Term Bioassays in the Fractionation and Analysis of Complex Environmental Mixtures,* M.D. Waters, S. Nesnow, J.L. Huisingh, S.S. Sandhu, and L.D. Claxton, Eds., Environmental Science Research, Vol. 15, Plenum Press, New York, 269–290.

Esancy, J.F., H.S. Freeman, and L.D. Claxton (1990a), The effect of alkoxy substitutents on the mutagenicity of some phenylenediamine-based diazo dyes, *Mutat. Res.,* 238, 23–38.

Esancy, J.F., H.S. Freeman, and L.D. Claxton (1990b), The effect of alkoxy substitutents on the mutagenicity of some aminoazobenzene dyes and their reductive-cleavage products, *Mutat. Res.,* 238, 1–22.

Eskenazi, B., A.J. Wyrobek, L. Fenster, D.F. Katz, M. Sadler, J. Lee, M. Hudes, and D.M. Rempel (1991), DMA study of the effect of perchloroethylene exposure on semen quality in dry cleaning workers, *Am. J. Ind. Med.,* 20, 575–591.

Evans, H.J. (1976), Cytological methods for detecting chemical mutagens, in *Chemical Mutagens: Principles and Methods for Their Detection,* Vol. 4, A. Hollaender, Ed., Plenum Press, New York, 1–30.

Evans, H.J. and D.C. Lloyd, Eds. (1978), *Mutagen-Induced Chromosome Damage in Man,* University Press, Edinburgh, Scotland.

Evans, H.J., K.E. Buckton, G.E. Hamilton, and A. Carothers (1979), A radiation-induced chromosome aberrations in nuclear-dockyard workers, *Nature (London),* 277, 531–534.

Everson, R.B., E. Randerath, R.M. Santella, R.C. Cefalo, T.A. Avitts, and K. Randerath (1986), Detection of smoking-related covalent DNA adducts in human placenta, *Science,* 231, 54–57.

Everson, R.B., C.M. Wehr, G.L. Erexson, and J.T. MacGregor (1988a), Association of marginal folate depletion with increased human chromosomal damage *in vivo:* demonstration by analysis of micronucleated erythrocytes, *J. Natl. Cancer Inst.,* 80, 525–529.

Everson, R.B., E. Randerath, R.M. Santella, T.A. Avitts, I.B. Weinstein, and K. Randerath (1988b), Quantitative associations between DNA damage in human placenta and maternal smoking and birth weight, *J. Natl. Cancer Inst.,* 80(8), 567–576.

Fenech, M. and A.A. Morley (1986), Cytokinesis-block micronucleus method in human lymphocytes: effect of *in vivo* aging and low dose X-irradiation, *Mutat. Res.,* 161, 193–198.

Fichtinger-Schepman, A.M.J., P.H.M. Lohman, and J. Reediji (1982), Detection and quantification of adducts formed upon interaction of diamminedichloroplatinum (II) with DNA, by anion-exchange chromatography after enzymatic degradation, *Nucleic Acids Res.,* 10, 5345–5356.

Fichtinger-Schepman, A.M.J., J.L. van der Veer, P.H.M. Lohman, and J. Reediji (1984), A simple method for the inactivation of monofunctionally DNA-bound *cis*-diamminedichloroplatinum(ll), *J. Inorg. Biochem.,* 21, 103–112.

Fichtinger-Schepman, A.M.J., J.L. van der Veer, J.H.J. den Hartog, P.H.M. Lohman, and J. Reediji (1985), Adducts of the antitumor drug *cis*-diamminedichloroplatinum(ll) with DNA: formation, identification, and quantitation, *Biochemistry,* 24, 707–713.

Fichtinger-Schepman, A.M.J., A. T. van Oosterom, P.H.M. Lohman, and F. Berends (1987a), *cis*-Diamminedichloroplatinum (II)-induced DNA adducts in peripheral leukocytes from seven cancer patients: quantitative immunochemical detection of the adduct induction and removal after a single dose of *cis*-diamminedichloroplatinum (II), *Cancer Res.,* 47, 3000–3004.

Fichtinger-Schepman, A.M.J., A.T. van Oosterom, P.H.M. Lohman, and F. Berends (1987b), Interindividual human variation in cisplatinum sensitivity, predictable in an *in vitro* assay?, *Mutat. Res.,* 190, 59–62.

Floyd, R.A., M.S. West, W.E. Hogsett, and D.T. Tingey (1989), Increased 8-hydroxyguanine content of chloroplast DNA from ozone-treated plants, *Plant Physiol.,* 91, 644–647.

Ford, C.E. and J.L. Hamerton (1956), A colchicine hypotonic citrate squash sequence for mammalian chromosomes, *Stain Technol.,* 31, 247.

Freeman, H.S., J.F. Esancy, and L.D. Claxton (1990), An approach to the design of nonmutagenic azo dyes: analogs of the mutagen CI direct black 17, *Dyes Pigm.,* 13, 55–70.

Gentile, J.M., P. Johnson, and S. Robbins (1990), Activation of aflatoxin B1 and benzo[a]pyrene by tobacco cells in the plant cell/microbe coincubation assay, in *Plants for Toxicity Assessment,* Vol. 2, J. Gorsuch, W. Lower, M. Lewis, and W. Wang, Eds., ASTM 1115, American Society for Testing and Materials, Philadelphia, 318–325.

Gibson, T.L, P.E. Korsog, and G.T. Wolff (1985), Evidence for the transformation of polycyclic organic matter in the atmosphere, *Atmos. Env.,* 20, 1575–1578.

Gilbert, R.O. (1987), *Statistical Methods for Environmental Pollution Monitoring,* Van Nostrand Reinhold, New York, 320.

Gledhill, B.L. and F. Mauro (1991), *New Horizons in Biological Dosimetry,* Wiley-Liss, New York, 619.

Glickman, L.T., L.M. Domanski, T.G. Maquire, R.R. Dubielzig, and A. Churg (1983), Mesothelioma in pet dogs associated with exposure of their owners to asbestos, *Environ. Res.,* 32, 305–313.

Graedel, T.E., D.T. Hawkin, and L.D. Claxton (1986), *Atmospheric Chemical Compounds: Sources, Occurrence, and Bioassay,* Academic Press, Orlando, Florida, 732.

Grant, W.F. and K.D. Zura (1982), Plants are sensitive *in situ* detectors of atmospheric mutagens, in *Mutagenicity: New Horizons in Genetic Toxicology,* J.A. Heddle, Ed., Academic Press, New York, 407–434.

Green, R.H. (1984), Some guidelines for the design of biological monitoring programs in the marine environment, in *Concepts in Marine Pollution Measurements,* H.H. White, Ed., University of Maryland, College Park, MD, 647–655.

Gupta, R.C., M.V. Reddy, and K. Randerathy (1982), ^{32}P-Postlabeling analysis of nonradioactive aromatic carcinogen-DNA adducts, *Carcinogenesis,* 3, 1081–1092.

Han, T.L., G.C. Webb, S.P. Flaherty, A. Correll, C.D. Matthews, and J.H. Ford (1992), Detection of chromosome 17 and x-bearing human spermatozoa using fluorescence *in situ* hybridization, *Mol. Reprod. Dev.,* 33, 189–194.

Harris, C.C., K. Vahäkängas, M.J. Newman, G.E. Triers, A. Shamsuddin, N. Sinopoli, D.L. Mann, and W.E. Wright (1985), Detection of benzo(a)pyrene diol epoxide DNA adducts in peripheral blood lymphocytes and antibodies to the adducts in serum from coke oven workers, *Proc. Natl. Acad. Sci. U.S.A.,* 82, 6672–6676.

Harrison, F.L. and I.M. Jones (1982), An *in vivo* sister chromatid exchange assay BXB in the larvae of the mussel *Mytilus edulis:* response to 3 mutagens, *Mutat. Res.,* 105, 235–242.

Harrison, F.L., D.W. Rice Jr., and D.H. Moore (1984), Induction of Chromosomal Aberrations and Sister Chromatid Exchanges in the Benthic Worm, *Neanthes arenaceodentata* Exposed to Ionizing Radiation, UCRL 53524, Lawrence Livermore National Laboratory, Livermore, California, 57.

Harrison, F., D.W. Rice Jr., D.H. Moore, and M. Varela (1987), Effects of radiation on frequency of chromosomal aberrations and sister chromatid exchange in *Neathes arenaceodentata,* in *Oceanic Processes in Marine Pollution, Vol. I, Biological Processes and Wastes in the Ocean,* J.M. Capuzzo and D.R. Kester, Eds., Robert E. Krieger, Malabar, FL, chap. 14.

Haugen, A., G. Becher, C. Benestad, K. VaHäkängas, G.E. Trivers, M.J. Newman, and C.C. Harris (1986), Determination of polycyclic aromatic hydrocarbons in the urine, benzo(a)pyrene diol-epoxide DNA adducts in lymphocyte DNA, and antibodies to the adducts in sera from coke oven workers exposed to measured amounts of polycyclic aromatic hydrocarbons in the work atmosphere, *Cancer Res.*, 46, 4178–4183.

Hayes, H.M., Jr., R. Hoover, and R.E. Tarone (1981), Bladder cancer in pet dogs: A sentinel for environmental cancer? *Am. J. Epidemiol.*, 114, 229–233.

Hemminki, K., E. Grzybrowska, M. Chorazy, K. Twardowska-Saaucha, J.W. Sroczynski, K.L. Putman, K. Randerath, D.H. Phillips, A. Hewer, R.M. Santella, T.L. Young, and F.P. Perera (1990), DNA adducts in humans environmentally exposed to aromatic compounds in an industrial area of Poland, *Carcinogenesis*, 11, 1229–1231.

Hemsworth, B.N and A.A. Wardhaugh (1978), The induction of dominant lethal mutations in *Tilapia mossambica* by alkane sulfonic esters, *Mutat. Res.*, 58, 263–268.

Higashi, K., K. Ikeuchi, Y. Karasaki, and M. Obara (1983), Isolation of immunochemically distinct forms of cytochrome from microsomes of tulip bulbs, *Biochem. Biophysics. Res. Commun.*, 115, 46–52.

Higashi, K., K. Ikeuchi, M. Obara, Y. Karasaki, H. Hirano, S. Totoh, and Y. Koga (1985), Purification of a single form of microsomal cytochrome P-450 from tulip bulbs (*Tulipa Gesneriana* L.), *Agric. Biol. Chem.*, 49, 2399–2405.

Holzberg, S. and J.H. Schroeder (1972), Behavioral mutagenesis in the convict cichlid fish, *Cichlasoma nigrofasciatum* Guenther. I. The reduction of male aggressiveness in the first postirradiation generation, *Mutat. Res.*, 16, 289.

Honghua L., X. Cui, and N. Arnheim (1990), Analysis of DNA sequences in individual gametes: application to human genetic mapping, *Prog. Clin. Biol. Res.*, 340C, 207–211.

Hongslo, J.K., A. Brogger, C. Bjorge, and J.A. Holme (1991), Increased frequency of sister-chromatid exchange and chromatid breaks in lymphocytes after treatment of human volunteers with therapeutic doses of paracetomolm, *Mutat. Res.*, 261, 1–8.

Hooftman, R.N. and W.K. de Raat (1982), Induction of nuclear anomalies (micronuclei) in the peripheral blood erythrocytes of the eastern mudminnow *Umbra pygmaea* by ethyl methanesulphonate, *Mutat. Res.*, 104, 147–152.

Hopke, P.K., N.S. Plewa, J.B. Johnston, D. Weaver, S.G. Wood, B.A. Larson, and T. Hinesly (1982), Multitechnique screening of Chicago municipal sewage sludge for mutagenic activity. *Environ. Sci. Technol.*, 16, 140–147.

Hose, J.E. (1985), Potential uses of sea urchin embryos for identifying toxic chemicals: description of a bioassay incorporating cytologic, cytogenetic and embryologic endpoints, *J. Appl. Toxicol.*, 5(4), 245–254.

Hose, J.E. and S.S. Smith (1986), Hematological abnormalities in fish from highly contaminated areas of southern California. Abstract in Society of Environmental Toxicology and Chemistry, Seventh Annual Meeting, 66.

Hose, J.E. J.N. Cross, S.G. Smith, and D. Diehl (1987), Elevated circulating erythrocyte micronuclei in fishes from contaminated sites off southern California, *Mar. Environ. Res.*, 22, 167–176.

Houk, V.S. (1992), The genotoxicity of industrial wastes and effluents: a review, *Mutat. Res.*, 277, 91–138.

Houk, V.S. and D.M. DeMarini (1987), Induction of prophage lambda by chlorinated pesticides, *Mutat. Res.,* 182, 193–201.

Houk, V.S. and D.M. DeMarini (1988), Use of the Microscreen phage-induction assay to assess the genotoxicity of 14 hazardous industrial wastes, *Environ. Mol. Mutagen.,* 11, 13–29.

Houk, V.S., S. Schalkowsky, and L.D. Claxton (1989), Development and validation of the spiral Salmonella assay: an automated approach to bacterial mutagenicity testing, *Mutat. Res.,* 223, 49–64.

Houk, V.S., L.D. Claxton, and G. Early (1991), Use of the spiral Salmonella assay to detect the mutagenicity of complex environmental mixtures, *Environ. Mol. Mutagen.,* 17, 112–121.

Hsu, T.C. (1952), Mammalian chromosomes *in vitro.* I. The karyotype of man., *J. Hered.,* 43, 167–172.

Ireland, F., W.R. Lower, O. Sehgal, and B. Judy (1990), DNA adducts in plants as a bioassay. Proc., 43rd Southern Weed Science Society, 366–372.

Jensen, R.H., W.L. Bigbee, and R.G. Langlois (1990), Multiple endpoints for somatic mutations in humans provide complementary views for biodosimetry, genotoxicity, and health risks, *Prog. Clin. Biol. Res.,* 340C, 81–92.

Jensen, R.H., W.L. Bigbee, R.G. Langlois, S.G., Grant, P.G, Pleshanov, A.A. Chirkov, and M.A. Pilinskaya, M.A. (1990), Laser-based flow cytometric analysis of genotoxicity of humans exposed to ionizing radiation during the Chernobyl accident, in (Proceedings), Laser Applications in Life Sciences, *Proc. Soc. Photo-Optical Inst. Eng.,* 1403, 372–380.

Jungers, R., R. Burton, L. Claxton, and J. Lewtas Huisingh (1981), Evaluation of collection and extraction methods for mutagenesis studies on ambient air particulate, in *Short-Term Bioassays in the Analysis of Complex Environmental Mixtures II,* M.D. Waters, S.S. Sandhu, J. Lewtas Huisingh, L. Claxton, and S. Nesnow, Eds., Plenum Press, New York, 45–65.

Kada, T. and S. Watanabe (1983), *Bacillus subtilis* rec-assay with and without activation: improvements and applications, in *In Vitro Toxicity Testing of Environmental Agents: Current and Future Possibilities,* Kolber, A.R., T.K. Wong, L.D. Grant, R.S. DeWoskin, and T.J. Hughes, Eds., Plenum Press, New York, 41–60.

Kaderlik, K.R., G. Talaska, D. Gayle DeBord, A.M. Osorio, and F.F. Kadlubar, (1993), 4,4′-Methylene-bis(2-chloroaniline)-DNA adduct analysis in human exfoliated urothelial cells by ^{32}P-postlabeling, *Cancer Epidemiol. Biomarker Prev.,* 2, 63–69.

Kado, N.Y., G.N. Guirguis, C.P. Flessel, R.C. Chan, K. Chang, and J.J. Wesolowski (1986), Mutagenicity of fine (.5 µm) airborne particles: diurinal variation in community air determined by a salmonella micro pre-incubation (microsuspension) procedure, *Environ. Mutagen.,* 8, 53–66.

Kado, N.Y., S.J. Tesluk, S.K. Hammond, S.R. Woskie, S.J. Samuels, and M.B. Schenker (1987), Use of a Salmonella micro preincubation procedure for studying personal exposure to mutagens in environmental tobacco smoke: pilot study of urine and airborne mutagenicity from passive smoking, in *Short-Term Bioassays in the Analysis of Complex Environmental Mixtures V,* S.S. Sandhu, D.M. DeMarini, M.J. Moss, M.M. Moore, and J.L. Mumford, Eds., Environmental Science Research, 36, Plenum Press, New York, 375–390.

Kallioniemi, A., O.-P. Kallioniemi, F. Waldman, L.-C. Chen, L.-C. Yu, Y. Fung, H. Smith, D. Pinkel, and J.W. Gray (1992a), Detection of retinoblastoma gene deletions in metaphase chromosomes and interphase nuclei by fluorescence in situ hybridization, *Cytogenet. Cell Genet.*, 60, 190–193.

Kallioniemi, A., O.-P. Kallioniemi, D. Sudar, D. Rutovitz, J. Gray, F. Waldman, and D. Pinkel, (1992b), Comparative genomic hybridization for molecular cytogenetic analysis of solid tumors, *Science*, 258, 818–821.

Kamens, R.M., D. Bell, A. Dietrich, J. Perry, L. Claxton, and S. Tejada (1984), Mutagenic transformation of dilute wood smoke systems in the presence of O_3 and NO_2. Analysis of selected high-pressure liquid chromatography fractions from wood smoke particle extracts, *Environ. Sci. Technol.*, 19, 63–69.

Kanter, P.M. and H.S. Schwartz (1982), A fluorescence enhancement assay for cellular DNA damage, *Mol. Pharmacol.*, 22, 145–151.

Kligerman, A.D. (1980), The use of aquatic organisms to detect mutagens that cause cytogenetic damage, in *Radiation Effects on Aquatic Organisms*, N. Egami, Ed., Japan Science Society Press, Tokyo, 241–252.

Kligerman, A.D. (1982), Fishes as biological detectors of the effects of genotoxic agents, in *Mutagenicity: New Horizons in Genetic Toxicology*, Academic Press, New York, 435.

Kligerman, A.D. and S.E. Bloom (1975), A cytogenetics model for the study of chromosome aberrations in fishes, *Mutat. Res.*, 31, 334–335.

Kligerman, A.D. and S.E. Bloom (1976), Sister-chromatid differentiation and exchanges in adult mudminnows *(Umbra limi)* after *in vivo* exposure to 5-bromodeoxyuridine, *Chromosoma*, 56, 101–109.

Kligerman, A.D., S.E. Bloom, and W.M. Howell (1975), *Umbra limi:* A model for the study of chromosome aberrations in fishes, *Mutat. Res.*, 31, 225–233.

Kocan, R.M., K.M. Sabo, and M.L. Landolt (1985), Cytotoxicity/genotoxicity: the application of cell culture techniques to the measurement of marine sediment pollution, *Aquat. Toxicol.*, 6, 165.

Kronberg, L. (1990), Characterization of mutagenic compounds formed during disinfection of drinking water, in *Genetic Toxicology of Complex Mixtures*, M.D. Waters, F.B. Daniel, J. Lewtas, M.M. Moore, and S. Nesnow, Eds., Plenum Press, New York, 173–183.

Kucerova, M. (1982), Chromosomal aberrations induced in occupationally exposed persons, in *Mutagenicity: New Horizons in Genetic Toxicology*, J.A. Heddle, Ed., Academic Press, New York, 242–266.

Kushiro, J., Y. Hirai, Y. Kusunoki, S. Kyoizumi, Y. Kodama, A. Wakisaka, A. Jeffreys, J.B. Cologne, K. Dohi, and N. Nakamura, et al. (1992), Development of a flow-cytometric HLA-A locus mutation assay for human peripheral blood lymphocytes, *Mutat. Res.*, 272(1), 17–29.

Kyoizumi, S., S. Umeki, M. Akiyama, Y. Hirai, Y. Kusunoki, N. Nakamura, K. Endoh, J. Konishi, M.S. Sasaki, and T. Mori (1992), Frequency of mutant T lymphocytes defective in the expression of the T-cell antigen receptor gene among radiation-exposed people, *Mutat. Res.*, 265(2), 173–180.

Landolt, M.L. and R.M. Kocan (1987), The sea surface microlayer: A complex mixture which causes genotoxic damage to fish cells and embryos, in *Short Term Bioassays in the Analysis of Complex Environmental Mixtures, V.* S.S. Sandhu, D.M. DeMarini, M.J. Mass, M.M. Moore, and J.L. Mumford, Eds., Plenum Press, New York. 225–236.

Langlois, R.G., B.A. Nisbet, W.L. Bigbee, D.N. Ridinger, and R.H. Jensen (1990), An improved flow cytometric assay for somatic mutations at the glycophorin A locus in humans, *Cytometry,* 11, 513–521.

Langlois, R.G., W.L. Bigbee, R.H. Jensen, and J. German (1989), Evidence for elevated *in vivo* mutation and somatic recombination in Bloom's syndrome, *Proc. Natl. Acad. Sci., U.S.A.,* 86, 670–674.

Latt, S.A., J. Allen, S.E. Bloom, A. Carrano, E. Falke, D. Kram, E. Schneider, R. Schreck, R. Tice, B. Whitfield, and S. Wolff (1981), Sister chromatid exchange: a report of the Gene Tox Program, *Mutat. Res.,* 87, 17.

Latt, S. (1974), Sister chromatid exchanges, indices of human chromosome damage and repair: detection by fluorescence and induction by mitomycin C, *Proc. Natl. Acad. Sci. U.S.A.,* (Wash.), 71, 3162–3166.

Lave, L.B. and G.S. Omenn (1986), Cost effectiveness of short term tests for carcinogenicity, *Nature (London),* 324, 29–34.

Lewis, C. W., R. E. Baumgardner, R. K. Stevens, L. D. Claxton, and J. Lewtas (1988), Contribution of woodsmoke and motor vehicle emissions to ambient aerosol mutagenicity, *Environ. Sci. Technol.,* 22, 968–971.

Lewtas, J., Ed. (1982), *Toxicological Effects of Emissions from Diesel Engines,* Elsevier, New York, 380.

Lewtas, J. (1983), Evaluation of the mutagenicity and carcinogenicity of motor vehicle emissions in short-term bioassays, *Environ. Health Perspect.,* 47, 141–152.

Lewtas, J. (1985), Combustion Emissions: Characterization and Comparison of Their Mutagenic and Carcinogenic Activity, in *Carcinogens and Mutagens in the Environment, Volume V, the Workplace: Sources of Carcinogens,* H.F. Stich, Ed., CRC Press, Boca Raton, FL, 59–74.

Lewtas, J. (1988), Genotoxicity of complex mixtures: strategies for the identification and comparative assessment of airborne mutagens and carcinogens from combustion sources, *Fundam. Appl. Toxicol.,* 10, 571–589.

Lewtas, J. (1990), Experimental evidence for the carcinogenicity of air pollutants, in *Air Pollution and Human Cancer,* L. Tomatis, Ed., Springer-Verlag, Berlin, 49–61.

Lewtas, J. (1991), Environmental monitoring using genetic bioassays, in: *Genetic Toxicology: A Treatise,* A.P. Li and R.H. Heflich, Eds., Telford Press, Caldwell, N.J., 357–372.

Lewtas, J., A. Austin, L. Claxton, R. Burton, and R. Jungers (1982), The relative contribution of PNAs to the microbial mutagenicity of respirable particles from urban air, in *Polynuclear Aromatic Hydrocarbons: Physical and Biological Chemistry,* M. Cooke, A.J. Dennis, and G.L. Fisher, Eds., Battelle Press, Columbus, OH, 603–613.

Lewtas, J., S. Goto, K. Williams, J.C. Chuang, B.A. Petersen, and N.K. Wilson (1987), The mutagenicity of indoor air particles in a residential pilot field study: application and evaluation of new methodologies, *Atmos. Environ.,* 21, 443–449.

Lewtas, J, L.D. Claxton, J. Mumford, and G. Lofroth (1993), Bioassay of complex mixtures of indoor air pollutants, in *Environmental Carcinogens: Methods of Analysis and Exposure Measurement, Indoor Air,* B. Seifert and I. O'Neill, Eds., Vol. 12, International Agency for Research on Cancer, Lyon, France, (109) 85–95.

Li, A.P., A.L. Brooks, C.R. Clark, R.W. Shimizu, R.L. Hanson, and J.S. Dutcher (1983), Mutagenicity testing of complex environmental mixtures with Chinese Hamster Ovary Cells, in *Short-Term Bioassays in the Analysis of Complex Environmental Mixtures III,* M.D. Waters, S.S. Sandhu, J. Lewtas, L. Claxton, N. Chernoff, and S. Nesnow, Eds., Plenum Press, New York, 183–196.

Liquori, V.M. and M.L. Landolt (1985), Anaphase aberrations: An *in vivo* measure of genotoxicity, in *Short Term Genetic Bioassays in the Analysis of Complex Environmental Mixtures IV.* M.D. Waters, S.S. Sandhu, J. Lewtas, L. Claxton, G. Strauss, and S. Nesnow, Eds., Plenum Press, New York, 87–98.

Littlefield, L.G., R.A. Kleinerman, A.M. Sayer, R. Tarone, and J.D. Boice, Jr. (1991), Chromosome aberrations in lymphocytes-biomonitors of radiation exposure, *Prog. Clin. Biol. Res.,* 372, 387–397.

Livingston, G.K., R.N. Reed, B.L. Olson, and J.E. Lockey (1990), Induction of nuclear aberrations by smokeless tobacco in epithelial cells of human oral mucosa, *Environ. Mol. Mutagen.,* 15(3), 136–144.

Lloyd, D.C., R.J. Purrott, and E.J. Reeders (1980), The incidence of unstable chromosome aberrations in peripheral blood lymphocytes from unirradiated and occupationally exposed people, *Mutat. Res.,* 72, 523–532.

Lohman, P.H.M., R. Laauwerys, and M. Sorsa (1984), Methods of monitoring human exposure to carcinogenic and mutagenic agents, in *Monitoring human exposure to carcinogenic and mutagenic agents,* A. Berlin, M. Draper, K. Hemminki, and H. Vainio, Eds., International Agency for Research of Cancer, Scientific Publications, Lyon, France, 59, 423–427.

Lohman, P.H.M., R.A. Baan, A.M.J. Fichtinger-Shepman, M.A.A. Muysken-Schoen, M.J. Lansbergen, and F. Berends (1985), Molecular dosimetry of genotoxic damage: biochemical and immunochemical methods to detect DNA-damage *in vitro* and *in vivo,* TIPS-FEST supplement, Elsevier, New York, 1–7.

Longwell, A.C. (1976), Chromosome mutagenesis in developing mackerel eggs sampled from the New York Bight. NOAA Tech. Memo. ERL MESA 7, U.S. Department of Commerce, Washington, D.C., 61.

Longwell, A.C. and J.B. Hughes (1980), Cytologic, cytogenetic, and developmental state of Atlantic mackerel eggs from sea surface waters of the New York Bight, and prospects for biological effects monitoring with ichthyoplankton, *Rapp. P.V. Reun. Cons. Int. Explor. Mer.,* 179, 275.

Longwell, A.C. and J.B. Hughes (1981), Cytologic, cytogenetic, and embryologic state of Atlantic mackerel eggs from surface waters of the New York Bight in relation to pollution, *Rapp. P.V. Reun. Cons. Int. Explor. Mer.,* 178, 76.

Loper, J.C. (1980), Mutagenic effects of organic chemicals in drinking water, *Mutat. Res.,* 76, 241–268.

Loper, J.C. and M.W. Tabor (1983), Isolation of mutagens from drinking water: something old, something new, in *Short-Term Bioassays in the Analysis of Complex Environmental Mixtures III,* M.D. Waters, S.S. Sandhu, J. Lewtas, L.D. Claxton, N. Chernoff, and S. Nesnow, Eds., Environmental Science Research, Vol. 27, Plenum Press, New York, 165–181.

Lower, W.R., V.K. Drobney, B.J. Aholt, and R. Politte (1983a), Mutagenicity of the environment in the vicinity of an oil refinery and petrochemical complex, *Teratogen. Carcinogen. Mutagen.,* 3, 65–73.

Lower, W.R., W.A. Thompson, V.R. Drobney, and A.F. Yanders (1983b), Mutagenicity in the vicinity of a lead smelter, *Teratogen. Carcinog. Mutagen.*, 3, 231–253.

Lower, W.R., D.D. Hemphill, K. Roberts, T.E. Clevenger, A.G. Underbrink, and A.F. Yanders (1984), New methodology for assessing mutagenicity of water and water-related sediments, in Proc. 2nd Int. Conf. Groundwater Quality Research, Tulsa, OK, 194–196.

Lower, W.R. and R.J. Kendall (1990), Sentinel species and sentinel bioassay, in *Biological Markers of Environmental Contamination*, J. McCarthy and L. Shugart, Eds., Lewis Publishers, Boca Raton, FL, 309–331.

Lower, W.R., S.S. Sandhu, and M.W. Thomas (1990), Utility of *in situ* assays for detecting environmental pollutants, in *Waste Testing and Quality Assurance*, D. Freedman, Ed., ASTM STP 1062, American Society for Testing and Materials, Philadelphia, 163–175.

Lower, W.R., F.A. Ireland, and B.M. Judy (1991), ^{32}P-postlabeling for DNA adduct determination in plants, in *Plants for Toxicity Assessment*, J. Gorsuch, W. Lower, M. Lewis, and W. Wang, Eds., ASTM STP 1115, American Society for Testing and Materials, Philadelphia, 297–308.

Lucas, J.N., T. Tenjin, T. Straume, D. Pinkel, D. Moore, II, M. Litt, and J.W. Gray (1989), Rapid human chromosome aberration analysis using fluorescence *in situ* hybridization, *Int. J. Radiat. Biol.*, 56, 35–44.

Lucas, J., A. Awa, M. Poggensee, Y. Kodama, M. Nakano, K. Ohtaki, T. Straume, U. Weier, D. Pinkel, and J. Gray (1992), Rapid frequency analysis in humans decades after exposure to ionizing radiation, *Int. J. Radiat. Biol.*, 62, 53–63.

Ma, T.-H. (1979), Micronuclei induced by X-ray and chemical mutagens in meiotic pollen mother cells of Tradescantia — a promising mutagen test system, *Mutat. Res.*, 64, 307–313.

Ma, T.-H. (1982), Vicia cytogenetic tests for environmental mutagens. A report of the U.S. EPA's GENE-TOX program, *Mutat. Res.*, 99, 257–271.

Ma, T.-H., V.A. Anderson, and S.S. Sandhu (1981), A preliminary study of clastogenic effects of diesel exhaust fumes using Tradescantia-Micronucleus bioassay, in *Short-Term Bioassays in the Analysis of Complex Environmental Mixtures II*, M.D. Waters, S.S. Sandhu, J.L. Huisingh, L. Claxton, and S. Nesnow, Eds., Plenum Press, New York, 351–358.

Ma, T.-H, W.R. Lower, F.D. Harris, J. Poku, V.A. Anderson, M.M. Harris, and J.L. Bare (1983), Evaluation by the Tradescantia-Micronucleus test of the mutagenicity of internal combustion engine exhaust fumes from diesel and diesel-soybean oil mixed fuels, in *Short-Term Bioassays in the Analyses of Complex Environmental Mixtures*, M.D. Waters, S.S. Sandhu, J. Lewtas, L. Claxton, N. Chernoff, and S. Nesnow, Eds., Plenum Press, New York, 89–99.

Ma, T.-H. and M.M. Harris (1985), *In situ* monitoring of environmental mutagens, *Hazard Assess. Chem.*, 4, 77–105.

MacGregor, J.T., C.M. Wehr, and D.H. Gould (1980), Clastogen-induced micronuclei in peripheral blood erythrocytes: The basis of an improved micronucleus test, *Environ. Mutagen,* 2(4), 509–514.

Maddock, M.B. and J.J. Kelly (1980), A sister chromatid exchange assay for detecting genetic damage to marine fish exposed to mutagens and carcinogens, in *Water Chlorination: Environmental Impact and Health Effects,* R.L. Jolley, W.A. Brungs, R.B. Cummings, and V.A. Jacobs, Eds., Vol. 3, Ann Arbor Science Publishers, Ann Arbor, Michigan.

Maron, D. and B.N. Ames (1983), Revised methods for the Salmonella mutagenicity test, *Mutat. Res.,* 113, 173–212.

Martin, R.H., K. Hildebrand, J. Yamamoto, A.W. Rademaker, M. Barnes, G. Douglas, K. Arthur, T. Ringrose, and I.S. Brown (1986), An increased frequency of human sperm chromosomal abnormalities in normal men, *Am. J. Hum. Genet.,* 41, 484–492.

Mathews, J.G., J.B. Favor, and J.W. Crenshaw (1978), Dominant lethal effects of triethylenemelamine in the guppy *Poecilia reticulata, Mutat. Res.,* 54, 149–157.

Matsushita, H., O. Endo, S. Goto, H. Shimizu, H. Matsumoto, K. Tamakawa, T. Endo, Y. Sakabe, H. Tokiwa, and M. Ando (1992), Collaborative study using the preincubation *Salmonella typhimurium* mutation assay for airborne particulate matter in Japan. A trial to minimize interlaboratory variation, *Mutat. Res.,* 271, 1–12.

McBee, K. (1987), Chromosomal aberrations in native small mammals, *Peroymscus leucopus* and *Sigmodon hispidus,* at a petrochemical waste disposal site I. Standard Caryo. *Arch. Environ. Contam. Toxicol.,* 16, 681–688.

McCarron, M.A., A. Kutlaca, and A.A. Morley (1989), The HLA-A mutation assay: improved technique and normal results, *Mutat. Res.,* 225, 189–193.

McGeorge, L.J., J.B. Louis, T.B. Atherholt, and G.J. McGarrity (1985), Mutagenicity analyses of industrial effluents: results and considerations for integration into water pollution control programs, in *Short-Term Bioassays in the Analysis of Complex Environmental Mixtures IV,* M.D. Waters, S.S. Sandhu, J. Lewtas, L.D. Claxton, G. Strauss, and S. Nesnow, Eds., Environmental Science Research, Vol. 32, Plenum Press, New York, 247–268.

McKelvey, V.J., M. Butler, and L.H. Stewart (1992), Analysis of DNA content and integrity in cells extracted from bladder washing and voided urine specimens, in bladder cancer patients, using the comet assay, *Mutat. Res.,* 271, 163.

Meier, J. (1988), Genotoxic activity of organic chemicals in drinking water, *Mutat. Res.,* 196, 211–245.

Meier, J.R., A.B. DeAngelo, F.B. Daniel, K.M. Schenck, J.U. Doerger, et. al. (1990), Genotoxic and carcinogenic properties of chlorinated furanones: important byproducts of water chlorination, in *Genetic Toxicology of Complex Mixtures,* M.D. Waters, F.B. Daniel, J. Lewtas, M.M. Moore, and S. Nesnow, Eds., Plenum Press, New York, 185–195.

Metalli, P. and E. Ballardin (1962), First results on X ray induced genetic damage in *Artemia salina* Leach, *Atti. Assoc. Genet. Ital.,* 7, 219.

Metalli, P. and E. Ballardin (1972), Radiobiology of Artemia: Radiation effects and ploidy, *Curr. Top. Radiat. Res.,* 2, 7, 181.

Meyers, L.J., L.R. Shugart, and B.T. Walton (1988), Freshwater turtles as indicators of contaminated aquatic environments. Paper presented at 9th Annual Meet. Society of Environmental Toxicology and Chemistry, Arlington, VA, November 15, 1988.

Migliore, L., F. Di Marino, R. Scarpato, R. Barale, and G. Cognetti-Alfinito (1990), Detection of mutagenic/carcinogenic compounds in the marine environment, in *Advances in Applied Biotechnology Series. Vol. 5 Carcinogenic, Mutagenic, and Teratogenic Marine Pollutants: Impact on Human Health and the Environment,* Gulf Publishing, Houston, 111–120.

Mitchell, A.D., M.L. Casciano, D.E. Metz, R.H.C. Robinson, G.M. San, G.M. Williams, and E.S. Von Halle (1983), Unscheduled DNA synthesis tests: A report of the "Gene Tox" program, *Mutat. Res.,* 123, 363–410.

Mix, M.C. (1986), Cancerous diseases in aquatic animals and their association with environmental pollutants: a critical literature review, *Mar. Environ. Res.,* 20, 1–141.

Mohrenweiser H.W., R.D. Larsen, and J.V. Neel (1989), Development of molecular approaches to estimating germinal mutation rates. I. Insertion/deletion/rearrangement variants in the human genome, *Mutat. Res.,* 212, 241–252.

Mong, S.J. and T.M. Berra (1979), The effects of increasing dosages of X-irradiation on the chromosomes of the central mudminnow, *Umbra limi* (Kirkland) (Salmoniformes:Umbridae), *J. Fish. Biol.,* 14, 523–527.

Moorehead, P.S., et al. (1960), Chromosome preparations of leukocytes from human peripheral blood, *Exp. Cell Res.,* 20, 613–616.

Morley, A.A., S.A. Grist, D.R. Turner, A. Kutlaca, and G. Bennett (1990), Molecular nature of *in vivo* mutations in human cells at the autosomal HLA-A locus, *Cancer Res.,* 50, 4584–4587.

Moruzzi, J.F., A.J. Wyrobek, B.H. Mayall, and B.L. Gledhill (1988), Quantification and classification of human sperm morphology by computer assisted image analysis, *Fertil. Steril.,* 50, 142–152.

Mulvihill J.J. (1990), Sentinel and other mutational effects in offspring of cancer survivors, *Prog. Clin. Biol. Res.,* 340C, 179–186.

Mulvihill J.J. and A. Czeizel (1983), Perspectives in mutation epidemiology: a 1983 view of sentinel phenotypes, *Mutat. Res.,* 123, 345–361.

Mumford, J.L. , R.S. Chapman, S. Nesnow, C.T. Tucker Helmes, and X. Li (1990), Mutagenicity, Carcinogenicity, and Human Cancer Risk from Indoor exposure to coal and wood combustion in Xuan Wei, China, in *Genetic Toxicology of Complex Mixtures,* M.D. Waters, F.B. Daniel, J. Lewtas, M.M. Moore, and S. Nesnow, Eds., Plenum Press, New York, 157–165.

Mumm, R.O. and R.H. Hamilton (1979), *Xenobiotic Metabolism: In Vitro Methods,* D.D. Paulson, D.S. Frear, and E.P. Marks, Eds., ACS Symp. Ser. No. 97, American Chemical Society, Washington, DC, 35–76.

Mumm, R.O. and R.H. Hamilton (1983), Advances in pesticide metabolite identification through use of plant tissue, *J. Toxicol. Clin. Toxicol.,* 19, 535–555.

Nacci, D. and E. Jackim (1989), Using the DNA alkaline unwinding assay to detect DNA damage in laboratory and environmentally exposed cells and tissues, *Mar. Environ. Res.,* 28, 333–337.

Nair, U., G. Obe, J. Nair, G.B. Maru, S.V. Bhide, R. Pieper, and H. Bartsch (1991), Evaluation of frequency of micronucleated oral mucosa cells as a marker for genotoxic damage in chewers of betel quid with or without tobacco, *Mutat. Res.,* 261, 163–168.

Nakamura, N., S. Umeki, Y. Hirai, S. Kyoizumi, J. Kushiro, Y. Kusunoki, and M. Akiyama (1991), Evaluation of four somatic mutation assays for biological dosimetry of radiation-exposed people including atomic bomb survivors, *Progr. Clin. Biol. Res.,* 372, 341–50.

Nascimbeni, B., M.D. Phillips, D.K. Croom, P.W. Andrews, R.R. Tice, and C.H. Nauman (1991), Evaluation of DNA damage in golden mice inhabiting a hazardous waste site, *Environ. Mol. Mutagen.,* 17 (S17), 192.

Nayak, B.N. and M.L. Petras (1985), Environmental monitoring for genotoxicity: *in vivo* sister chromatid exchange in the house mouse *(Mus musculus), Can. J. Genet. Cytol.,* 27, 351–356.

Neel J.V. (1981), Genetic effects of the atomic bombs, *Science*, 213, 1205.

Neel, J.V. and S.E. Lewis (1990), The comparative radiation genetics of humans and mice, *Annu. Rev. Genet.*, 24, 327–362.

Neel J.V., C. Satoh, K. Goriki, J. Asakawa, M. Fujita, N. Takahashi, T. Kageoka, and T. Hazama (1988), Search for mutations altering protein charge and/or function in children of atomic bomb survivors: final report, *Am. J. Hum. Genet.*, 42, 663–676.

Nestman, E.R. (1983) Mutagenic activity of drinking water, in *Carcinogens and Mutagens in the Environment*, Vol. 3, H.F. Stich, Ed., CRC Press, Boca Raton, FL, 137–147.

Nicklas, J.A., J.P. O'Neill, and R.J. Albertini (1986), Use of T-cell receptor gene probes to quantify the *in vivo* hprt mutations in human T-lymphocytes, *Mutat. Res.*, 173, 65–72.

Nicklas, J.A., J.P. O'Neill, L.M. Sullivan, T.C. Hunter, M. Allegretta, B.F. Chastenay, B.L. Libbus, and R.J. Albertini (1988), Molecular analyses of *in vivo* hypoxanthine-guanine phosphoribosyltransferase mutations in human T-lymphocytes. II. Demonstration of a clonal amplification of hprt mutant T-lymphocytes *in vivo*, *Environ. Mol. Mutagen.*, 12, 271–284.

Nowell, P.C. (1960), Phytohemagglutinin: an initiator of mitosis in cultures of normal human leukocytes, *Cancer Res.*, 20, 462–466.

Nylund, L.E., E. Hakala, and M. Sorsa (1992), Application of a semi-automated SOS chromotest for measuring genotoxicities of complex environmental mixtures containing polycyclic aromatic hydrocarbons, *Mutat. Res.*, 276, 125–132.

Oda, Y., S. Nakamura, I. Oki, T. Kato, and H. Shinagawa (1985), Evaluation of the new system (umu-test) for the detection of environmental mutagens and carcinogens, *Mutat. Res.*, 147, 219–229.

O'Keefe, D.P. and K.J. Leto (1989), Cytochrome P450 from the mesocarp of avocado *(Persea americana)*, *Plant Physiol.*, 89, 1141–1149.

Oldereid, N.B., H. Rui, O.P. Clausen, and K. Purvis (1989), Cigarette smoking and human sperm quality by laser-Dopler spectroscopy and DNA flow cytometry, *J. Reprod. Fertil.*, 86, 731–736.

Olive, P.L., D. Wlodek, and J.D. Banath (1991), A double strand breaks measured in individual cells subjected to gel electrophoresis, *Cancer Res.*, 51, 4671–4676.

Ong, T., J.D. Stewart, J.D. Tucker, and W.-Z. Wong (1985), Development of an *in situ* test system for detection of mutagens in the workplace, in *Short-Term Bioassays in the Analysis of Complex Environmental Mixtures IV*, M.D. Waters, S.S. Sandhu, J. Lewtas, L.D. Claxton, G. Strauss, and S. Nesnow, Eds., Environmental Science Research, Vol. 32, Plenum Press, New York, 25–36.

Ong, T.M., J. Stewart, and W. Whong (1988), Detection of the genotoxicity of complex chemical mixtures with the SOS UMU/test, in *Short-Term Bioassays in the Analysis of Complex Environmental Mixtures V*, S.S. Sandhu, D.M. DeMarini, M.J. Mass, M.M. Moore, and J.L. Mumford, Eds., Plenum Press, New York, 181–189.

Pagano, G., P. Bove, M. de Angelis, A. Esposito, A. Rota, and G.G. Giordano (1982b), Mercury induced developmental defects and mitotic aberrations in sea urchin development, *Mutat. Res.*, 97, 210.

Pagano, G., A. Esposito, P. Bove, M. De Angelis, A. Rota, E. Vamvakinos, and G.G. Giordano (1982a), Arsenic induced developmental defects and mitotic abnormalities in sea urchin development, *Mutat. Res.*, 104, 351–354.

Park, J-W., K.C. Cundy, and B.N. Ames, (1989), Detection of DNA adducts by high-performance liquid chromatography with electrochemical detection, *Carcinogenesis*, 10(5), 827–832.

Park, J-W., M.K. Shigenaga, P. Degan, T.S. Korn, J.W. Kitzler, C.M. Wehr, P. Kolachana, and B.N. Ames (1992), Assay of oxidative DNA lesions: isolation of 8-oxoguanine and its nucleoside derivatives from biological fluids with monoclonal antibody column, *Proc. Natl. Acad. Sci. U.S.A.*, 89, 3375–3379.

Parry, J.M., D.J. Tweats, and M.A. Al Mossawi (1976), Monitoring the marine environment for mutagens, *Nature (London)*, 264, 538.

Payne, J.F. (1976), Field evaluation of benzopyrene hydroxylase induction as a monitor for marine petroleum pollution, *Science*, 191, 945.

Pereira, M.A. (1982), Genotoxicity of diesel exhaust emissions in laboratory animals, in *Toxicological Effects of Emissions from Diesel Engines*, J. Lewtas, Ed., Elsevier, New York, 265–276.

Perera, F.P., K. Hemminki, T.L. Young, R.M. Santella, D. Brenner, and G. Kelly (1988), Detection of polycylic aromatic hydrocarbon-DNA adducts in white blood cells of foundry workers. *Cancer Res.*, 48, 2288–2291.

Perry, P. and H.J. Evans (1975), Cytological detection of mutagen carcinogen exposure by sister chromatid exchange, *Nature (London)*, 258, 121–125.

Pesch, G. (1990), Sister chromatid exchange and genotoxicity measurements using polychaete worms, *Rev. Aquat. Sci.*, 2(1), 19–25.

Pesch, G.G., C.E. Pesch, and A.R. Malcolm (1981), *Neanthes arenaceodentata*, a cytogenetic model for marine genetic toxicology, *Aquat. Toxicol.*, 1, 301–311.

Pesch, G.G., C. Mueller, C.E. Pesch, A.R. Malcolm, P.F. Rogerson, W.R. Munns, Jr., G.R. Gardner, J. Heltshe, T.C. Lee, and A. Senecal (1987), Sister chromatid exchange in a marine polychaete exposed to a contaminated harbor sediment, in *Short Term Bioassay in the Analysis of Complex Environmental Mixture V*, S.S. Sandhu, D.M. De Marini, M.J. Mass, M.M. Moore, and J.L. Mumford, Eds., Plenum Press, New York, 237–253.

Phillips, D.H., K. Hemminki, A. Alhonen, A. Hewer, and P.L. Grover (1988), Monitoring occupational exposure to carcinogens: detection by ^{32}P-postlabeling of aromatic DNA adducts in white blood cells from iron foundry workers, *Mutat. Res.*, 204, 531–541.

Pierson, W.R., R.A. Gorse, A.C. Szkarlat, W.W. Brachaczek, S.M. Japar, F.S.C. Lee, R.B. Zweidinger, and L. Claxton (1983), Mutagenicity and chemical characterization of carbonaceous particulate matter from vehicles on the road, *Environ. Sci. Technol.*, 17(1), 31–44.

Piper, J., S. Bayley, S. Boyle, J.A. Fantes, D.K. Green, J. Gordon, W. Hill, L. Ji, P. Malloy, P. Perry, D. Rutovitz, M. Stark, and D. Whale (in press), Automatic aberration scoring using whole chromosome FISH, *Cytometry*.

Pittinger, C.A., A.L. Buikema, Jr., and J.O. Falkinham, III (1987), *In situ* variations in oyster mutagenicity and tissue concentrations of polycyclic aromatic hydrocarbons, *Environ. Toxicol. Chem.*, 6, 51.

Pitts, J.N., K.A. Van Cauwenberghe, D. Grosjean, J.P. Schmid, D.R. Fitz, et. al. (1979), Chemical and microbiological studies of mutagenic pollutants in real and simulated atmospheres, in *Application of Short-Term Bioassays the Fractionation and Analysis of Complex Environmental Mixtures*, Environmental Science Research, Vol. 15, M.D. Waters, S. Nesnow, J. Lewtas-Huisingh, S.S. Sandhu, and L. Claxton, Eds., Plenum Press, New York, 354–379.

Plewa, J.J. and J.M. Gentile (1976), Mutagenicity of atrazine: a maize-microbe bioassay, *Mutat. Res.*, 38, 287–296.

Prein, A.E., G.M. Thie, G.M. Alink, J.H. Koeman, and C.L.M. Poels (1978), Cytogenetic changes in fish exposed to water of the river Rhine, *Sci. Total Environ.*, 9, 287–291.

Prosser, J.S., J.E. Moquet, D.C. Lloyd, and A.A. Edwards (1988), Radiation induction of micronuclei in human lymphocytes, *Mutat. Res.*, 199, 37–45.

Prosser, J.S., D.C. Lloyd, and A.A. Edwards (1989), A comparison of chromosomal and micronuclear methods for radiation accident dosimetry, in *Radiation Protection — Theory and Practice,* Session 6, 4th Int. Symp., Malvern, England, June, 1989.

Purdom, C.E. and D.S. Woodhead (1973), Radiation damage in fish, in *Genetics and Mutagenesis of Fish,* J.H. Schroeder, Ed., Springer-Verlag, Berlin, 67–73.

Quillardet, P., O. Huisman, R. D'Ari, and M. Hofnung (1982), SOS chromotest, a direct assay of induction of an SOS function in *Escherichia coli* K12 to measure genotoxicity, *Proc. Natl. Acad. Sci, U.S.A.*, 79, 5971–5975.

Randerath, K., E. Randerath, H.P. Agrawal, R.C. Gupta, M.E. Schurdak, and M.V. Reddy (1985), Postlabeling methods for carcinogen-DNA adduct analysis, *Environ. Health Perspect.*, 62, 57–65.

Randerath, K., R.H. Miller, D. Mittal, T.A. Avitts, H.A. Dunsford, and K. Randerath (1989), Covalent DNA damage in tissues of cigarette smokers as determined by ^{32}P-Postlabeling assay, *J. Natl. Cancer Inst.*, 81, 341–347.

Reddy, M.V. and K. Randerath (1986), Nucleasue P1-mediacted enhancement of sensitivity of ^{32}P-postlabeling test for structurally diverse DNA adducts, *Carcinogenesis,* 7, 1543–1551.

Redei, G.P. (1982), Mutagen assay with Arabidopsis. A report of the U.S. EPA's GENE-TOX program, *Mutat. Res.*, 99, 243–255.

Redei, G.P., W.W. Redei, W.R. Lower, and S.S. Sandhu (1980), Identification of carcinogens by mutagenicity of Arabidopsis, *Mutat. Res.*, 74, 469–475.

Reif, J.S. and D. Cohen (1970), II. Retrospective radiographic analysis of pulmonary disease in rural and urban dogs, *Arch. Environ. Health,* 20, 684–689.

Rether, B., A. Pfohl-Leszkorvick, P. Guillemut, and G. Keith (1990), Benzo[a]pyrene induces nuclear DNA adducts in plant cell suspension culture, *FEBS Lett.*, 763, 172–174.

Roesch, W.C., Ed., (1987) *Final Report of the US-Japan Joint Reassessment of Atomic Bomb Radiation Dosimetry in Hiroshima and Nagasaki,* Vol. 1 and 2, Radiation Effects Research Foundation, Hiroshima.

Rosenkranz, H.S. (1979), The use of microbial assay systems in the detection of environmental mutagens in complex mixtures, in *Application of Short-Term Bioassays in the Fractionation and Analysis of Complex Environmental Mixtures,* Environmental Science Research, Vol. 15, M.D. Waters, S. Nesnow, J.L. Huisingh, S.S. Sandhu, and L.D. Claxton, Eds., Plenum Press, New York, 3–41.

Rossman, G.G., L.W. Meyer, J.P. Butler, and J.M. Daisey (1985), Use of the microscreen assay for airborne particulate organic matter, in *Short-Term Bioassays in the Analysis of Complex Environmental Mixtures IV,* M.D. Waters, S.S. Sandhu, J. Lewtas, L. Claxton, G. Strauss, and S. Nesnow, Eds., Plenum Press, New York, 9–24.

Sala-Trepat, M., J. Cole, M.H.L. Green, O. Rigaud, J.R. Vilcoq, and E. Moustacchi (1990), Genotoxic effects of radiotherapy and chemotherapy on the circulating lymphocytes of breast cancer patients. III. Measurement of mutant frequency to 6-thioguanine resistance, *Mutagenesis,* 5, 593–598.

Sandhu, S.S. and W.R. Lower (1989), *In situ* assessment of genotoxic hazards of environmental pollution, *Toxicol. Indust. Health,* 5, 73–83.

Sandhu, S.S., D.M. DeMarini, M.J. Moss, M.M. Moore, and J.L. Mumford, Eds., (1987) *Short-Term Bioassays in the Analysis of Complex Environmental Mixtures V,* Environmental Science Research, Vol. 36, Plenum Press, New York, 407.

Sarto, F., R. Tomanin, L. Giacomelli, et al. (1990), The micronucleus assay in human exfoliated cells of the nose and mouth: application to occupational exposures to chromic acid and ethylene oxide, *Mutat. Res.,* 244, 345–351.

Savard, S., R. Otson, and G.R. Douglas (1992), Mutagenicity and chemical analysis of sequential organic extracts of airborne particulates, *Mutat. Res.,* 276, 101–115.

Schaeffer, D.J., E.W. Novak, W.R. Lower, A.F. Yanders, S. Kapila, and R. Wang (1987), Effects of chemical smokes on flora and fauna under field and laboratory exposures, *Ecotoxicol. Environ. Saf.,* 13, 301–315.

Schairer, L.A. (1983), Mutagenicity of ambient air at selected sites in the United States using *Tradescantia* as a monitor, in *In Vitro Toxicity Testing of Environmental Agents: Current and Future Possibilities,* A.R. Kolber, T.K. Wong, L.D. Grant, R.S. DeWoskin, and T.J. Hughes, Eds., Plenum Press, New York, 167–190.

Schairer, L.A., J. Hof Van't, C.G. Hayes, R.M. Burton, and F.J. deSerres (1978a), Measurement of biological activity of ambient air mixture using a mobil laboratory for *in situ* exposures: preliminary results from the Tradescantia plant test system, in *Application of Short-Term Bioassays in the Fractionation and Analyzing Complex Environmental Mixtures,* M.D. Waters, S. Nesnow, J.L. Huisingh, S.S. Sandhu, and L. Claxton, Eds., EPA 600/9-78-0277, U.S. Environmental Protection Agency, Washington, D.C.

Schairer, L.A., J. Van't, C.G. Hayes, R.M. Burton, and F.J. deSerres (1978b), Exploratory monitoring of air pollutants for mutagenicity activity with the Tradescantia stamen hair system, *Environ. Health Perspect,* 27, 51–60.

Schlegel, R., J.T. MacGregor, and R.B. Everson (1986), Assessment of cytogenetic damage by quantitation of micronuclei in human peripheral blood erythrocytes, *Cancer Res.,* 46, 3717–3721.

Schmid, W. (1976), The micronucleus test for cytogenetic analysis, in *Chemical Mutagens: Principles and Methods for Their Detection. Vol 4,* A. Hollaender, Ed., Plenum Press, New York, 31.

Schneider, R. (1970), Comparison of age, sex, and incidence rates in human and canine breast cancer, *Cancer,* 26, 419–426.

Schneider, R., C.R. Dorn, and D.O.N. Taylor (1969), Factors influencing canine mammary cancer development and postsurgical survival, *J. Natl. Cancer Inst.,* 43, 1249–1261.

Schoket, B., A. Hewer, P.L. Grover, and D.H. Phillips (1988), Covalent binding of components of coal-tar, creosote and bitumen to the DNA of the skin and lungs of mice following topical application, *Carcinogenesis,* 9, 1253–1258.

Schoket, B., D.H. Phillips, A. Hewer, and I. Vincze (1991), ^{32}P-Postlabeling detection of aromatic DNA adducts in peripheral blood lymphocytes from aluminum production plant workers, *Mutat. Res.,* 260, 89–98.

Schroeder, J.H. (1969), X ray induced mutations in the poecillid fish, *Lebistes reticulatus* Peters, *Mutat. Res.,* 7, 75–90.

Schroeder, J.H. (1979), Methods for screening radiation induced mutations in fish, in *Methodology for Assessing Impacts of Radioactivity on Aquatic Ecosystems,* Technical Report Series 190, International Atomic Energy Agency, Vienna, 371–402.

Schuetzle, D. and J. Lewtas (1986), Bioassay-directed chemical analysis in environmental research, *Anal. Chem.,* 58, 1060A–1075A.

Schull W.J., M. Otake, and J.V. Neel (1981), Genetic effects of the atomic bombs: a reappraisal, *Science,* 213, 1220–1227.

Schy, W.E. and M.J. Plewa (1989), Molecular dosimetry studies of forward mutation induced at the yg 2 locus in maize by ethyl methanesulfonate, *Mutat. Res.,* 211, 231–241.

Seiler, J.P. (1991), Pentachlorophenol, *Mutat. Res.,* 257, 27–47.

Sexton K. and P.B. Ryan (1988), Assessment of human exposure to air pollution: methods, measurements, and models, in *Air Pollution, the Automobile, and Public Health,* A.Y. Watson, R.R. Bates, and D. Kennedy, Eds., National Academy Press, Washington, D.C., 207–238.

Shepson, P., T.E. Kliendienst, E.O. Edney, G.R. Namie, J.H. Pittman, L.T. Cupitt, and L.D. Claxton (1985), The mutagenic activity of irradiated toluene/NO_x/H_2O/air mixtures, *Environ. Sci. Tech.,* 19, 249–255.

Shields, P.G., A. Weston, H. Sugimura, E.D. Bowman, N.E. Caporaso, D.K. Manchester, G.E. Trivers, S. Tamai, J.H. Resau, B.F. Trump, and C.C. Harris (1991), Molecular epidemiology: dosimetry, susceptibility, and cancer risk, in *Immunoassays for Trace Chemical Analysis,* M. Vanderlaan, et. al., Eds., American Chemical Society, Washington, D.C., 186–206.

Shigenaga, M.K., J-W. Park, K.C. Cundy, C.J. Gimeno, and B.N. Ames (1990), Assaying *in vivo* oxidative DNA damage: measurement of 8-hydroxydeoxyguanosine in DNA and urine by HPLC with electrochemical detection, in *Oxygen Radicals in Biological Systems, Part B: Oxygen Radicals and Antioxidants,* L. Packer, and A. Glazer, Eds., New York, Academic Press.

Shugart, L.R. (1988a), An alkaline unwinding assay for the detection of DNA damage in aquatic organisms, *Mar. Environ. Res.,* 24, 321–325.

Shugart, L.R. (1988b), Quantitation of chemically induced damage to DNA of aquatic organisms by alkaline unwinding assay, *Aquat. Toxicol.,* 13, 43–52.

Shugart, L.R., J.F. McCarthy, B.D. Jimenez, and J. Daniel (1987), Analysis of adduct formation in the Bluegill sunfish *(Lepomis macrochirus)* between benzo[a]pyrene and DNA of the liver and hemoglobin of the erythrocyte, *Aquat. Toxicol.,* 9, 319–325.

Singh, N.P., M.T. McCoy, R.R. Tice, and E.L. Schneider (1988), A simple technique for quantitation of low levels of DNA damage in individual cells. *Exp. Cell Res.,* 175, 184–191.

Singh, N.P., D.B. Danner, R.R. Tice, J.B. Pearson, L.J. Brant, and E.L. Schneider (1990), DNA damage and repair with age in individual human lymphocytes, *Mutat. Res.,* 237, 123–130.

Skipper, P.L. and S.R. Tannenbaum (1990), The role of protein adducts in the study of chemical carcinogenesis, *Prog. Clin. Biol. Res.,* 340C, 301–310.

Smith, D.F., J.T. MacGregor, R.A. Hiatt, N.K. Hooper, C.M. Wehr, B. Peters, L.R. Goldman, L.A. Yuan, P.A. Smith, and C.E. Becker (1990), Micronucleated erythrocytes as an index of cytogenetic damage in humans: demographic and dietary factors associated with micronucleated erythrocytes in splenectomized subjects, *Cancer Res.,* 50, 5049–5054.

Sorsa, M., K. Hemminki, and H. Vainio (1985), Occupational exposure to anticancer drugs — Potential and real hazards, *Mutat. Res.,* 154, 135–149.

Squire, R.D. (1973), The effects of acute gamma irradiation on the brine shrimp, *Artemia*. III. Male F 1 reproductive performance following paternal irradiation of mature sperm, *Biol. Bull.*, 144, 192.

Stich, H.F. (1987), Micronucleated exfoliated cells as indicators for genotoxic damage and as markers in chemoprevention trials, *J. Nutr. Growth Cancer*, 4, 9–18.

Stich, H.F., W. Stich, and B.B. Parida (1982a), Elevated frequency of micronucleated cells in the bucca mucosa of individuals at high risk for oral cancer: betel quid chewers, *Cancer Lett.*, 17, 125–134.

Stich, H.F., J.R. Curtis, and B.B. Parida (1982b), Application of the micronucleus test to exfoliated cells of higher cancer risk groups: tobacco chewers, *Int. J. Cancer*, 30, 553–559.

Stich, H.F. and M.P. Rosin (1983a), Micronuclei in exfoliated human cells as an internal dosimeter for exposures to carcinogens, in *Carcinogens and Mutagens in the Environment*, H.F. Stich, Ed., Vol. 2, CRC Press, Boca Raton, FL, 17–25.

Stich, H.F. and M.P. Rosin (1983b), Quantitating the synergistic effect of smoking and alcohol consumption with the micronucleus test on human buccal mucosa cells, *Int. J. Cancer*, 31, 305–308.

Stich, H.F., A.P. Hornby, and B.P. Dunn (1985), A pilot beta-carotene intervention trial with Inuits using smokeless tobacco, *Int. J. Cancer*, 36, 321–327.

Straume, T., R.G. Langlois, J. Lucas, R.H. Jensen, W.L. Bigbee, A.T. Ramalho, and C.E. Brandao-Mello (1991), Novel biodosimetry methods applied to the victims of the Goiania accident, *Health Phys.*, 60, 71–76.

Stromberg, P.T., M.L. Landolt, and R.M. Kocan (1981), Alterations in the Frequency of Sister Chromatid Exchanges in Flat Fish from Puget Sound, Washington, Following Experimental and Natural Exposure to Mutagenic Chemicals. NOAA Tech. Memo. OMPA-10, National Oceanic and Atmospheric Administration, Office of Marine Pollution Assessment, Washington, D.C., 43.

Sugatt, R.H. (1978), Chromosome aberrations in the eastern mudminnow *(Umbra pygmaea)* exposed *in vivo* to Trenimon or river water. Report MD-N and E 78/3, Central Laboratory TNO, Delft, Netherlands.

Tates, A.D., L.F. Bernini, A.T. Natarajan, J.S. Ploem, N.P. Verwoerd, J. Cole, N.H.L. Green, C.F. Arlett, and P.N. Norris (1989), Detection of somatic mutants in man: HPRT mutations in lymphocytes and hemoglobin mutations in erythrocytes, *Mutat. Res.*, 213, 73–82.

Taylor, J.H. (1958), Sister chromatid exchanges in tritium-labeled chromosomes, *Genetics*, 43, 515.

Thompson, R., G. Schroder, S. Pathak, and T. Connor (1986), A study of chromosomal aberrations in the cotton rat, *Sigmodon hispidus* exposed to hazardous waste. (Abstr.) in EPA Symp. Application of Short-Term Bioassays in the Analyses of Complex Environmental Mixtures, Durham, NC, October 20 to 23, 1986.

Thompson, R. A., G.D. Schroder, and T.H. Connor (1988), Chromosomal aberrations in the cotton rat, *Sigmodon hispidus*, exposed to hazardous waste, *Environ. Mol. Mutagen.*, 11, 359–367.

Tice, R.R., B.G. Ormiston, R. Boucher, C.A. Luke, and D.E. Paquette (1988), Environmental biomonitoring with feral rodent species, in *Short-Term Bioassays in the Analysis of Complex Environmental Mixtures V*, S.S. Sandhu, D.M. DeMarini, M.J. Mass, M.M. Moore, and J.L. Mumford, Eds. Plenum Press, New York, 175–180.

Tice, R.R., P.W. Andrews, O. Hirai, and N.P. Singh (1991), The single cell gel (SCG) assay: an electrophoretic technique for the detection of DNA damage in individual cells, in *Biological Reactive intermediates IV, Molecular and Cellular Effects and Their Impact on Human Health,* C.R. Witmer, R.R. Snyder, D.J. Jollow, G.F. Kalf, J.J. Kocsis, and I.G. Sipes, Eds., Plenum Press, New York, 157–164.

Tice, R.R., G.H.S. Strauss, and W.P. Peters (1992), High-dose combination alkylating agents with autologous bone marrow support in patients with breast cancer: preliminary assessment of DNA damage in individual peripheral blood lymphocytes using the single cell gel electrophoresis assay, *Mutat. Res.,* 271, 101–113.

Tjalma, R.A. (1966), Canine bone sacroma: Estimation of relative risk as a function of body size, *J. Natl. Cancer Inst.,* 36, 1137–1150.

Tjio, J.H. and A. Levan (1956), The chromosome number of man, *Hereditas,* 42, 1–6.

Tolbert, P.E., C.M. Shy, and J.W. Allen, (1991), Micronuclei and other nuclear anomalies in buccal smears: field test in snuff users, *Am. J. Epidemiol.,* 134(8), 840–850.

Tolbert, P.E., C.M. Shy, and J.W. Allen (submitted), Micronuclei and other nuclear anomalies in buccal smears: methods development, *Mutat. Res.*

Tough, I., K.E. Buckton, A.G. Baikie, and W.M. Court-Brown (1960), X-ray-induced chromosome damage in man, *Lancet,* 2, 849.

Tsoi, R.M. (1971), Effect of methyl sulfate on mutation frequency of genes S and in the carp (*Cyprinus carpio* L.) (trans.), *Dokl. Biol. Sci.,* 197, 197–200.

Tsoi, R.M., A.I. Men'shova, and Y.F. Golodov (1976), Frequency of spontaneous and induced mutations in genes determining carp scales, (transl.) *Sov. Genet.,* 10, 1368–1370.

Turner, D.R. and A.A. Morley (1990), Human somatic mutation at the HLA-A locus, *Prog. Clin. Biol. Res.,* 340C, 37–46.

Underbrink, A.G., L.A. Schairer, and A.H. Sparrow (1973), Tradescantia stamen hairs: A radiobiological test system applicable to chemical mutagenesis, in *Chemical Mutagens: Principals and Methods for Their Detection,* Vol. 3, A. Hollaender, Ed., New York, Plenum Press, 171–207.

U.S. Congress, Office of Technology Assessment, Technologies for Detecting Heritable Mutations in Human Beings, OTA-H-298, U.S. Government Printing Office, Washington, D.C., September 1986.

Vankerkom, J., M. Janowski, and L. Verschaeve (1993), Systematic investigations on the experimental parameters that influence the accuracy and reproducibility of the single cell gel electrophoresis assay (comet test) (Abstr.). 6th Int. Conf. Environmental Mutagens, Melbourne, Australia, February 21–26.

Varanasi, U., W.L. Reichert, B.L. Eberhart, and J.E. Stein (1989), Formation and persistence of benzo(a)pyrene-diolepoxide-DNA adducts in liver of English sole *(Parophrys vetulus), Chem.-Biol. Interact.,* 69, 203–216.

Wagner, E.D., M.A. Verdier, and M.J. Plewa (1990), The biochemical mechanisms of the plant activation of promutagenic aromatic amines, *Environ. Mol. Mutagen.,* 15, 236-244.

Wang, H., X. You, Y. Qu, W. Wang, D. Wang, Y. Long, and J. Ni (1984), Investigation of cancer epidemiology and study of carcinogenic agents in the Shanghai rubber industry. *Cancer Res.,* 44, 3101–3105.

Wasserman, S.S., M.M. Cohen, and S. Schwartz (1990), Factors underlying variation in spontaneous and clastogen-induced sister chromatid exchanges and chromosome breakage frequencies, *Eviron. Mol. Mutagen.*, 16(4), 255–259.

Waters, M.D., S. Nesnow, J. Lewtas-Huisingh, S. Sandhu, and L. Claxton, Eds. (1979), *Application of Short-Term Bioassays in the Fractionation and Analysis of Complex Environmental Mixtures,* Plenum Press, New York, 588.

Waters, M.D., S.S. Sandhu, J. Lewtas-Huisingh, L. Claxton, and S. Nesnow, Eds. (1981), *Short-Term Bioassays in the Analysis of Complex Environmental Mixtures II,* Plenum Press, New York, 524.

Waters, M.D., S.S. Sandhu, J. Lewtas, L. Claxton, N. Chernoff, and S. Nesnow, Eds. (1983a), *Short-Term Bioassays in the Analysis of Complex Environmental Mixtures III,* Plenum Press, New York, 589.

Waters, M.D., V.F. Simmon, A.D. Mitchell, T.A. Jorgenson, and R. Valencia (1983b), A phased approach to the evaluation of environmental chemicals for mutagenesis and presumptive carcinogenesis, in *In Vitro Toxicity Testing of Environmental Agents: Current and Future Possibilities, Part B,* A.R. Kolber, T.K. Wong, L.D. Grant, R.S. DeWoskin, and T.J. Hughes, Eds., Plenum Press, New York, 417–442.

Waters, M.D., S.S. Sandhu, J. Lewtas, L. Claxton, G. Strauss, and S. Nesnow, Eds. (1985), *Short-Term Bioassays in the Analysis of Complex Environmental Mixtures IV,* Plenum Press, New York, 384.

Waters, M.D., L.D. Claxton, H.F. Stack, A.L. Brady, and T.E. Graedel (1990a), Genetic activity profiles in the testing and evaluation of chemical mixtures, *Teratog. Carcinog. Mutagen.*, 10, 147–164.

Waters, M.D., F.B. Daniel, J. Lewtas, M.M. Moore, and S. Nesnow, Eds. (1990b), *Genetic Toxicology of Complex Mixtures,* Environmental Science Research, Vol. 39, Plenum Press, New York, 376.

Welch, L.S., S.M. Schrader, T.W. Turner, and M.R. Cullen (1988), Effects of exposure to ethylene glycol ethers on shipyard painters. II. Male reproduction, *Am. J. Ind. Med.*, 14, 509–526.

Whong, W.-Z, Y.-F. Wen, J. Steward, and T. Ong (1986), Validation of the SOS/Umu test with mutagenic complex mixtures, *Mutat. Res.*, 175, 139–144.

Wolff, S. (1991), Biological dosimetry with cytogenetic endpoints, in *New Horizons in Biological Dosimetry,* B.L. Gledhill and F. Mauro, Eds., Wiley-Liss, New York, 351–362.

Wyrobek, A.J., T. Alhborn, R. Balhorn, L. Stanker, and D. Pinkel (1990), Fluorescence *in situ* hybridization to Y chromosomes in decondensed human sperm nuclei, *Mol. Reprod. Dev.*, 27, 200–208.

Yager, J.W., M. Sorsa, and S. Slevin (1988), Micronuclei in cytokinesis-blocked lymphocytes as an index of occupational exposure to alkylating cytostatic drugs, in *Methods for detecting DNA damaging agents in humans: Applications in cancer epidemiology and prevention:* Vol. 89, International Agency for Research on Cancer, Lyon, France.

Yager, J.W., W.M. Paradisin, E. Symanski, and S.M. Rappaport (1990), Sister-chromatid exchanges induced in peripheral lymphocytes of workers exposed to low concentrations of styrene, *Prog. Clin. Biol. Res.*, 340C, 347-356.

Yamasaki, E. and B.N. Ames (1977), The concentration of mutagens from urine with the nonpolar resin XAD-2: cigarette smokers have mutagenic urine, *Proc. Natl. Acad. Sci. U.S.A.*, 74, 3555–3559.

Yan, P. and T.-H Ma (1990), Tradescantia sister-chromatid exchange (SCE) bioassay for mutagens, in *Plants for Toxicity Assessment,* ASTM STP 1091, American Society for Testing and Materials, Philadelphia, 319–323.

Zimmermann, F.K. (1975), Procedures used in the induction of mitotic recombination and mutation in the yeast *Saccharomyces cerevisiae, Mutat. Res.,* 31, 71–86.

Zimmermann, F.K. and R.E. Taylor-Mayer, Eds. (1985), *Mutagenicity Testing in Environmental Pollution Control,* Halsted Press, New York, 195.

INDEX

A

ADAPT program, 14–16
Aging, 6, 160
Air, as contact medium, 35–36
Air monitoring, 202–203
Ames test, see *Salmonella* assay
Amino acid substitution assay, 196
Aneuploidy, 11, 17
Aquatic species assays, 48–49, 179–184, 189–190, 206–208, see also Bioassay techniques
Ashby structural alert method, 15
Assessment, see Hazard identification, Risk assessment
Atherosclerotic plaque, 6
AUC (area under curve) as dose surrogate, 89–91

B

Background mutation rate, 159–160
BCF correlation models, 47–48
Bioassay techniques
  in aquatic organisms
    cytogenetic: anaphase, 179–180, 182
    cytogenetic: metaphase, 180, 182–183
    DNA adduct analysis, 182
    DNA-damage-related synthesis, 182
    DNA strand breaks, 181–182, 184
    dominant-lethal, 181, 184
    micronucleus, 180–181
    sister-chromatid exchange, 180, 183–184
    specific locus, 181, 184
  in human germ cells
    protein variants, 198
    sperm cells, 198–199
  in human monitoring, 191–199
  in human somatic cells
    blood, urine, or tissues, 197–198
    cytogenetic damage, 191–194
    direct measurement of mutation, 194–196
    DNA damage, 196–197
  for *in situ* environmental monitoring
    aquatic species, 189–190
    mammals, 191
    microorganisms, 187
    terrestrial plants, 187–189

mammalian
  in vivo, 185
  cell culture, 184–185
microbial assays
  gene mutation, 176–177
  mitotic recombinations, 177
  repair, 177
plant assays
  cytogenetic, 178
  DNA adduct, 178–179
  multilocus, 178
  specific-locus, 178
sampling and analytic methods, 185
Bioconcentration factors (BCFs), 47–48
Biota, hazard identification in nonhuman, 136–138
Biotransfer factors (BTFs), 49
Blood tests, 197–198

C

Carcinogenesis, 5–6
Carcinogenicity
  of ethylene/ethylene oxide, 75–76
  of 2-nitropropane, 72–74
  relative and GAPs, 100–104
  risk assessment for, 141–142
  of vinyl chloride/vinyl bromide, 72
Cardiovascular disease, 6
CASE program, 14–16
Cataracts, senile, 6
Cell replication and risk, 160
Cells
  germ, 3–4, 127–128
  somatic, 4–6, 127
Chironomus tentans assay, 183, 207
Chromosomal mutations
  of number (aneuploidy), 3
  of structure, 3
Computational tools, 14, 19–20
Consumer products, see Contact media
Contact media
  air (indoor and outdoor), 35–36
  cross-media transfers and multimedia models, 44–46
  environmental landscape properties, 46–47
  ingested substances, 36
  physicochemical properties, 47
  soils, 36

transformation processes in, 42–44
transport processes in, 38–41
water, 36
Corn, see *Zea maize* studies
Cross-media transfers, 44–45
Cytogenetic assays, 178
   of aquatic organisms, 179–180, 182–183
   in humans, 191–194

D

Data sources, 8–14
Dermal uptake exposure, 52–53
Diet and risk, 160–161
Dioxyribonucleic acid (see DNA)
Direct extrapolation, 93–94
Direct risk-estimation method, 152–153
DNA
   damage-related synthesis assays, 182
   strand break assays, of aquatic organisms, 181–182, 184
DNA adducts, see also Molecular dosimetry
   assay methods, 77–78, 178–179, 196–197
   correlation of formation with mutagenic effects, 81–83
   damage assay, 196–197
   formation, 78–79
   removal/repair, 79–80, 177
dN/dt (amount of change per time unit), 68–69
Dominant-lethal assays, 181, 184
Dose and effect assessment
   of comparative mutagenesis
      carcinogenetic potency ranking, 97–100, 153
      extrapolation, 104–107
      potency and GAPs, 100–104
      qualitative vs. quantitative approaches, 95–97
   conclusions, 107–108
   of germ cell mutagenesis
      dose response in, 91–92
      extrapolation, 95
      of female cells, 94–95
      and genetic endpoints, 92
      mammalian mutagenetic mechanisms, 88–91
      and risk estimation, 92–94
   and molecular dosimetry
      DNA adduct formation, 78–79
      DNA adduct removal/repair, 79–80
      exposure vs. mutation fixation, 81–84
      methodology, 77–78

   perspectives, 108–109
   and pharmacokinetics
      general principles, 67–68
      principles and models, 68–76
   of somatic cell mutagenesis
      human mutation testing, 86–88
      short-term assays, 84–86
Dose-response extrapolation, 132
   comparative methods
      comparative potency method, 152–153
      radiation equivalence, 152
   doubling dose and relative mutagenic effect (RME), 92–93, 146–148, 151
   human risk methods, 150–151
   LED method, 85–86, 148–149
   low-dose methods, 150
   potency estimation methods, 145–146
   probabilistic methods, 151–152
   slope, 146
Doubling dose approaches, 92–93, 146–148, 151
Drosophila assays, 83–84, 103–104
Duration of exposure, 161–162

E

E. coli assay, 83–84, 102, see also Bioassay techniques, microbial
Echinacea purpurea assay, 178
Economic risk factors, 161
Effluent testing, 201–202
EMIC (Environmental Mutagen Information Center), 8–9
Emission testing, 201–202
Endogenous risk factors, 159
Endpoint determination, 92
Environmental landscape modeling, 46–47
Environmental monitoring, see Monitoring
Epidemiology of genetic disease, 4
Erythrocyte assays, 86–88
Ethylene/ethylene oxide, 75–76
Ethylnitrosurea (ESU) assay, 88–89
Experimental design, 199
Exposure
   defined, 33
   duration and risk, 161–162
   vs. mutation fixation, 81–84
Exposure assessment, 133, see also Hazard identification
   concentration
      cross-media transfers and multimedia models, 44–46

INDEX

environmental/physicochemical
properties, 46–47
transformation processes and, 42–44
transport processes and, 38–41
exposure pathways
for biota, 47–49
for humans, 49–53
methods, 53–55
population risk estimation, 33–34
quantitation and validation, 55–56
recommendations, 56–57
in risk assessment and management, 31–33
source identification
indoor and outdoor air, 35–36
ingested substances, 37
soils, 37
water, 37
Exposure commitment model (Bennett), 54
Exposure monitoring, 203–204
Exposure pathways
for biota, 47–49
for humans
dermal uptake, 52–53
ingestion, 50–51
inhalation, 51–52
Extraction sampling, 186
Extrapolation
in comparative mutagenesis, 104–107
direct, 93–94
of dose response curves, 76
doubling dose method, 92–93
in germ cell mutation, 15, 92–84
human risk methods, 150–151
LED method, 85–86, 145–146, 150
of molecular dosimetry, 84

F

Fractionation sampling, 186
Fugacity model (Mackay), 45, 53

G

GAPs, see Genetic activity profiles (GAPs)
Gastrointestinal metaplasias, 6
Gene mutations, defined, 3
Genetic activity profiles (GAPs), 9–11
carcinogenic potency and, 100–104
Gene-Tox database, 8–11
Genotoxicity, see also Genotoxicity tests; Hazard identification
classification of
by endpoints, 140–141
organism and relevance to humans, 139–140
weight-of-evidence, 139, 142
Genotoxicity Score, 149
Genotoxicity tests
NTP comparative study, 17–18
types and selection, 17
vs. mutagenic potency, 19
GEOTOX model, 41, 45
Germ cell mutagenesis, 88–95, see also Mutagenesis, germ cell
Germ cells, 3–4
Glycophorin A (GPA) assay, 194, 195, 210
Groundwater, see Water, ground

H

Haber's rule, 69–70
Hansch-type quantitative SAR methods (QSAR), 15–16
Hazard identification, 2–18, 131–132; see also Exposure assessment
future research, 20–22
general principles of
definition of terms, 2–3
germ cells, 4
mutation, 3
somatic cells, 4–6
mutagen identification, 6–18
chemical and physical properties, 14
computational methods and structure-activity studies, 14–16
data generation, 16–18
information sources/databases, 7–14
ranking of hazards, 18–20
HazardExpert program, 15
HID/LED systems, 85–86, 148–149, 151
HLA assay, 195–196, 210
HRPT assay, 194–195, 210
Human exposure pathways, 49–53, see also Exposure pathways
Human risk extrapolation methods, 150–151
Hydrolysis, 43

I

ICPEMC dose response system, 85–86
ICPEMC Genotoxicity Score, 149
ICPEMC rem-equivalent chemical unit (REC), 149–151

Incidence, see Epidemiology
Inhalation exposure, 51–52
International Commission for Protection against Environmental Mutagens and Carcinogens, see ICPEMC entries

L

Laboratory vs. human studies, 20–21
LED/HID systems, 85–86, 148–149, 151
Locus-specific assays, 86–88, 181, 184
Low-dose extrapolation methods, 150
Lowest effective dose (LED), see LED/HID systems
Lymphocyte assays, 86–88, 196

M

Mammalian assays, 91, 184–185, see also Bioassay techniques
MCM model (Cohen et al.), 45
Metaplasias, of gastrointestinal tract, 6
Michaelis-Menten equations, 69
Microbial assays, 176–177, see also Bioassay techniques
Microbial transformation, 44
Microenvironmental monitoring, 203
Micronucleus test, 180–181, 183–184
Microorganism monitoring, 187
Mitotic recombination assays, 177
Molecular dosimetry
  DNA adduct formation, 78–79
  DNA adduct removal/repair, 79–80
  exposure vs. mutation fixation, 81–84
  extrapolation, 84
  methodology, 77–78
Monitoring
  future needs, 218
  human
    experimental design and interpretation, 199
    germ cells, 198–199
    somatic cells, 191–198
  in situ applications
    aquatic species studies, 189–190, 206–208
    human studies, 209–211
    mammalian studies, 208–209
    microorganism studies, 187
    plant studies, 187–189, 191, 204–206
      laboratory applications
        ambient air and water, 202–203
        emissions and effluents, 201–202
        microenvironments, 203
        mutagen transport and transformation, 204
        personal exposure monitoring, 203–204
  laboratory bioassays
    of aquatic organisms, 179–184
    mammalian, 184–185
    microbial, 175–178
    of plants, 178–179
    sampling and analysis for, 186
  objectives
    data management, 213–215
    endpoint selection, 215–218
    sampling design and statistical analysis, 212–213
  principles and applications, 173–175
  in risk assessment, 134–136
Mudminnow (*Umbra*) assay, 190
Multilocus assay method, 178
Multimedia models, 44–47
Mussel (*Mytilus*) assay, see *Mytilus* (mussel) assay
Mutagenesis
  germ cell
    dose response and, 91–92
    endpoint determination, 92
    ethylnitrosurea-induced, 88–89
    extrapolation, 92–94, 95
    in female cells, 94–95
    mammalian mechanisms, 88–91
    specific locus method (Russell), 89–91
  somatic cell
    human mutation testing, 86–88
    short-term assays, 84–86
Mutagenicity, see also Genotoxicity; Genotoxicity testing; Mutagens
  comparative
    qualitative vs. quantitative, 95–97
    vs. exposure, 81–84
  extrapolation, 104–107
  ranking procedures, 18–20, 97–100, 153
  GAPs and, 100–104
Mutagens, see also Mutagenicity
  classification of, 142–143
  identification of, 6–18
    of chemical properties, 14
    computational methods, 14
    data sources for, 7–14
    new data generation, 16–18
    of physical properties, 14
    structure-activity (SAR) studies, 14–16

INDEX

Mutations
    aneuploidy, 3
    chromosomal, 3
    gene (point), 3
Mytilus (mussel) assay, 183, 190, 207–208

N

NIOSH (National Institute of Safety and Health), 8–9
2-Nitropropane, 72–74
NTP (U.S. National Toxicology Program), 9
    distribution of GAP tests, 10–11
    range of tests available, 13

O

Ochrotomys nuttalli assay, 209
Oncogenes defined, 5
Oocyte studies, 94–95
Organ specificity, 81
Osmunda regalis (royal fern) assay, 189, 206
Oxidation and reduction, 43–44

P

Pathway exposure factors (PEFs), 54–55, 133, 154–157
PATHWAY model, 48
PCB, transfer and PEF comparison, 56
Peromyscus leucopus assay, 188, 208
Pharmacokinetics
    dose response curves
        of ethylene and ethylene oxide, 75–76
        extrapolation issues, 76
        of 2-nitropropane, 72–75
        of vinyl chloride and vinyl bromide, 72
    general principles, 67–68
    molecular dosimetry and, 77–84
    principles and models
        and area under curve (AUC), 69–72
        basic considerations, 68–69
Phenotypes, sentinel, 198
Photolysis, 42–43
Physicochemical properties modeling, 47
Plant assays, 178–179
    in situ, 205–206
    of emissions, 201–202
Plant monitoring, 187–189
Point mutations, see Gene mutations
Protein variant tests, 198
Pseudopleuronectes assay, 207

Q

QSAR methods, 15–16
Qualitative risk characterization, 143–144
Quantification, 55
Quantitative risk assessent
    dose-response extrapolation, 145–153
    exposure assessment, 153–156
    risk quantitation, 156–158

R

Radiation dose equivalence: ICPEMC rem-equivalent chemical unit (REC), 149–151
Radiation exposure, 87
Ranking of mutageneic hazards, 18–20, 153
Redox potentials, 43–44
Relative mutagenic effect (RME), 146–147
Repair assays, 177
Repair processes, 79–80
Research trends and priorities, 20–22
Risk
    as compared with exposure, 33–34
    factors affecting, 158–162
    prediction formula, 34
Risk assessment, see also Exposure assessment; Hazard identification
    characterization of risk, 136
    dose-response assessment, 132
    exposure assessment and, 31–33, 133
    general principles, 126–128
    hazard identification, 2–18, 131–132
    for human exposure, 138–158, see also Dose-reponse extrapolation; Quantitative risk assessment
        classification of genotoxicity, 139–143
        qualitative methods, 143–144
        quantitative methods, 144–153
    monitoring, 134–136
    for nonhuman biota, 136–138
    significance of comparative, 162
    stages and techniques, 128–131
Risk categorization, 143–144
RME (relative mutagenic effect), 146–147
RTECS (Registry of Toxic Effects of Chemical Substances), 9

S

Salmonella reversion mutation assay (Ames test), 11–13, 17–18, 35, 85, 101,

201–202, see also Bioassay techniques
Sampling, 186
SAR (structure-activity relationship) modeling, 14–16
Saturation kinetics, 68–69
SCG alkaline electrophoresis assay, 211
Sentinel phenotypes, 198
Sigmodon hispidus assay, 208
Single-gel electrophoresis, 209
Sister-chromatid exchange analysis, of aquatic organisms, 180, 183, 194
Skin contact, see Dermal uptake exposure
Socioeconomic risk factors, 161
Soil(s)
 as contact medium, 37
 fugacity model of exposure, 53
 transport in, 41
Somatic cell mutagenesis, 84–88, see also Mutagenesis, somatic cell
Somatic cells, 4–6
SOS/phase induction assays, 177
Specific-locus assays
 in aquatic organisms, 181, 184
 in male mice, 105
 method, 86–91
 in plants, 10
Sperm analysis, 198–199
Structural alert method, 15
Structure-activity relationship (SAR) modeling, 14–16

T

TCDD, transfer and PEF comparison, 56
Terrestrial species transfers, 48–49
Threshold calculation, 33–34, see also Exposure assessment
Tissue testing, 197–198
T lymphocyte assays, 196
TOPKAT program, 14–16
TOXLINE database, 8
TOXNET database, 8–9

Tradescantia assay, 178, 187–189, 178, 202, 205–206
Transformation
 environmental monitoring of, 204
 by hydrolysis, 43
 microbial, 44
 by oxidation and reduction, 43–44
 by photolysis, 42–43
Transport
 environmental monitoring of, 204
 in groundwater, 40
 in indoor and outdoor air, 38–39
 in soil, 41
 in surface waters, 39–40
Tumor suppressor genes, 5–6

U

Umbra (mudminnow) assay, 190
United Nations Environment Programme Study, 131–162, see also Risk assessment
Urine tests, 197–198
U.S. National Toxicology Program, 17–18

V

Vinyl chloride/vinyl bromide, 72
Volatilization, 35–36

W

Water
 as contact medium, 37
 ground, 40
Water monitoring, 202–203
Waters (LED/HID) method, 85–86
Weight-of-evidence classification, 139, 141

Z

Zea maize bioassays, 178–179, 189, 205–206

THE LIBRARY
UNIVERISTY OF CALIFORNIA, SAN FRANCISCO
(415) 476-2335

THIS BOOK IS DUE ON THE LAST DATE STAMPED BELOW

Books not returned on time are subject to fines according to the Library Lending Code. A renewal may be made on certain materials. For details consult Lending Code.

INTERLIBRARY LOAN DUE
14 DAYS AFTER RECEIPT

RETURNED
JUN - 3 1994

Series 4128